COMMUNICATION PROTOCOL ENGINEERING

COMMUNICATION PROTOCOL ENGINEERING

Miroslav Popovic

Taylor & Francis
Taylor & Francis Group
Boca Raton   London   New York

CRC is an imprint of the Taylor & Francis Group,
an informa business

CRC Press
Taylor & Francis Group
6000 Broken Sound Parkway NW, Suite 300
Boca Raton, FL 33487-2742

© 2006 by Taylor and Francis Group, LLC
CRC Press is an imprint of Taylor & Francis Group, an Informa business

No claim to original U.S. Government works
Printed in the United States of America on acid-free paper
10 9 8 7 6 5 4 3 2 1

International Standard Book Number-10: 0-8493-9814-2 (Hardcover)
International Standard Book Number-13: 978-0-8493-9814-8 (Hardcover)
Library of Congress Card Number 2005036572

This book contains information obtained from authentic and highly regarded sources. Reprinted material is quoted with permission, and sources are indicated. A wide variety of references are listed. Reasonable efforts have been made to publish reliable data and information, but the author and the publisher cannot assume responsibility for the validity of all materials or for the consequences of their use.

No part of this book may be reprinted, reproduced, transmitted, or utilized in any form by any electronic, mechanical, or other means, now known or hereafter invented, including photocopying, microfilming, and recording, or in any information storage or retrieval system, without written permission from the publishers.

For permission to photocopy or use material electronically from this work, please access www.copyright.com (http://www.copyright.com/) or contact the Copyright Clearance Center, Inc. (CCC) 222 Rosewood Drive, Danvers, MA 01923, 978-750-8400. CCC is a not-for-profit organization that provides licenses and registration for a variety of users. For organizations that have been granted a photocopy license by the CCC, a separate system of payment has been arranged.

Trademark Notice: Product or corporate names may be trademarks or registered trademarks, and are used only for identification and explanation without intent to infringe.

Library of Congress Cataloging-in-Publication Data

Popovic, Miroslav.
    Communication protocol engineering / Miroslav Popovic.
        p. cm.
    Includes bibliographical references and index.
    ISBN-13: 978-0-8493-9814-8 (alk. paper)
    ISBN-10: 0-8493-9814-2 (alk. paper)
    1. Computer network protocols. 2. Computer networks--Standards. I. Title.

TK5101.55.P67 2006
621.382'12--dc22   2005036572

Visit the Taylor & Francis Web site at
http://www.taylorandfrancis.com

and the CRC Press Web site at
http://www.crcpress.com

Preface

I wrote this book as a textbook for postgraduate students, but it might also be used by people in the industry to update specific knowledge in their lifelong learning processes. The book partly covers the actual postgraduate course on computer communications and networks undertaken during the first semester of studies for the M.Sc. degree in computer engineering. Since nowadays we are witnessing the convergence of the Internet and public telephone network, this book might also be useful to engineers with B.Sc. degrees in telecommunications.

The prerequisite for this book is the knowledge of the first order logic (predicate calculus), operating systems, and computer network fundamentals. The reader should also be familiar with C++ and Java programming languages.

My approach in writing this book was to provide all the details that the reader may need. I assumed that nothing is obvious. However, if you the reader find something obvious while reading the book, you are encouraged to skip ahead. If something is not clear later on, you may always return to what you skipped. Communication protocol engineering is a very interesting combination of abstraction and practice that requires a lot of details. It starts from a vision that gradually materializes in the real-world artifacts. This happens through a typical engineering process. This book covers all aspects of the communication protocol engineering, including requirements and analysis, design, implementation, and test and verification.

Many people helped me in writing this book. My gratitude goes to all of them. I thank my family for their continuous support, my niece Silvia Likavec for her valuable text corrections, and B.J. Clark, Nora Konopka, and Helena Redshaw, of Taylor & Francis, for their professional support. Special thanks go to my colleagues from the University of Novi Sad, Prof. Vladimir Kovacevic for giving his blessing to this book, Ph.D. student Ivan Velikic for the excellent cooperation (in his M.Sc. thesis we actually developed the FSM Library, one of the anchors of this book), Ph.D. student Ilija Basicevic (for helping me in preparation of examples in Sections 3.10.5, 4.5.2, and 5.5.2), Sonja Vukobrat (for helping me in preparation of the example in Section 3.7), Laslo Benarik and Aleksander Stojicevic (for helping me in preparation of Chapter 6), Milan Savic, Aleksander Stojicevic, and Cedomir Rebic (for helping me in preparation of examples in Sections 3.10.1 and 3.10.2), and Nenad Cetic (for helping me in preparation of the example in Section 4.5.1). Thank you all!

Miroslav Popovic
Novi Sad

The Author

Miroslav Popovic, Ph.D. received all his degrees from the University of Novi Sad. He defended his diploma thesis, "An Intelligent System Restart," in 1984, his M.Sc. thesis, "An Efficient Virtual Machine System," in 1988, and his Ph.D. thesis, "A Contribution to Standardization of ISO OSI Presentation Layer," in 1990. He began working at the University of Novi Sad immediately after graduating in 1984, and since then he has been teaching students operating systems and intercomputer communications in various forms. The research area he is mainly interested in today is engineering of computer-based systems. He is a member of the program committee of the IEEE Annual Conference on Engineering of Computer Based Systems. He is also a member of IEEE (both Computer and Communications societies) and ACM. In the last 20 years he has published approximately 100 papers and he has supervised many real-world projects for the industry.

Contents

Chapter 1 Introduction 1
1.1 The Notion of the Communication Protocol 5
References 8

Chapter 2 Requirements and Analysis 9
2.1 Use Case Diagrams 13
2.2 Collaboration Diagrams 21
2.3 Requirements and Analysis Example 30
     2.3.1 SIP Domain Specifics 30
     2.3.2 SIP Softphone Requirements Model 34
     2.3.3 SIP Softphone Analysis Model 39
References 42

Chapter 3 Design 45
3.1 Class Diagrams 50
3.2 Object Diagrams 60
3.3 Sequence Diagrams 64
3.4 Activity Diagrams 71
3.5 Statechart Diagrams 85
3.6 Deployment Diagrams 99
3.7 Specification and Description Language 103
     3.7.1 Telephone Call Processing Example 116
3.8 Message Sequence Charts 120
3.9 Tree and Tabular Combined Notation 124
3.10 Examples 134
     3.10.1 Example 1 134
     3.10.2 Example 2 139
     3.10.3 Example 3 144
     3.10.4 Example 4 149
     3.10.5 Example 5 154
References 164

Chapter 4 Implementation 165
4.1 Component Diagrams 167
4.2 The Spectrum of FSM Implementations 172
4.3 State Design Pattern 194

4.4 Implementation Based on the FSM Library 197
    4.4.1 Using the FSM Library 203
    4.4.2 FSM Library Internals 204
        4.4.2.1 FSMSystem Internals 205
        4.4.2.2 FiniteStateMachine Internals 207
        4.4.2.3 Kernel Internals 214
    4.4.3 Writing FSM Library-Based Implementations 216
4.5 Examples 217
    4.5.1 Example 1 217
    4.5.2 Example 2 235
References 245

Chapter 5 Test and Verification 247
5.1 Unit Testing 251
5.2 Conformance Testing 261
5.3 Formal Verification Based on Theorem Proving 265
5.4 Statistical Usage Testing 277
5.5 Examples 291
    5.5.1 Example 1 291
    5.5.2 Example 2 299
5.6 Further Reading 305
References 306

Chapter 6 FSM Library 307
6.1 Introduction 307
6.2 Basic FSM System Components 308
    6.2.1 Class FSMSystem 308
        6.2.1.1 FSM System Initialization 309
        6.2.1.2 FSM System Startup 312
    6.2.2 Class FiniteStateMachine 313
6.3 Time Management 316
6.4 Memory Management 317
6.5 Message Management 318
6.6 TCP/IP Support 323
    6.6.1 Class FSMSystemWithTCP 324
    6.6.2 Class NetFSM 325
6.7 Global Constants, Types, and Functions 326
6.8 API Functions 327
    6.8.1 FSMSystem 329
    6.8.2 Add(ptrFiniteStateMachine, uint8, uint32, bool) 340
    6.8.3 Add(ptrFiniteStateMachine, uint8) 342
    6.8.4 InitKernel 342
    6.8.5 Remove(uint8) 343

6.8.6	Remove(uint8, uint32)	344
6.8.7	Start	344
6.8.8	StopSystem	344
6.8.9	FSMSystemWithTCP	345
6.8.10	InitTCPServer	345
6.8.11	FiniteStateMachine	346
6.8.12	AddParam	347
6.8.13	AddParamByte	348
6.8.14	AddParamDWord	348
6.8.15	AddParamWord	348
6.8.16	CheckBufferSize	349
6.8.17	ClearMessage	349
6.8.18	CopyMessage()	350
6.8.19	CopyMessage(uint*)	350
6.8.20	CopyMessageInfo	351
6.8.21	Discard	351
6.8.22	DoNothing	351
6.8.23	FreeFSM	352
6.8.24	GetAutomata	352
6.8.25	GetBitParamByteBasic	352
6.8.26	GetBitParamWordBasic	353
6.8.27	GetBitParamDWordBasic	353
6.8.28	GetBuffer	354
6.8.29	GetBufferLength	355
6.8.30	GetCallId	355
6.8.31	GetCount	356
6.8.32	GetGroup	356
6.8.33	GetInitialState	356
6.8.34	GetLeftMbx	356
6.8.35	GetLeftAutomata	357
6.8.36	GetLeftGroup	357
6.8.37	GetLeftObjectId	357
6.8.38	GetMbxId	358
6.8.39	GetMessageInterface	358
6.8.40	GetMsg()	359
6.8.41	GetMsg(uint8)	359
6.8.42	GetMsgCallId	359
6.8.43	GetMsgCode	360
6.8.44	GetMsgFromAutomata	360
6.8.45	GetMsgFromGroup	360
6.8.46	GetMsgInfoCoding	360
6.8.47	GetMsgInfoLength()	361
6.8.48	GetMsgInfoLength(uint8*)	361
6.8.49	GetMsgObjectNumberFrom	361
6.8.50	GetMsgObjectNumberTo	362
6.8.51	GetMsgToAutomata	362

6.8.52	GetMsgToGroup	362
6.8.53	GetNewMessage	362
6.8.54	GetNewMsgInfoCoding	363
6.8.55	GetNewMsgInfoLength	363
6.8.56	GetNextParam	363
6.8.57	GetNextParamByte	364
6.8.58	GetNextParamDWord	364
6.8.59	GetNextParamWord	365
6.8.60	GetObjectId	366
6.8.61	GetParam	366
6.8.62	GetParamByte	367
6.8.63	GetParamDWord	367
6.8.64	GetParamWord	368
6.8.65	GetProcedure	368
6.8.66	GetRightMbx	369
6.8.67	GetRightAutomata	369
6.8.68	GetRightGroup	370
6.8.69	GetRightObjectId	370
6.8.70	GetState	370
6.8.71	IsBufferSmall	370
6.8.72	Initialize	371
6.8.73	InitEventProc	371
6.8.74	InitTimerBlock	372
6.8.75	InitUnexpectedEventProc	373
6.8.76	IsTimerRunning	373
6.8.77	NoFreeObjectProcedure	373
6.8.78	NoFreeInstances	374
6.8.79	ParseMessage	374
6.8.80	PrepareNewMessage(uint8)	375
6.8.81	PrepareNewMessage(uint32, uint16, uint8)	375
6.8.82	Process	376
6.8.83	PurgeMailBox	376
6.8.84	RemoveParam	377
6.8.85	Reset	377
6.8.86	ResetTimer	377
6.8.87	RestartTimer	378
6.8.88	RetBuffer	378
6.8.89	ReturnMsg	378
6.8.90	SetBitParamByteBasic	379
6.8.91	SetBitParamDWordBasic	379
6.8.92	SetBitParamWordBasic	380
6.8.93	SetCallId()	380
6.8.94	SetCallId(uint32)	380
6.8.95	SetCallIdFromMsg	381
6.8.96	SetDefaultFSMData	381
6.8.97	SetDefaultHeader	381

6.8.98	SetGroup	382
6.8.99	SetInitialState	382
6.8.100	SetKernelObjects	382
6.8.101	SetLeftMbx	383
6.8.102	SetLeftAutomata	383
6.8.103	SetLeftObject	383
6.8.104	SetLeftObjectId	383
6.8.105	SetLogInterface	384
6.8.106	SendMessage(uint8)	384
6.8.107	SendMessage(uint8, uint8*)	384
6.8.108	SetMessageFromData	385
6.8.109	SetMsgCallId(uint32)	385
6.8.110	SetMsgCallId(uint32, uint8*)	385
6.8.111	SetMsgCode(uint16)	386
6.8.112	SetMsgCode(uint16, uint8*)	386
6.8.113	SetMsgFromAutomata(uint8)	386
6.8.114	SetMsgFromAutomata(uint8, uint8*)	387
6.8.115	SetMsgFromGroup(uint8)	387
6.8.116	SetMsgFromGroup(uint8, uint8*)	388
6.8.117	SetMsgInfoCoding(uint8)	388
6.8.118	SetMsgInfoCoding(uint8, uint8*)	388
6.8.119	SetMsgInfoLength(uint16)	389
6.8.120	SetMsgInfoLength(uint16, uint8*)	389
6.8.121	SetMsgObjectNumberFrom(uint32)	389
6.8.122	SetMsgObjectNumberFrom(uint32, uint8*)	390
6.8.123	SetMsgObjectNumberTo(uint32)	390
6.8.124	SetMsgObjectNumberTo(uint32, uint8*)	390
6.8.125	SetMsgToAutomata(uint8)	391
6.8.126	SetMsgToAutomata(uint8, uint8*)	391
6.8.127	SetMsgToGroup(uint8)	391
6.8.128	SetMsgToGroup(uint8, uint8*)	392
6.8.129	SendMessageLeft	392
6.8.130	SendMessageRight	392
6.8.131	SetNewMessage	393
6.8.132	SetObjectId	393
6.8.133	SetRightMbx	393
6.8.134	SetRightAutomata	394
6.8.135	SetRightObject	394
6.8.136	SetRightObjectId	394
6.8.137	SetState	394
6.8.138	StartTimer	395
6.8.139	StopTimer	395
6.8.140	SysClearLogFlag	395
6.8.141	SysStartAll	396
6.8.142	NetFSM	396
6.8.143	convertFSMToNetMessage	397

	6.8.144 convertNetToFSMMessage	397
	6.8.145 establishConnection	397
	6.8.146 getProtocolInfoCoding	397
	6.8.147 sendToTCP	398
6.9	A Simple Example with Three Automata Instances	398
6.10	A Simple Example with Network-Aware Automata Instances	422

Index **439**

Dedication

To my wife Vlasta and our sons Marko and Andrej

1
Introduction

Originally, the term *protocol* was related to the customs and regulations dealing with diplomatic formality, precedence, and etiquette. A protocol is actually the original draft, minutes, or record from which a document, especially a treaty, is prepared, e.g., an agreement between states. Today, in the context of computer networks, the term *protocol* is interpreted as a set of rules governing the format of messages that are exchanged between computers. Sometimes, especially if we want to be more specific, we use the term *communication protocol* instead.

The title of this book, *Communication Protocol Engineering*, is used to emphasize the process of developing communication protocols. Like other engineering disciplines, communication protocol engineering typically comprises the following phases (Figure 1.1):

- Requirements and analysis
- Design
- Implementation
- Test and verification

The process as described in this book is ideally the union of the UML (Unified Modeling Language)-driven unified development process (Booch et al., 1998), Cleanroom engineering (formal system design verification and statistical usage testing), and some elements of Agile programming (particularly unit testing based on JUnit). Of course, each organization should adapt and tune the process to its own needs and goals. For example, one organization may stick to the UML-driven unified development process, another may prefer Cleanroom engineering, yet another may use the combination of both, and so forth.

Because this book is written for the process in which all the existing state-of-the-art methods and techniques in the area are applied, it is independent of any particular engineering process; this is as far as we will go in discussions on processes in this book. This book is not about managing processes. Rather, this book is intended for engineers. It provides the knowledge that

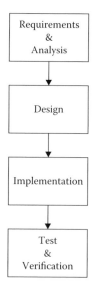

FIGURE 1.1
Typical communication protocol engineering phases.

an engineer needs to work in a modern organization involved in communication protocol engineering.

The chapters are named by typical process phases: requirements and analysis, design, implementation, and test and verification. These chapters are actually used to classify various methods and techniques (and the accompanying tools). As already stated, the attitude in the selection of methods and techniques included in this book was making a union rather than falling into a trap of separatism. The methods and techniques introduced here originate from the following methodologies:

- UML methodology
- ITU-T system specification and description methodology
- Agile unit testing methodology
- Cleanroom engineering methodology

UML methodology is based on various kinds of graphs, also referred to as diagrams. This book covers all of them, namely:

- Use case diagrams (Section 2.1)
- Collaboration diagrams (Section 2.2)
- Class diagrams (Section 3.1)
- Object diagrams (Section 3.2)
- Sequence diagrams (Section 3.3)
- Activity diagrams (Section 3.4)

- Statechart diagrams (Section 3.5)
- Deployment diagrams (Section 3.6)
- Component diagrams (Section 4.1)

ITU-T system specification and description methodology is based on three domain-specific languages, which this book also covers. These languages are:

- Specification and description language (SDL) (Section 3.7)
- Message sequence charts (MSC) (Section 3.8)
- Three and tabular combined notation (TTCN) (Section 3.9)

Agile unit testing methodology assumes writing the test cases before the code. Today, it is supported by the following two open-source packages (both are covered in this book):

- JUnit, a package for automated unit testing of Java packages (Section 5.1)
- CppUnit, a library for automated unit testing of C++ modules (Section 5.5.1)

Cleanroom engineering methodology is based heavily on two main methods, both covered in this book. These methods are:

- Formal system design verification. Today, more approaches exist to formal system design verification. This book covers formal verification based on automated theorem proving (Section 5.3).
- Statistical usage testing (Section 5.4).

The text of the book is organized as follows. At the end of this chapter, in Section 1.1, we introduce the notion of the communication protocol and related definitions.

Chapter 2 is devoted to the requirements and analysis phase of communication protocol engineering. The first part of that chapter introduces UML use case and collaboration diagrams (Section 2.1 and Section 2.2, respectively). The former is used for capturing both functional and nonfunctional system requirements, whereas the latter is used for making system analysis models. The second part of that chapter presents a real-world example — requirements and analysis of an SIP (Session Initiation Protocol, RFC 3261) Softphone. The example starts with the presentation of the domain-specific information related to SIP, continues with the SIP Softphone requirements model (in the form of the corresponding use case diagram), and ends with the SIP Softphone analysis model (in the form of the corresponding collaboration diagram).

Chapter 3 covers the design phase of communication protocol engineering. In this chapter, we will see that communication protocols are actually modeled as finite state machines (FSM). The first part of the chapter introduces UML diagrams related to the design phase: class, object, sequence, activity, statechart, and deployment diagrams (Section 3.1, Section 3.2, Section 3.3, Section 3.4, Section 3.5, and Section 3.6, respectively). The second part of Chapter 3 covers domain-specific languages originated at ITU-T; namely SDL, MSC, and TTCN (Sections 3.7, Section 3.8, and Section 3.9, respectively). The third part consists of design examples, with the first three examples rather academic. The fourth example shows the design of the sliding window concept. The fifth example is a real-world design example — the design of the SIP INVITE client transaction, a part of the SIP protocol stack.

Chapter 4 is devoted to the implementation phase of communication protocol engineering. At the beginning of this chapter, we introduce the UML component diagrams (Section 4.1). The second part of Chapter 4 presents various implementation approaches. Section 4.2 presents three examples of approaches that can be used. The main goal of this study is to provoke dilemmas by studying three different concepts of implementation and to promote creative thinking about a spectrum of possible implementation paradigms before restricting ourselves to a single one. This short overview includes the implementations as nested switch-case statements, the implementation based on the interpretation of protocol messages using a protocol definition data structure, and the implementation based on a class hierarchy and state transition map. The second part of Chapter 4 ends with the introduction of the state design pattern (Section 4.3), a catalogued FSM implementation approach.

The third part of Chapter 4 (Section 4.4) introduces one concrete, industrial-strength implementation paradigm based on the FSM Library, a library of C++ classes used for modeling communication protocols as FSM. This paradigm has been successfully used on a series of real-world projects, such as SS7, DSS1, V5.2, H.323, SIP, and so on. This part of the book covers FSM Library features and internals as well as the rules for writing FSM Library-based implementations. The last part of Chapter 4 contains two real-world examples of the FSM Library-based implementations. The first is the implementation of the POP3 communication protocol, the TCP/IP Internet protocol for receiving e-mail messages. The second is the SIP INVITE client transaction, a part of the SIP protocol stack.

Chapter 5 deals with the testing and verification phase of communication protocol engineering. The first part starts with the introduction of unit testing based on JUnit, the open-source testing framework for unit testing Java programs, originally developed by Erich Gamma and Kent Beck (Section 5.1). Next, we introduce conformance testing (Section 5.2), actually the first stage of communication protocol acceptance testing. Conformance testing is typically based on the TTCN test suite specification. We then introduce formal verification of both system design and implementation based on automated theorem proving (Section 5.3). In this book, we use the theorem prover Theo for this purpose.

The first part of Chapter 5 ends with the introduction of statistical usage testing (Section 5.4) based on product operational profiles. The second part of Chapter 5 consists of two real-world examples. The first example shows the unit testing of the SIP INVITE client transaction based on the usage of the CppUint, the library for unit testing C++ modules. The second example demonstrates the integration testing of the SIP INVITE client transaction.

Chapter 6 is written as a programmer's reference manual for the FSM Library. The first part starts with the introduction of two main classes, *FSMSystem* and *FiniteStateMachine* (Section 6.2). Next, we introduce three main groups of basic functions supported by the FSM Library: time, memory, and message management functions (Sections 6.3, Sections 6.4, and Sections 6.5, respectively). We then introduce two classes that support the communication of FSMs over the TCP/IP Internet (Section 6.6), namely the classes *FSMSystemWithTCP* and *NetFSM*. The first part of Chapter 6 ends with the introduction of global constants, types, and functions (Section 6.7).

The second part of Chapter 6 contains detailed descriptions of the individual FSM Library Application Programming Interface (API) functions (Section 6.8). The third part of Chapter 6 consists of two examples. The first is a simple example with three automata (FSM) instances (Section 6.9), and the second is a simple example with TCP/IP network-aware automata instances (Section 6.10).

1.1 The Notion of the Communication Protocol

What is a communication protocol? A wide range of definitions are available in the literature today, for example: "An established set of conventions by which two computers or communication devices validate the format and content of the messages exchanged;" "A set of defined interfaces that permits the computers to communicate with each other;" "A method by which two computers coordinate their communication;" "Common agreed rules followed in order to interconnect and communicate between computers;" "The rules governing the exchange of information between devices on a data link;" "The set of rules governing how information is exchanged on a network;" and so on.

In this book, we begin with a wider informal definition. A **protocol** is a set of conventions and rules governing their use that regulates the communication of an entity under observation with its environment. Such a definition enables the study of any communication, e.g., an agenda for a technical meeting of representatives of two companies. The subject of this book is one special class of protocols, referred to as **communication protocols**, that regulate the communication of geographically distributed program objects. The communicating program objects are deployed on different processors in the network. We will sometimes use the term **protocol** as an abbreviated form of the phrase **communication protocol** to save space.

A **process**, as generally defined in the theory of operating systems, is a program in execution or prepared for execution. A process may be specialized for data processing, communication, or some other special task (e.g., I/O control, time management). Traditionally, a data processing algorithm is specified by the flow chart. What the flow chart means for the data processing process, the protocol means for the communication process.

The flow chart specifies the program control flow by the use of graphic symbols related to the series of sequential calculations, selection, iteration, procedure/function call, and input/output operation needed to read input data or write output data. On the other hand, the formal specification of a communication protocol is based on messages and consists of the following three parts:

- The message format specification.
- The message-processing procedures specification. This is essentially a formal description of process reactions to input stimuli (i.e., messages).
- The error processing specification. This is the formal description of process reactions to exceptional events (i.e., corrupted data, timeouts).

The **message format** completely defines the structure of the message, i.e., it defines the set of fields that constitute the message by defining the width of individual fields (most commonly in bits, bytes, or words), the applied coding scheme (e.g., binary, ASCII, Unicode, ASN.1), and optionally legal values (e.g., constants in binary or some symbolic form, value intervals).

Therefore, a **message** is a series of bits logically divided into various fields. Typically, a message consists of a message header, which most commonly comprises more subfields, and useful data referred to as a payload. The **payload** contains data interpreted by the communicating program objects. The message header contains data added for supervision and control purposes in accordance with the established conventions.

The **message-processing procedure** (i.e., the process reaction) begins with the message reception and is described as a series of primitive operations that define the rules of the communication, which are the essential parts of a protocol. Typical **primitive operations** include timer-start operations, timer-stop operations, message-send operations, message-receive operations, and message-data processing operations (e.g., cyclic redundancy checking of message data, calculating expected order number of the next message to be received).

In terms of software implementation, message processing is performed by a message processing routine. Depending on the selected working environment, this routine can be a subroutine that consists of a series of machine instructions in a symbolic form (assembly language) or a function comprising a series of statements in a higher-level programming language, such as C/C++ or Java.

The error-processing specification defines a set of error reactions. An **error reaction** is a special protocol reaction to exceptional events or, in other words, a reaction to unexpected situations, i.e., conditions. Typical examples of unexpected events are: the reception of a message that contains corrupted data, the reception of a message that is out of the original order (e.g., after receiving the messages numbered 1, 2, and 3, we receive the message numbered 7 instead of the message numbered 4), timer expiration (e.g., the receiver has not acknowledged the reception of a message to its sender within a certain interval of time, determined by the value of the corresponding timer), and so on.

Note that a protocol can be described informally or formally. The informal description of a protocol is referred to as its **informal specification** and has the following characteristics:

- It frequently has the form of a combination of textual and graphical description of the most common scenarios of communication.
- It may state nothing about the order of the activities to be conducted in the course of the communication.
- It is always incomplete. Most frequently, missing parts are specifications of timers, which determine time limits over individual phases of communication.

Let us forget for a moment the communication protocols and use the old example of informal specification of a group of tasks to get a feeling about the issues stated above. While leaving the house, the mother says to her daughter:

"Do not forget to finish your homework."

"Have your breakfast when you get hungry."

"Before you go to school, throw the garbage out."

Obviously, this specification does not say anything about the order of the individual tasks. For example, the daughter may complete the tasks in any order without interrupting the individual tasks (e.g., task order may be 1, 2, 3, or 1, 3, 2), or she may complete them in any order and switch between them (e.g., she starts with task 1, then before completing it, she switches to task 2, completes tasks 2 and 3, and at the end finishes task 1). An essential question here is how to organize the task executions in time, i.e., how to allocate time to them. Clearly, a need exists to limit task duration, i.e., to control the task execution time. What happens if the daughter gets preoccupied with her homework and forgets to have breakfast before it is time to go to school?

The example above might appear to be an exaggeration of the problems we face in reality, but its goal is to show that informal systems specification is insufficient, and that we need a formal systems specification to make a precise and correct system implementation. Formal specification in the area of communication protocols is based on modeling a protocol as a **finite state machine** (FSM). A single FSM is often referred to by the term **automata,** and we will use these two terms interchangeably in this book.

The formal specification of an FSM defines all its states and state transitions, including transitions initiated by expiration of timers, in a unique and detailed way. Today, we may make formal protocol specifications in either UML or ITU-T domain-specific languages. Once we have a formal protocol specification, we can implement it in Java or C++. Finally, we must test and verify it. This procedure is basically what this book is all about.

References

Booch, G., Rumbaugh, J., and Jacobson, I., *The Unified Modeling Language User Guide*, Addison-Wesley, Reading, MA, 1998.

Booch, G., Rumbaugh, J., and Jacobson, I., *The Unified Software Development Process*, Addison-Wesley, Reading, MA, 1998.

2
Requirements and Analysis

At the beginning of any project, engineers face the fundamental question, "What must be done and how do we verify (deliver) the solution (system, device, products, service, hardware or software)?" Answering this question leads to what are called **requirements**. To simplify the matter, the process of answering this question — i.e., the corresponding engineering phase — is also commonly called *requirements*. So both the working phase and the resulting documents have the same name, but the meaning is easily deduced from the context.

The previous question actually consists of the following two questions:

1. What must be done?
2. How can the solution be verified?

Answering the former question leads to a set of functional requirements, most frequently adorned by non-functional requirements. **Functional requirements** describe the desired system behavior, while **nonfunctional requirements** can be imagined as the additional attributes to the behavior related to time restrictions, performance, and so on. To answer the latter question, we must quantify the behavior of the system. Normally, we would say, "For this input, the system should produce this output." Such thinking implies the existence of a test setup that enables automated (most preferably automatic) testing, referred to as a **test bed**. A test bed provides a **test harness** by generating the input to the system and capturing its output.

The ordered pair of the given input and the expected output informally stated in the text above is called a **test case**. To verify complex systems, we need many test cases. A set of test cases packed in a suitable form is referred to as a **test suite**. Ideally, we would like the test suite to completely cover the systems behavior (i.e., the functional requirements), which are adorned with their non-functional requirements. Typically, one or more test cases will be derived from each functional requirement. Clearly for any nontrivial system, the number of test cases needed to verify the system may be huge.

However, while thinking about the desired behavior of the system and its verification, we inevitably think about the question, "How can we make it?" Actually, we are trying to make a concept of the system or, more precisely, its architecture. This engineering phase is called an **analysis**. Obviously, it is tightly coupled with the requirements. These two phases have a highly interactive relation.

Typically, work on the definition of the system architecture yields the refinement of system functional requirements, and vice versa. This is especially true for communication protocol engineering. Therefore, we think of these two phases, the requirements and the analysis, as one indivisible front-end phase of communication protocol engineering. This is the reason they are covered together in this chapter.

As already mentioned, the area of communication protocol engineering is very well founded, and many standards, recommendations, and well-known experiences exist — hence, this chapter is rather short compared to the others. Unlike other areas of engineering, here a vast majority of engineers will be faced with the task of implementing some already defined standards, such as IETF RFC, ITU-T/ETSI recommendations, and so on. A very few engineers will be in a position to create a completely new protocol, and even then they will have many existing protocols for reference and as starting points.

Many existing standards actually represent very detailed designs accompanied by the corresponding test suites, but others are rather informal and bring nothing more than the message syntax and encoding together with some textual explanations of the message handling procedures. However, most of the standards can be viewed at least as rather good starting functional requirements that must be further formalized and analyzed. This chapter tries to help the reader exactly in this direction. It tries to answer the question, "How can we deal with the requirements in a systematic way?" Or, in other words, "How do we capture the requirements and how do we proceed with forward engineering from there?"

An overall consensus seems to exist in both academia and industry today that the UML paradigm (Booch et al., 1998) can help in this respect. The behavior of the system is described with a set of use cases. Each **use case** captures one functional requirement adorned with its corresponding non-functional requirements. The requirements engineer models the system by specifying the individual actors and the corresponding use cases of the system. The result is referred to as a **requirements model** of the system. The means for making such models are **use case diagrams**, which will be introduced in the next section.

The next step in the UML paradigm is to transform the requirements model into the **analysis model**. Typically, a use case is viewed as a collaboration of classifiers. In the analysis model, three different **stereotypes** of classes are used: <<*boundary class*>>, <<*control class*>>, and <<*entity class*>>. The means of specifying the collaborations in UML are **collaboration diagrams**, which will be introduced in a following section.

Sometimes the analysts describe the static structure of the system — in addition to its behavior — with **class diagrams**. This practice can be helpful in really complex systems. In this chapter, we will present the collaboration diagrams sufficient for the examples at hand, therefore the introduction to class diagrams is postponed until the next chapter. The next chapter deals with the communication protocol design phase in which class diagrams are essential to show the static relations among classes.

Further on, in accordance with the UML paradigm, the requirements model should be transformed into the **test model** to facilitate the system verification (the test model is actually the test suite needed for the system verification). Essentially, the use cases should be translated into the corresponding test cases described by test scripts of some kind. UML is not specific in that respect. Of course, a few scripting languages are popular today, such as TCL/TK, Perl, and Payton, but being general purpose languages, these might be inappropriate for some of the projects.

To close this gap, we will introduce a domain-specific language known as **tree and tabular combined notation** (TTCN). The TTCN tables are used for specifying the test suites for communication protocols once the software architecture is rather well known. Therefore, we will postpone the introduction to the TTCN language until the next chapter, which deals with the design phase of communication protocol engineering.

A general problem when transforming use cases to test cases is that the transformation is typically done manually, i.e., it is semiautomatic. Such an approach is both time consuming and prone to error. However, the main conceptual problem is the test coverage of the system behavior. In practice, the number of possible scenarios and all possible combinations of message parameters can be impossible to cover manually. Therefore, testing at least the most frequently used system scenarios and message parameter combinations should somehow be possible.

Clearly, more detailed UML models made during the system design phase (e.g., statecharts, to be introduced in the next chapter) can be used later for the automatic generation of test cases. However, the problem with this approach is that if an error exists in the UML model, it will be propagated into the test suite and the test suite will not be able to detect the error. A well-known principle from mathematical logic is that negation of negation leads to affirmation, so the bug will remain undiscovered. No matter how large test suite we generate, it will not be able to detect the bug.

The former problem can be solved by the application of **statistical usage testing**, also referred to as **behavior testing**. This paradigm is based on the operational profile model of the system, which describes the statistics of the system usage. It enables the practitioners to thoroughly test the system and even estimate the system or software reliability. This practice is recognized as a *de facto* standard by the industry (Broekman and Notenboom, 2003) and it will be covered in detail in Chapter 5 (test and verification phase of communication protocol engineering).

The latter problem can be solved by using one model as a source for the software implementation generated with forward engineering and a completely different model for the system test suite generation. Also highly desirable is that these two models are made by two separate individuals or teams. For example, the well-known Cleanroom engineering paradigm is conducted by three completely separate teams. The design team makes the design and does its formal verification, the implementation team just does the coding, and the test team makes the operational profile of the system and conducts the statistical usage testing. Cleanroom engineering will be described together with statistical usage testing in Chapter 5.

Before proceeding further to the introduction of the mainstream approach to requirements and analysis, which is based on UML, worth mentioning is that until recently, many opponents to this paradigm existed. Some ongoing doubts still exist as to if this is the correct choice. For example, in his article, "Use-Cases Are Not Requirements" (Meyer and Apfelbaum, 1999), Meyer argues that a better approach to requirements and analysis is transforming the functional requirements into the behavior model that takes the form of a finite state machine (FSM). He sees use cases as just walks across the FSM and claims it is possible to generate them automatically rather than writing them manually.

According to the methodology proposed by Meyer, after creating the behavior model, two parallel streams of activities are started. The first stream covers the analysis, the design, and the implementation, and yields the implementation. The second stream covers the operational profile and the performance analysis, as well as the automatic test suite generation. These two streams merge at the automated testing phase.

This approach is very similar to the one used in this book. A slight difference is that the latter promotes separation of concerns between design and implementation, and promotes test teams, including the models they make, very much like the Cleanroom engineering model does. Also, it gives more credit to the UML use cases. If we go back to the original ideas of the UML authors (Booch et al., 1998) and try to think of a single use case as a family of closely related collaborations among the same set of objects, clearly a use case really captures a part of the traditional **list of functional requirements**. Use cases help us group simple and closely related functional requirements, as will be illustrated by the examples in this chapter.

As already mentioned, use cases are the starting point of the software development in the unified software development process (Booch et al., 1998). The requirements model, essentially a set of use cases, is used to develop all the models that correspond to the engineering phases of the process, namely, the analysis model (result of the analysis phase), the design and deployment models (results of the design phase), the implementation model (result of the implementation phase), and the test model (result of the test preparation phase). The focus of this chapter is on requirements and analysis modeling.

The rest of the chapter is organized as follows: use case and collaboration diagrams are introduced in the next two sections. The last section of this chapter illustrates the requirements and analysis phases of communication protocol engineering by presenting the case of the session initiation protocol (SIP), RFC 3261 (Rosenberg et al., 2002). That last section is divided into three

Requirements and Analysis 13

subsections: SIP domain-specifics, the SIP requirements model, and the SIP analysis model.

2.1 Use Case Diagrams

Use case diagrams are special kinds of graphs whose vertices are connected with arcs. Two types of vertices are found in use case diagrams, namely, actors and use cases. The **actors** represent humans, machines, or software components that are the users of the software under development. They are rendered as stick figures. **Use cases** represent possible uses of the software under development and are rendered as ellipses. As already mentioned, we think of use cases as collaborations between the corresponding objects that constitute the part of the software under development. Clearly, they have different roles in the requirements and the analysis phases.

In the requirements phase, we concentrate on the functional requirements and use the use cases to capture them ("What must be done?"). At that time, how these requirements will be fulfilled does not matter. The only important concern is to build, together with the customer, a vision of the future system. This vision is expressed as a desirable behavior of the system and modeled by drawing the use case diagram and writing down the descriptions of the individual use cases as they are added to the diagram.

In other words, we concentrate on the client's perspective of the system. The requirements engineer tries to define the services that the system under development should provide. They also try to define an interface to these services. Later, the main problems that the requirements engineer must face are:

- Structuring the set of use cases by establishing the relationships among them
- Prioritizing the set of use cases by assigning different priorities to the individual use cases (especially important for the evolving systems)

Use cases have another role in the analysis phase. The job of the analyst is to realize the use cases by the corresponding collaborations between objects. The analyst reads the descriptions of the use cases and uses domain-specific knowledge to identify the individual objects (horizontal structuring) and to establish a hierarchy among them (vertical structuring). This process will be described in the next section.

Both actors and use cases are classifiers and, normally, they are connected by associations. The association between the actor and the use case shows the communication between the user and the part of the system modeled

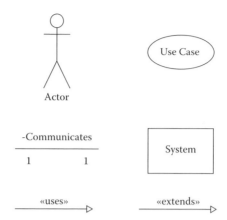

FIGURE 2.1
The basic set of graphical symbols available for rendering use case diagrams.

by the use case. Using associations enables us to indicate explicitly the points of connection between the users and the system.

Because both actors and use cases are classifiers, we can define general actors and general use cases and then specialize them using the generalization relationship. For example, we may specify the general actor *Client* and its specializations *SIP Client* and *H.323 Client* (Figure 2.2). Or, we can specify the general use case *Make a connection* and its specializations *Make a local connection* and *Make a long distance connection* (Figure 2.3).

Furthermore, while capturing the individual use cases, it may become obvious that a certain use case extends another use case or that a certain use case includes some other use cases. In such circumstances, the requirements engineer may structure the use cases using <<*extends*>> and <<*includes*>> stereotyped relationships. Especially important things can be indicated by using the sticky notes. Invariants, preconditions, and postconditions can be specified by the corresponding constraints. In more complex use case diagrams, we may need to indicate the packages and the interfaces.

Use case diagrams are normally rendered using the appropriate graphical tools, e.g., Microsoft® Visio. This tool provides the set of graphical symbols that are placed on the working sheet by the drag-and-drop paradigm. The basic set of graphical symbols is shown in Figure 2.1. The requirements engineer must specify the properties for each instance of a symbol in the drawing.

Five categories of actor properties are found: general information, table of attributes, table of operations, table of constraints, and tagged values. The general information includes name, full path, stereotype, visibility (private, protected, or public), and the indicators for *Root*, *Leaf*, and *Abstract* types of actors. The table of attributes includes columns for the attribute name, type, visibility, multiplicity (1, *, 0..1, 0..*, 1..1, or 1..*), and its initial value. The table of operations comprises columns for the operation name, return type, visibility, scope (classifier or instance), and the indicator for the polymorphic

Requirements and Analysis

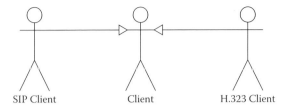

FIGURE 2.2
An example of the generalization and specialization of actors.

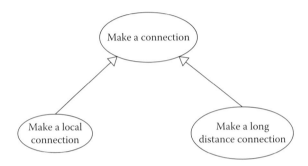

FIGURE 2.3
An example of structuring use cases.

operations. The table of constraints consists of four columns: the constraint name, stereotype (precondition, postcondition, or invariant), language type (OCL, text, pseudocode, or code), and body of the constraint. The tagged values include notes for the documentation, location, persistence, responsibility, and semantics.

A use case — being a classifier like an actor — has the same five categories of properties as the actor, as well as the additional sixth category. The sixth category of the use case properties contains the notes about the extension points that are used to describe the <<extends>> stereotyped relations.

An association between an actor and a use case has three categories of properties: general information about the association, table of constraints, and tagged values. The general information includes the association name, full path, stereotype, direction (none, forward, and backward), association end count (default 2), and the attributes for each end of the association. The attributes of the association end are its name, aggregation (none, composite, or shared), visibility, multiplicity, and navigability indicator (navigable or not). The graphical symbol *System* is used to show the system boundaries, i.e., to group the use cases that constitute the system under development. It has no properties.

All the relations between the use cases have three categories of properties: general information, table of constraints, and tagged values. The general information includes the relation name, full path, stereotype (extends,

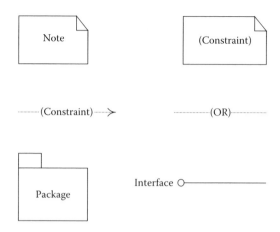

FIGURE 2.4
The additional graphical symbols available for rendering use case diagrams.

inherits, private, protected, subclass, subtype, or uses), and discriminator. The table of constraints is the same as the table of constraints for the actors and use cases. The tagged values are notes for the documentation.

The additional graphical symbols available for drawing use case diagrams are shown in Figure 2.4. These symbols include notes, general constraints, two-element constraints, OR constraints, packages, and interfaces. The notes have two categories of properties: general properties and tagged values. The general properties include the note name and its stereotype (none or requirement). The tagged values are notes for the documentation.

All the constraints, including general, two-element, and OR constraints, have the same categories of properties: general properties and tagged values. The general properties include the constraint name, full path, stereotype (precondition, postcondition, or invariant), language type (OCL, code, pseudocode, or text), and constraint body.

Four categories of package properties exist, including general properties, table of events, table of constraints, and tagged values. The general properties are the package name, full path, stereotype (facade, framework, stub, or system), visibility (private, protected, or public), and the indicators for *Root*, *Leaf*, and *Abstract* types of packages. The table of events contains an entry for each event. The attributes of individual events are the event name and event type (call event, signal event, change event, or time event). The table of constraints has the same format as the table of constraints for the actors, and use cases and tagged values are just the notes for the documentation.

The interface has four categories of properties, actually a subset of the actor properties. These are general properties, table of operations, table of constraints, and tagged values. All of them are the same as the corresponding actor properties.

The requirements engineer renders the use case diagram along as they talk to the customer about the desired behavior of the system to be developed.

Requirements and Analysis

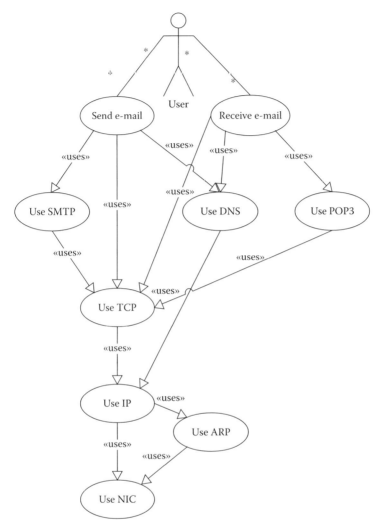

FIGURE 2.5
The use case diagram of the simple program for sending and receiving e-mails.

The use case diagram is intented as a medium to communicate the requirements between the customer and the system provider. Drawing use case diagrams is simple: the right graphical symbol is selected, dragged-and-dropped to the working sheet, the corresponding properties are filled in, and it is connected to the other symbols in the sheet.

As an illustration of a use case diagram, consider a simple program for sending and receiving electronic mail messages over the Internet. The use case diagram for such a program might look like the one shown in Figure 2.5. A single actor is found in this diagram, who is the user of the program (named *User*). On the highest level of abstraction, this program has two main use cases, *Send e-mail* and *Receive e-mail*.

Both of these highest-level use cases make use of the use cases *Use DNS* (Domain Name System) and *Use TCP* (Transmission Control Protocol). The DNS service provides the mapping of the e-mail server domain name into its IP (Internet Protocol) address. The TCP provides reliable data delivery service. Other than that, the use case *Send e-mail* uses the use case *Use SMTP* (Simple Mail Transfer Protocol) and the use case *Receive e-mail* uses the use case *Use POP3* (Post Office Protocol, Version 3). Normally, an e-mail client uses SMTP to send an e-mail message to the e-mail server. Similarly, a user uses POP3 to read the e-mail messages from their mailbox.

The use case *Use DNS* uses the use case *Use IP* to send a DNS requests to the DNS server and to receive DNS responses from it. The use case *Use TCP* uses the use case *Use IP* to send and receive segments of data and control information over the Internet. The use case *Use IP* uses the use case *Use ARP* (Address Resolution Protocol) to map the IP address of the destination host to its physical (e.g., Ethernet) address. Alternatively, the use case *Use IP* uses the use case *Use NIC* (Network Interface Controller) to send and receive IP datagrams over the Internet. Finally, the use case *Use ARP* uses the use case *Use NIC* to send an ARP request to the ARP server and to receive an ARP response from it.

This hierarchy of use cases actually follows the hierarchy of protocols in the TCP/IP protocol stack. As already mentioned, the concept of layered software architecture, which is traditionally explained by the ISO OSI, was actually invented to enable the separation of functions and the corresponding functional requests, which are referred to as *use cases* in UML.

After creating the skeleton of the use case model, the requirements engineer must fill in the descriptions of the individual use cases. The descriptions in this example are simplified for the sake of clarity. The description of the use case *Send e-mail* in plain text is the following:

Precondition:
```
The user has issued the send mail command.
```

Main flow of events:
```
Extract the recipient's e-mail address from the e-mail message header (defined by
the RFC 822).
Extract the e-mail server domain name from the recipient's e-mail address (string
after the character "@").
Use the use case Use DNS to map the server domain name into its IP address.
Use the use case Use TCP to open the TCP connection.
Use the use case Use SMTP to send the e-mail message to the e-mail server.
Use the use case Use TCP to close the TCP connection.
Prompt the user for the next command.
```

Exceptional flow of events:
```
The user may cancel the use case at any time by issuing the cancel command.
```

Requirements and Analysis

Exceptional flow of events:
```
If the use case Use SMTP indicates the problem in the mail delivery, this use case
should report it to the actor User.
```

The use case *Receive e-mail* is identical to the use case *Send e-mail* with the difference being that the former uses the use case *Use POP3* instead of the use case *Use SMTP*. The following description of the use case *Use DNS* is rather simple (actually, this is the description of the behavior of the DNS client):

Main flow of events:
```
Send the recursive DNS request by using Use IP.
Receive the DNS response by using Use IP.
```

The use case Use TCP is the active (initiator's) side of the TCP. It is defined as follows:

Main flow of events:
```
The procedure to open the TCP connection:
  Send SYN data segment.
  Receive SYN + ACK data segment.
  Send ACK data segment.
  Indicate that the connection is established.
The data transmission procedure:
  Send and receive the data segments using the sliding window.
The procedure to close the TCP connection:
  Send FIN data segment.
  Receive ACK data segment.
  Receive FIN + ACK data segment.
  Send ACK data segment.
  Indicate that the connection is closed both ways.
```

Exceptional flow of events:
```
The use case Send e-mail may close the TCP connection at any time.
```

The use case *Use SMTP* is actually the client side of the SMTP (defined by IETF RFC 821 and RFC 788) and can be described as follows (for simplicity, only one exceptional flow of events is given):

Main flow of events:
```
Receive the message 220 READY FOR MAIL.
Send the message HELLO.
Receive the message 250 OK.
Send the message MAIL FROM: <recipient's e-mail address>.
Receive the message 250 OK.
Send the message RCPT TO: <sender's e-mail address>.
Receive the message 250 OK.
Send the message DATA.
Receive the message 354 START MAIL INPUT.
Send the body of the e-mail message terminated with <CR><LF>.<CR><LF>.
Receive the message 250 OK.
```

```
Send the message QUIT.
Receive the message 221.
```

Exceptional flow of events:

```
If a use case receives the message 550 NO SUCH USER HERE, as a reply to its RCPT
TO: message, it indicates the problem to the use case Send e-mail.
```

The use case *Use POP3* is the client side of the POP3 protocol, similar to the use case *Use SMTP*. The use case *Use IP* is actually the IP protocol, which is described as follows:

Main flow of events:

```
The procedure that is used to receive the datagrams:
  Receive a datagram by using the Use NIC.
  Send the received datagram to the use case Use TCP.
The procedure that is used to send the datagrams:
  Decrement the contents of the time-to-live field of the IP datagram.
  Extract the destination IP address from the datagram header.
  Extract the destination network id from the destination IP address.
  If the destination network is local the network:
    Use the use case Use ARP to determine the physical address.
    Deliver the datagram by using the Use NIC.
  Else, route the datagram.
```

Exceptional flow of events:

```
If the datagram has been corrupted during the transmission, drop it.
```

Exceptional flow of events:

```
If the time-to-live field of the datagram counts down to 0, drop it.
```

The use case *Use ARP* is an ARP client and the use case *Use NIC* is a network card driver. The former is defined as follows:

Main flow of events:

```
Send an ARP request by using the use case Use NIC.
Receive the ARP response by using the use case Use NIC.
```

The example above, especially the use cases *Use TCP* and *Use SMTP*, should help the reader understand that a use case is a set of event sequences, not just a single sequence. To keep use cases simple, separating the main and the alternative flows of events is always desirable. Usually, we start by just writing the main flow of events for each use case and later refine them by adding the exceptional flow of events.

After this example, it should be clear that a use case captures the intended behavior of the part of the system (subsystem, class, or interface). Of course, after specifying the intended behavior, we must create a set of classes that work together to implement that behavior. The means of modeling both static and dynamic structures of the society of objects in UML are the collaboration diagrams.

2.2 Collaboration Diagrams

As already mentioned, we think of use cases as collaborations between objects. Actually, in UML we realize a use case as a collaboration of a set of objects. This concept can be explicitly shown in UML by connecting the use case with the corresponding collaboration using the realization relationship.

A **collaboration diagram** is a special kind of graph consisting of a set of vertices interconnected by a set of arcs. Basically, the vertices are the objects and the arcs are the links that carry the messages between the interconnected objects. Additional vertices and arcs are the notes and the constraints (general, two-element, and OR constraints).

Collaboration diagrams are normally rendered using the appropriate graphical tools, e.g., Microsoft Visio. This tool provides the set of graphical symbols that are placed on the working sheet by the drag-and-drop paradigm. The basic set of graphical symbols is shown in Figure 2.6. The engineer that renders the diagram must specify the properties for each instance of a symbol in the drawing.

Three categories of object properties exist: general properties, table of constraints, and tagged values. The general properties include the object name, full path, classifier name, and multiplicity. The table of constraints and the tagged values contain the same properties as the corresponding categories for the use cases (see the previous section of this chapter).

While adding objects to the collaboration diagram, we are forced to introduce the corresponding classifiers and to specify their properties (at least the classifiers names, for a start). The classifiers have eight categories of properties, including general properties, table of attributes, table of operations, table of receptions, table of template parameters, list of the components, table of constraints, and tagged values. The general properties, the

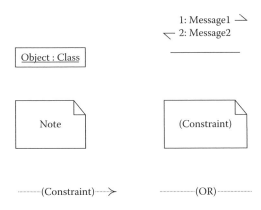

FIGURE 2.6
The set of graphical symbols available for rendering collaboration diagrams.

table of attributes, the table of operations, the table of constraints and the tagged values contain the same properties as the corresponding categories for the use cases (see the previous section of this chapter).

The table of receptions has five columns, which contain the reception name, signal name, visibility (private, protected, or public), polymorphic indicator (false or true), and scope (classifier or instance). The table of template parameters includes the columns for the parameter name and its type. The list of components is just a list of components that implement this class.

The links in collaboration diagrams have four categories of properties, including general properties, table of messages, table of constraints, and tagged values. The general properties are the link name, its full path, and the table of link ends roles, which has two columns, the end name and its stereotype (none, association, global, local, parameter, self). The table of link messages has four columns, including the message name, its direction (forward or backward), flow kind (procedure call, flat, or asynchronous), and sequence expression. The table of constraints contains the same properties as the corresponding category of object (and classifier) properties. The tagged values are just the notes for the documentation. The notes and the constraints have the same properties as in the use case diagrams (see the previous section of this chapter).

Most frequently, we model sequential flow of control with collaboration diagrams. In this case, a message sequence expression takes the simple form of a message sequence number. However, collaboration diagrams allow modeling of more complex flows, such as iteration and branching. Iteration is modeled by prefixing the message sequence number with the iteration expression

$$*[<control\ variable> := <start\ value>..<end\ value>]$$

$$e.g., *[j := 1..m].$$

Branching is modeled by prefixing the message sequence number with the condition clause [<condition>], e.g., [$i > 10$]. Alternate paths of the branch have the same message sequence number prefixed by the unique nonoverlapping condition, where the set of conditions must cover all the possibilities.

Next, we illustrate the use of collaboration diagrams in the example of a simple program for sending and receiving electronic mail messages over the Internet, which was introduced and modeled in the previous section of this chapter. The use case diagram for this program is shown in Figure 2.5. We start by making the real collaboration between objects that is a realization of the use case model, and continue with the study of virtual collaborations, which correspond to the peer-to-peer protocols present in this example.

To start, imagine that we are provided with the classifier FSM for modeling finite state machines. Clearly a single object of this class could be a realization of a single use case, as shown in Figure 2.5. The assumption that each use

Requirements and Analysis

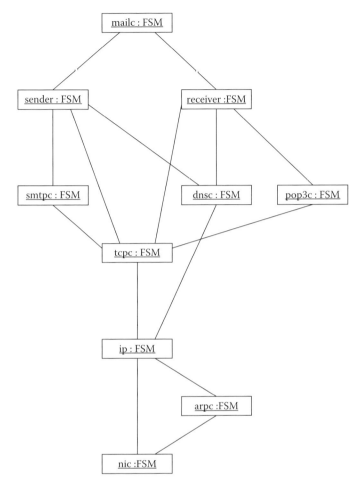

FIGURE 2.7
The collaboration diagram of the simple program for sending and receiving e-mails.

case is materialized by a single FSM object leads to a real collaboration between objects, shown in Figure 2.7.

In this class diagram, the object *mailc* (abbreviation for a mail client) is the <<boundary class>> object. All other objects are the <<control class>> objects. The e-mail message itself would be the <<entity object>>, but it is not shown in Figure 2.7. Obviously, the realization of the individual use cases is as follows:

- The object *sender* is a realization of the use case *Send e-mail*.
- The object *receiver* is a realization of the use case *Receive e-mail*.
- The object *dnsc* (abbreviation for a DNS Client) is a realization of the use case *Use DNS*.

- The object *tcpc* (abbreviation for a TCP Client, i.e., the side that initiates the establishment of the TCP connection) is a realization of the use case *Use TCP*.
- The object *smtpc* (abbreviation for an SMTP Client) is a realization of the use case *Use SMTP*.
- The object *pop3c* (abbreviation for a POP3 Client) is a realization of the use case *Use POP3*.
- The object *ip* is a realization of the use case *Use IP*.
- The object *arpc* (abbreviation for an ARP Client) is a realization of the use case *Use ARP*.
- The object *nic* is a realization of the use case *Use NIC*.

Figure 2.7 shows general collaboration among the relevant objects, i.e., it just shows the links between objects. Essentially, it shows the software architecture. We may think of it as a family of particular collaborations. For example, the user of the program might select the use case *Send e-mail* and this would lead to a particular collaboration, or the user might select the use case *Receive e-mail* and that would lead to another particular collaboration.

Another important thing to notice and remember is that Figure 2.7 shows only the objects of the system under development. In this case, it is a program that runs on a computer connected to the Internet over its network interface card. If we want the overall picture, we can also add the models of the systems with which our system under development would normally communicate. By adding the models of these external systems, we are modeling end-to-end collaborations.

The system under development communicates with external servers, including the ARP server, the DNS server, and the e-mail server. If we assume that all of these servers run on the same computer, the model of the external environment of the system under development is rather simple (Figure 2.8). The external objects are as follows:

- The object *smtps* is the SMTP server.
- The object *pop3s* is the POP3 server.
- The object *tcps* is the TCP server, i.e., the side that accepts the establishment of the TCP connection.
- The object *dnss* is the DNS server.
- The object *arps* is the ARP server.
- The object *ips* is an instance of IP.
- The object *nics* is an instance of NIC.

The overall collaboration that corresponds to the main flow of events of the use case *Send e-mail*, up to the point when the SMTP client receives the

Requirements and Analysis

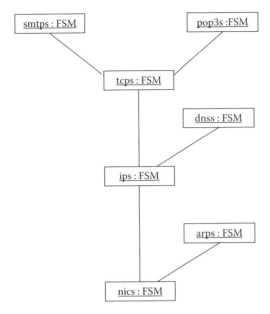

FIGURE 2.8
The collaboration diagram of the e-mail and DNS server.

message 220 READY FOR MAIL, is shown in Figure 2.9. The flow of events is as follows:

1: The object *mailc* sends the signal *sendMail(msg)* to the object *sender*. The signal parameter *msg* is the e-mail message itself.
2: The object *sender* sends the signal *domainToIP(domain)* to the object *dnsc*. The signal parameter *domain* is the domain name of the e-mail server.
3: The object *dnsc* sends the signal *dnsReq(domain)* to the object *ip*. The signal *dnsReq* is actually the DNS service request message.
4: The object *ip* sends the signal *data(dnsReq)* to the object *nic*. The general signal *data* is an IP datagram. Together with the parameter *dnsReq*, it represents the datagram carrying the DNS service request message.
5: The object *nic* sends the signal *frame(dnsReq)* to the object *nics*. The general signal *frame* is a data frame from the underlying physical network (e.g., Ethernet). The signal *frame(dnsReq)* is the data frame carrying the datagram that encapsulates the DNS service request message.
6: The object *nics* sends the signal *data(dnsReq)* to the object *ips*.
7: The object *ips* sends the signal *dnsReq(domain)* to the object *dnss*.

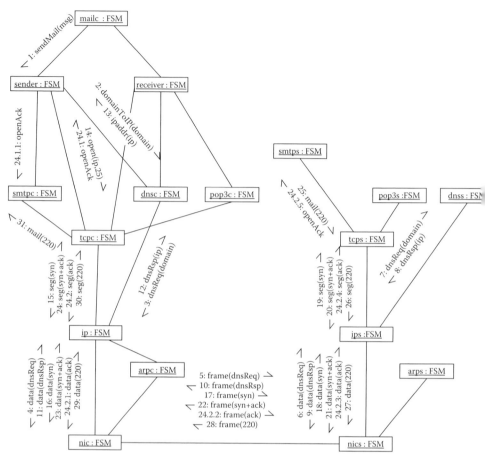

FIGURE 2.9
The overall real collaboration of the simple program for sending and receiving e-mails and its environment.

    8: The object *dnss* sends the signal *dnsRsp(ip)* to the object *ips*. The signal *dnsRsp* is the DNS service response message and its parameter *ip* is the IP address of the target e-mail server.
    9: The object *ips* sends the signal *data(dnsRsp)* to the object *nics*.
    10: The object *nics* sends the signal *frame(dnsRsp)* to the object *nic*.
    11: The object *nic* sends the signal *data(dnsRsp)* to the object *ip*.
    12: The object *ip* sends the signal *dnsRsp(ip)* to the object *dnsc*.
    13: The object *dnsc* sends the signal *ipaddr(ip)* to the object *sender*.
    14: The object *sender* sends the signal *open(ip,25)* to the object *tcpc*. The signal *open* is an active open request to TCP (TCP should send the SYN segment to initiate the TCP connection establishment procedure). Its parameters, *ip* and 25, are the IP address of the target e-mail

Requirements and Analysis

sever and the well-known TCP port number reserved for the SMTP, respectively.

15: The object *tcpc* sends the signal *seg(syn)* to the object *ip*. The general signal *seg* is a TCP segment. The signal *seg(syn)* is a SYN (synchronization) TCP segment (i.e., it has the SYN bit set in the code field).

16: The object *ip* sends the signal *data(syn)* to the object *nic*.

17: The object *nic* sends the signal *frame(syn)* to the object *nics*.

18: The object *nics* sends the signal *data(syn)* to the object *ips*.

19: The object *ips* sends the signal *seg(syn)* to the object *tcps*.

20: The object *tcps* sends the signal *seg(syn+ack)* to the object *ips*. The signal *seg(syn+ack)* is a SYN+ACK (synchronization and acknowledgment) TCP segment (i.e., it has both SYN and ACK bits set in the code field).

21: The object *ips* sends the signal *data(syn+ack)* to the object *nics*. The signal *data(syn+ack)* is the IP datagram that encapsulates the SYN+ACK TCP segment.

22: The object *nics* sends the signal *frame(syn+ack)* to the object *nic*. The signal *frame(syn+ack)* is the data frame carrying the IP datagram that encapsulates the SYN+ACK TCP segment.

23: The object *nic* sends the signal *data(syn+ack)* to the object *ip*.

24: The object *ip* sends the signal *seg(syn+ack)* to the object *tcpc*. (The event flow now forks into two parallel flows.)

 24.1: The object *tcpc* sends the signal *openAck* to the object *sender*. (The first flow begins here.)

  24.1.1: The object *sender* sends the signal *openAck* to the object *smtpc* (The first flow ends here.)

 24.2: The object *tcpc* sends the signal *seg(ack)* to the object *ip*. (The second flow begins here.)

  24.2.1: The object *ip* sends the signal *data(ack)* to the object *nic*.

  24.2.2: The object *nic* sends the signal *frame(ack)* to the object *nics*.

  24.2.3: The object *nics* sends the signal *data(ack)* to the object *ips*.

  24.2.4: The object *ips* sends the signal *seg(ack)* to the object *tcps*.

  24.2.5: The object *tcps* sends the signal *openAck* to the object *smtps*.

25: The object *smtps* sends the signal *mail(220)* to the object *tcps*. The general signal *mail* is the SMTP message. The particular signal *mail(220)* is actually the message 220 READY FOR MAIL, where the first three digits are mandatory and the rest of the message is a human-readable comment. (Note: We have restarted the message numbering here for brevity.)

26: The object *tcps* sends the signal *seg(220)* to the object *ips*.

27: The object *ips* sends the signal *data(220)* to the object *nics*.
28: The object *nics* sends the signal *frame(220)* to the object *nic*.
29: The object *nic* sends the signal *data(220)* to the object *ip*.
30: The object *ip* sends the signal *seg(220)* to the object *tcpc*.
31: The object *tcpc* sends the signal *mail(220)* to the object *smtpc*. (The example ends here.)

What we have just described is the real collaboration between objects within the system under development as well as with the relevant objects in its surroundings. The real collaboration for any nontrivial system could be rather complex. This behavior should be clear from the previous example, where we intentionally stopped at the certain point of the event flow, which was selected as a compromise between showing enough complexity and maintaining clarity.

The complete list of events for the use case *Send e-mail* is much longer than the one given above. For modeling the transfer of the rest of the SMTP messages (12 of them), we would need additional 84 (12 × 7) UML events, almost three times more than already in the list above. This complexity is why we try to break the system down into its parts and analyze them in detail later.

One important aspect of the simplification is the definition of the Application Programming Interfaces (API). For example, we may define the API between the *sender* and the hierarchically lower level objects (*dnsc*, *smtpc*, and *tcpc*), or the API between *tcpc* and *ip*, and so on. Other important items are the virtual collaborations that are governed by the peer-to-peer protocols. Consider for example the virtual collaboration between *dnsc* and *dnss* (Figure 2.10). The corresponding flow comprises only two events, *dnsReq(domain)* and *dnsRsp(ip)*.

The virtual collaboration between *tcpc* and *tcps* is governed by the TCP. It is slightly more complex and comprises the following flow of events (Figure 2.11):

1: The object *tcpc* sends the signal *seg(syn)* to the object *tcps*.
2: The object *tcps* sends the signal *seg(syn+ack)* to the object *tcpc*.
3: The object *tcpc* sends the signal *seg(ack)* to the object *tcps*.
4: The object *tcpc* sends the signal *seg(data)* to the object *tcps*. (Data transmission phase)

FIGURE 2.10
The virtual collaboration between the DNS client and the DNS server.

Requirements and Analysis

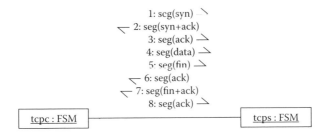

FIGURE 2.11
The virtual collaboration between two TCP entities.

5: The object *tcpc* sends the signal *seg(fin)* to the object *tcps*.
6: The object *tcps* sends the signal *seg(ack)* to the object *tcpc*.
7: The object *tcps* sends the signal *seg(fin+ack)* to the object *tcpc*.
8: The object *tcpc* sends the signal *seg(ack)* to the object *tcps*.

Finally, the virtual collaboration between *smtpc* and *smtps* (in accordance with SMTP) is of the same order of complexity (Figure 2.12; note that only the first eight events are shown in the figure). The corresponding flow of events is the following:

1: The object *smtps* sends the signal *mail(220)* to the object *smtpc*.
2: The object *smtpc* sends the signal *mail(HELO)* to the object *smtps*.
3: The object *smtps* sends the signal *mail(250_OK)* to the object *smtpc*.
4: The object *smtpc* sends the signal *mail(MAIL_FROM:)* to the object *smtps*.
5: The object *smtps* sends the signal *mail(250_OK)* to the object *smtpc*.
6: The object *smtpc* sends the signal *mail(RCPT_TO:)* to the object *smtps*.
7: The object *smtps* sends the signal *mail(250_OK)* to the object *smtpc*.
8: The object *smtpc* sends the signal *mail(DATA)* to the object *smtps*.
9: The object *smtps* sends the signal *mail(354_START_MAIL_INPUT)* to the object *smtpc*.

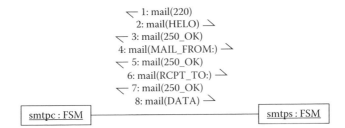

FIGURE 2.12
The virtual collaboration between the SMTP client and the SMTP server.

10: The object *smtpc* sends the signal *mail(MAIL_BODY)* to the object *smtps*.

11: The object *smtps* sends the signal *mail(250_OK)* to the object *smtpc*.

12: The object *smtpc* sends the signal *mail(QUIT)* to the object *smtps*.

13: The object *smtps* sends the signal *mail(221)* to the object *smtpc*.

2.3 Requirements and Analysis Example

This section of the book illustrates the requirements and analysis phases of communication protocol engineering with the example of a simple SIP softphone. Normally, the requirements phase starts by acquiring the relevant domain-specific knowledge and continues by the construction of the corresponding requirements model, which is the input for the analysis phase. As already mentioned, the output of the analysis phase is the corresponding analysis model. The next three sections cover a short overview of the domain-specific information, the requirements, and the analysis models of a simple SIP softphone.

2.3.1 SIP Domain Specifics

SIP is the application layer protocol used for creating, modifying, and terminating sessions, such as Internet telephone calls and multimedia distribution and conferences, with one or more participants. It has been standardized by the IETF RFC 3261 (Rosenberg et al., 2002) and related series of RFCs (RFC 3262, RFC 3263, RFC 3264, RFC 3265, RFC 3372, RFC 3428, RFC 3485, RFC 3487, and others). In contrast to the ITU-T H.323 family of protocols — which provide the whole protocol stack for multimedia communications — SIP is just the control and signaling component on the top of the multimedia architecture.

Aside from SIP, the multimedia architecture will typically include RTP (Real-Time Transfer Protocol, RFC 1889), RTSP (Real-Time Streaming Protocol, RFC 2326), MEGACO (Media Gateway Contol Protocol, RFC 3015), and SDP (Session Description Protocol, RFC 2327). SIP does not provide any service on its own. Instead of full services, it provides primitives for the services that are implemented in the overall architecture. These primitives are based on an HTTP-like (Hyper Text Transport Protocol) request and response transaction model.

The main SIP abstractions are the session, the dialog, and the transaction. A multimedia session is a set of multimedia senders and receivers, as well as data streams flowing from senders to receivers. A dialog is a peer-to-peer relationship between two user agents (end points in the communication) that persists for some time. A transaction is the collaboration between the client and the server, which comprises all the messages from the first request sent from the client to the server up to the final response sent from the server to the client. The requests are processed automatically, meaning that either all requested actions are conducted, if the request has been accepted, or none of the actions are conducted, if the request has not been accepted.

Two main transaction types exist, referred to as invite (officially written in capital letters, i.e., INVITE) and non-invite (or, more formally, non-INVITE) transactions. An invite transaction is a three-way handshake comprising the request, the response, and the acknowledgment. In contrast, a non-invite transaction is the two-way handshake starting with the request and ending with the corresponding response.

Notice that the roles of the user agents (communication end points) are not fixed, and they change on the transaction by transaction bases. The user agent that creates a new request becomes a user agent client (UAC), whereas the user agent that receives the request becomes the user agent server (UAS). Another important detail is that a new transaction (either invite or non-invite) may not be started while an invite transaction is in progress. Alternatively, a new invite transaction may be started while a non-invite transaction is in progress.

Besides user agents, the SIP standard defines three types of SIP servers, namely, the proxy server (stateful or stateless), the registrar, and the redirect server. A proxy server is the mediator that helps end points set up the session. Officially, it is an intermediary entity that acts as both a server and a client for the purpose of making requests on behalf of other clients. A registrar is a server that supports the registration of the user agents by maintaining the corresponding database for the domain it handles. This database is referred to as a location service. These two types of servers are most frequently collocated in the same physical machine. A redirect server can be viewed as a proxy server with limited capabilities. It is only capable of directing the client to contact an alternate set of Uniform Resource Identifications (URI).

Requests and responses between a server and a client are sent as SIP messages. The SIP message comprises the start line, one or more header fields, empty lines (carriage-return line-feed sequences, CRLF), and an optional message body. The start line is different in requests and in responses. In the former case, it is referred to as a request line, and in the latter as a status line. The request line comprises the method name (six methods are available in SIP: REGISTER, INVITE, ACK, CANCEL, BYE, and OPTIONS), the request URI, and the SIP version (currently "SIP/2.0"). The status line comprises the SIP version, the status code (a three-digit integer result code), and the reason phrase (textual status description).

The SIP protocol stack comprises four layers. Starting from the top and going down the hierarchy, these are the transaction user (TU) layer, the transaction layer, the transport layer, and the syntax and encoding of SIP messages. A transaction user is any SIP entity (client or server) except for the stateless proxy. The transaction layer supports transactions, which are the key component of SIP. The transport layer provides for the transfer of SIP messages across the Internet. SIP may use three types of transport services, including unreliable (UDP), reliable (TCP), and encrypted (Transport Layer Security, TLS) transport service. Much of SIP message and header field syntax is identical to HTTP/1.1. Although SIP is close to the HTTP philosophy, it is not an extension of HTTP.

As mentioned above, the SIP standard specifies six methods, including REGISTER for registering contact information, INVITE, ACK, and CANCEL for setting up sessions, BYE for terminating sessions, and OPTIONS for querying servers regarding their capabilities. Any INVITE after the initial invite to the same destination is called re-INVITE and is used for modifying the session and dialog parameters. The method INVITE starts the invite transaction; all other methods start non-invite transactions. Interestingly enough, six status code types are also found, depending on the value of status code first digit, as follows:

1xx: Provisional (the request has been received and its processing has been started)

2xx: Success (the request has been successfully processed)

3xx: Redirection (further action by the client is needed)

4xx: Client error (the request contains an error or it may not have been fulfilled on this server)

5xx: Server error (the request is valid, but the server failed to fulfill it)

6xx: Global failure (the request cannot be fulfilled on any server)

As an example, consider the typical scenario of the SIP session setup in Figure 2.13. (Note: This figure is actually a UML sequence diagram. Sequence diagrams are intentionally introduced later in the next chapter. For the moment, it is enough to assume that the rectangular symbols are the communicating entities and that the arrows are the messages they exchange. Time advances downwards.). Two user agents *ua1* and *ua2*, together with their corresponding proxy servers *p1* and *p2*, constitute the SIP trapezoid (imagine the trapezoid by "drawing" the lines that connect *ua1*, *p1*, *p2*, and *ua2*).

Suppose that *ua1* wants to set up a session with *ua2*. It starts by sending an invite request to the proxy server that is responsible for its domain, and that is *p1*. Proxy *p1* locates the proxy server responsible for the destination *ua2*, namely *p2*, and forwards the invite request to it. At the same time, *p1* sends back the response 100 TRYING to *ua1*. Proxy *p2* locates the destination user agent, *ua2*, forwards the invite request to it, and sends back the response 100 TRYING to the proxy *p1*. *ua2* receives the invite request and sends back the response 180 RINGING, which is forwarded by the proxies *p2* and *p1* to *ua1*.

Requirements and Analysis

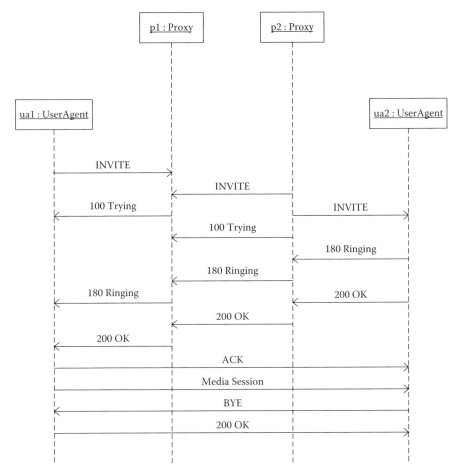

FIGURE 2.13
The example of SIP session setup (with SIP trapezoid).

At this point, *ua2* indicates the incoming invite request to its user. The user accepts the request and *ua2* sends back the response 200 OK, which is forwarded by the proxies *p2* and *p1* to *ua1*. The dialog between *ua1* and *ua2* is successfully established. Further on, *ua1* sends the ACK request to *ua2* directly (the end of the three-way handshake). The session is successfully established at this point. The communicating user agents may now exchange the media.

The session may be terminated by either *ua1* or *ua2*. Suppose that *ua2* wants to terminate the session. It sends the BYE request to *ua1* directly, which in its turn sends back the response 200 OK. The session is successfully closed. This is an example of the non-invite transaction.

This simplified explanation hides one rather important aspect of the invite three-way handshake, and that is the application of the offer-answer procedure. This procedure is used by *ua1* and *ua2* to determine the session parameters in accordance with SDP. The first offer must be carried either by the invite request or by the response 200 OK. If the offer is carried by the invite request (*ua1* makes the first offer), the answer must be included in the response 200 OK. If the offer is carried by the response 200 OK (*ua2* makes the first offer), the answer must be included in the ACK request (the last message in the three-way handshake). The session is successfully established only after the offer-answer procedure is successfully ended.

2.3.2 SIP Softphone Requirements Model

SIP softphone is the application that normally runs on some computer — for example, a desktop PC — and enables its user to set up multimedia sessions and to communicate with other SIP users or entities over the Internet. Such an application would typically use some type of graphical user interface (GUI) and device drivers for the sound card and the web camera, typically provided by the local operating system (out of scope for this book) and, of course, the SIP protocol stack.

This section shows how to construct the requirements model for the SIP protocol stack in a simple SIP softphone. As mentioned previously, the SIP protocol stack comprises the transaction user layer, the transaction layer, and the transport layer. In terms of use cases, the user uses the application (softphone), which in turn uses both the transaction layer and the transport layer. The transaction layer also uses the transport layer. The use case diagram shown in Figure 2.14 is a simple requirements model that captures these relations.

We can refine this simple model by taking into account the details of the individual layers of the SIP protocol stack. To start, the transaction user (TU) layer dynamically creates and uses the user agent clients (UAC) and the user agent servers (UAS) entities to support outgoing and incoming invite requests. Both UAC and UAS use the transaction layer (TAL), as well as the transport layer, which is accessible through the transport layer interface (TLI). TAL and TLI are abbreviations introduced here (they have not been taken from the RFC 3261).

Similar to TU, TAL dynamically creates and uses invite client transactions (INVITE CT), non-invite client transactions (non-INVITE CT), invite server transactions (INVITE ST), and non-invite server transactions (non-INVITE ST). TAL and all transactions use TLI, but they are all also used by TU. Finally, TLI uses UDP, TCP, or TLS. The detailed use case diagram of the simple SIP softphone is shown in Figure 2.15.

Before proceeding further, two important points must be emphasized. The first is that the direct relations between TU and TLI are strictly in accordance with the RFC 3261, although this may seem to be an error because it violates the ISO OSI ideal of a strictly layered architecture (no direct communication between layer $i + 1$ and layer i). The second point is that the relations between TU and transactions, and transactions and TLI, are not prescribed by the

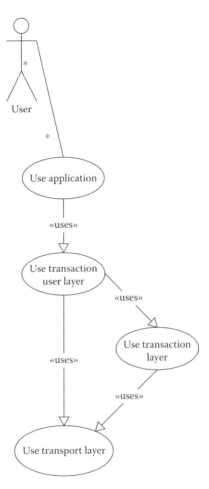

FIGURE 2.14
The use case diagram of the simple SIP softphone.

RFC 3261 but they are also not forbidden. These relations are introduced to minimize the message paths at the expense of the increased relations complexity.

To complete the requirements model, we need to describe the individual use cases. The use case *Use application* is actually the main program that interacts with the user and makes use of the SIP protocol stack and is out of the scope of this book. The use case *Use TU* is responsible for dispatching TU messages (coming from the application and the lower layers and going to the user agent clients and servers and to the application), as well as for dynamic creation of user agent clients and servers.

The use case *Use UAC* provides a set of procedures for the client side of the transactions. The high-level description of these procedures follows:

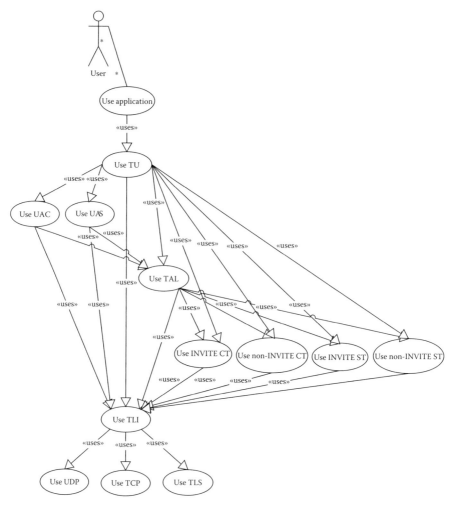

FIGURE 2.15
The detailed use case diagram of the simple SIP softphone.

Main flow of events:

```
Receive the request from the application.
Dispatch it to the corresponding procedure.
Registration procedure:
  Create and send REGISTER request.
  Receive the response.
  Indicate the response to the application.
Session setup procedure:
  Create and send INVITE request.
  Receive provisional responses (1xx), if any.
  Receive the final response (not 1xx).
  Indicate the final response to the application.
  If the final response is 2xx,
    Send ACK request.
```

Requirements and Analysis

```
Cancel session setup procedure:
  If the final response has not been received,
    Create and send CANCEL request.
    Receive the response.
    Indicate the response to the application.
Modify session/dialog procedure:
  Perform session setup procedure.
Query server capabilities procedure:
  Create and send OPTIONS request.
  Receive the response.
  Indicate the response to the application.
Terminate session procedure:
  Create and send BYE request.
  Receive the response.
  Indicate the response to the application.
```

The use case above includes only the main flow of events. A more detailed version would also include the exceptional flow of events that would describe the time management and the retransmissions of the unacknowledged SIP messages. These are skipped here for brevity (in reality, we also start from a very simple version of use cases and refine them later). The same is true for all the other use cases given in this subsection.

The use case *Use UAS* provides the set of procedures for the server side of the transactions. The high-level description of these procedures follows (the implementation is rather simple and it takes the passive and goodwill approach):

Main flow of events:

```
Receive the request from the TU dispatcher (i.e., remote SIP entity).
Dispatch it to the corresponding procedure.
Session setup service procedure:
  Receive the incoming INVITE request.
  Indicate INVITE request to the application.
  Send the provisional response, e.g., 180 RINGING.
  If the user accepts the call,
    Send the final response 200 OK.
    Receive ACK request.
Cancel session setup service procedure:
  Receive CANCEL request.
  Send the final response 200 OK.
  Report the outcome to the application.
Modify session/dialog service procedure:
  Receive INVITE request.
  Send the final response 200 OK.
  Report the outcome to the application.
Query server capabilities service procedure:
  Receive OPTIONS request.
  Send the final response 200 OK.
  Report the outcome to the application.
Terminate session service procedure:
  Receive BYE request.
  Send the final response 200 OK.
  Report the outcome to the application.
```

The use case *Use TAL* is responsible for dispatching TAL messages (coming from TU, UAC, UAS, and TLI and going to the TAL transactions), as well as for dynamic creation of TAL transactions. The use case *Use INVITE CT* is an invite client transaction. Its description follows:

Main flow of events:
```
Receive INVITE request from TAL.
Forward INVITE request to TLI.
Receive 1xx response from TAL.
Forward 1xx response to TU.
Receive the final response from TAL.
Forward the final response to TU.
If the final response is 3xx-6xx,
  Send ACK request to TLI.
```

The use case *Use INVITE ST* is an invite server transaction. Its description follows:

Main flow of events:
```
Receive INVITE request from TAL.
Forward INVITE request to TU.
Receive 1xx response from TAL.
Forward 1xx response to TLI.
Receive the final response from TAL.
Forward the final response to TLI.
```

The use case *Use non-INVITE CT* is a non-invite client transaction, whose description follows:

Main flow of events:
```
Receive the request from TAL.
Forward the request to TLI.
Receive the response from TAL.
Forward the response to TU.
```

The use case *Use non-INVITE ST* is a non-invite server transaction, which is defined as follows.:

Main flow of events:
```
Receive the request from TAL.
Forward the request to TU.
Receive the response from TAL.
Forward the response to TLI.
```

The use case *Use TLI* is responsible for dispatching transport messages. It routes the requests from upper layers toward its remote peer in a forward direction, and routes the responses received from its remote peer toward the upper layers in a backward direction (non-ACK responses are sent to TAL, whereas ACK responses are sent to TU). It may use UDP, TCP, or TLS for the communication with its peers over the Internet. The description of this use case follows:

Requirements and Analysis 39

Main flow of events:
```
Receive a request from upper layers.
Send the request to the remote peer.
Receive the response from the remote peer.
If the response is ACK,
  Send it to TU,
Else,
  Send it to TAL.
```

Now that we have completed the use case diagram, we can proceed to the next engineering phase. This phase is the analysis, whose main goal is the definition of the software architecture.

2.3.3 SIP Softphone Analysis Model

Generally, the analysis model is constructed by defining the collaboration in a set of objects for each use case in the source requirements model. This process becomes obvious when considering the rough use case diagram shown in Figure 2.14. However, by refining the use cases, we may reach a point when a single class can realize a single use case. Figure 2.15 is an example of exactly such a use case diagram. Each use case is rather simple, so that a single class can realize it. Along this approach, assume the following mapping:

- The instance of the class *FSM* named *app* realizes the use case *Use application*.
- The instance of the class *TUDisp* named *tud* realizes the use case *Use TU*.
- An unnamed instance of the class *UAClient* realizes the use case *Use UAC*.
- An unnamed instance of the class *UAServer* realizes the use case *Use UAS*.
- The instance of the class *TALDisp* named *tald* realizes the use case *Use TAL*.
- An unnamed instance of the class *InClientT* realizes the use case *Use INVITE CT*.
- An unnamed instance of the class *NIClientT* realizes the use case *Use non-INVITE CT*.
- An unnamed instance of the class *InServerT* realizes the use case *Use INVITE ST*.
- An unnamed instance of the class *NIServerT* realizes the use case *Use non-INVITE ST*.

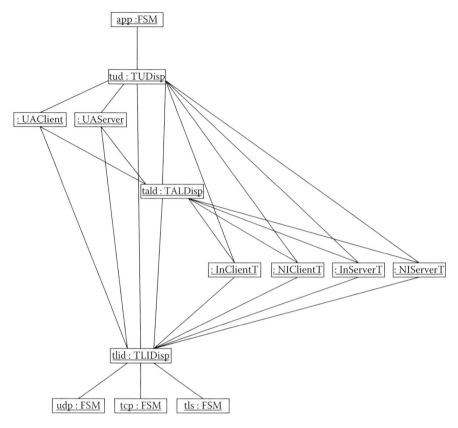

FIGURE 2.16
The general collaboration diagram of the simple SIP softphone.

- The instance of the class *TLIDisp* named *tlid* realizes the use case *Use TLI*.
- The instance of the class *FSM* named *udp* realizes the use case *Use UDP*.
- The instance of the class *FSM* named *tcp* realizes the use case *Use TCP*.
- The instance of the class *FSM* named *tls* realizes the use case *Use TLS*.

The mapping given above translates the use case diagram shown in Figure 2.15 into the general collaboration diagram shown in Figure 2.16. This diagram actually shows the software architecture, which defines the software objects that constitute the software system or product and the associations among them.

Requirements and Analysis

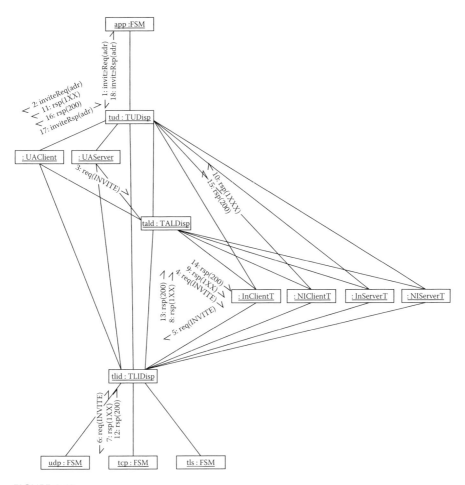

FIGURE 2.17
The collaboration diagram showing the part of the SIP session setup.

The software architecture can be used for the further study of particular object collaborations to check if the architecture is feasible and, if not, to refine the use case or collaboration diagram. An example of a particular collaboration is shown in Figure 2.17. This diagram shows the handling of the invite request initiated by the softphone user. The flow of events is as follows:

1: The object *app* sends the event *inviteReq(adr)* to the object *tud*.

2: The object *tud* sends the event *inviteReq(adr)* to an unnamed instance of the class *UAClient*.

3: The unnamed instance of the class *UAClient* sends the event *req(IN-VITE)* to the object *tald*.

4: The object *tald* sends the event *req(INVITE)* to an unnamed instance of the class *IClientT*.
5: The unnamed instance of the class *IClientT* sends the event *req(INVITE)* to the object *tlid*.
6: The object *tlid* sends the event *req(INVITE)* to its peer over the object *tcp*.
7: The object *tlid* receives the event *rsp(1xx)* from its peer over the object *tcp*.
8: The object *tlid* sends the event *rsp(1xx)* to the object *tald*.
9: The object *tald* sends the event *rsp(1xx)* to an unnamed instance of the class *IClientT*.
10: The unnamed instance of the class *IClientT* sends the even *rsp(1xx)* to the object *tud*.
11: The object *tud* sends the event *rsp(1xx)* to an unnamed instance of the class *UAClient*.
12: The object *tlid* receives the event *rsp(200)* from its peer over the object *tcp*.
13: The object *tlid* sends the event *rsp(200)* to the object *tald*.
14: The object *tald* sends the event *rsp(200)* to an unnamed instance of the class *IClientT*.
15: The unnamed instance of the class *IClientT* sends the event *rsp(200)* to the object *tud*.
16: The object *tud* sends the event *rsp(200)* to an unnamed instance of the class *UAClient*.
17: The unnamed instance of the class *UAClient* sends the event *inviteRsp(adr)* to the object *tud*.
18: The object *tud* sends the event *inviteRsp(adr)* to the object *app*.

Generally, *req()* and *rsp()* designate SIP requests and SIP responses in the flow of events shown above. For example, *req(INVITE)* is the SIP invite request, *rsp(1xx)* is the SIP provisional response, and *rsp(200)* is the SIP final response.

References

Booch, G., Rumbaugh, J., and Jacobson, I., *The Unified Modeling Language User Guide*, Addison-Wesley, Reading, MA, 1998.
Booch, G., Rumbaugh, J., and Jacobson, I., *The Unified Software Development Process*, Addison-Wesley, Reading, MA, 1998.
Broekman, B. and Notenboom, E., *Testing Embedded Software*, Addison-Wesley, London, 2003.

Meyer, S. and Apfelbaum, L., "Use Cases Are Not Requirements," http://www.geocities.com/model_based_testing/online_papers.htm, 1999.

Rosenberg, J. et al., "RFC 3261 – SIP: Session Initiation Protocol," http://www.faqs.org/rfcs/rfc3261.html, 2002.

3
Design

System **design** is a phase in engineering work that follows the system requirements and analysis phases. Its main goal is to synthesize a complete solution based on the result of the analysis phase (obtaining the analysis model of the system), which is actually a rough architecture — a skeleton — of the system. We can imagine the system synthesis as a process of creating the body of the system. This body is a reflection of the details related to the system structure and its behavior.

Note that the complete solution of the system mentioned above is not the system itself, but rather a detailed vision of the system that comprises all the details sufficient to construct the system. Technically, we refer to this vision as a **design model**. Therefore, the system synthesis is a process that takes an analysis model as its input and produces the design model as its output.

The design model defines the two most important system aspects:

- System structure
- System behavior

The **system structure** defines the elements of the system and their associations. Sometimes it is referred to as the **static structure** because it defines the static view of the system, i.e., a view without any respect to time. The system behavior defines the outputs of the systems as functions of time or their inputs. In the case of a family of communication protocols, which are most frequently modeled as groups of finite state machines (automata), the static structure defines the automata and the links between them whereas the system behavior defines the state transitions for the individual automata and the external messages.

Besides system synthesis, or system design, the communication protocol design phase described in this book includes two additional designs, namely **deployment design** and **test design**, which result in a **deployment model** and a **test model**, respectively. The main goals of the deployment design are identifying network nodes and configurations as well as identifying design subsystems and interfaces. The deployment model is especially important for the complex communication systems comprising many distributed components. For less complex systems, it is not as important, and for very simple systems it may not even be necessary.

Although the system design and deployment models make the complete vision of the system, they do not specify how the system can be verified. Therefore, the engineers conduct the test design by taking the requirements and design models and creating a test model. The test model actually defines the behavior of the testers, who emulate the environment of the system. As already mentioned in the previous chapter, the test model is most frequently referred to as a test suite, which comprises a set of test cases. Each test case specifies a series of test input values (events and messages) to the system and the corresponding output values (events and messages) that are expected at the system output as the results of correct system reactions to the given series.

To summarize, a communication protocol design is a process that takes the requirements and analysis as its input and provides the following models as its output:

- System design model
- System deployment model
- System test model

The means of making these models today are UML diagrams or some domain-specific languages, which are introduced in this chapter. The design engineer starts from the analysis model, essentially a collaboration of <<boundary>>, <<control>>, and <<entity>> classes, described in the corresponding collaboration diagram. The development model is made by mapping each class from the analysis model to a set of new classes in the development model. If the analysis model is well refined, this might even be a one-to-one mapping or close to it. For example, the analysis model of the SIP softphone given at the end of the previous chapter is detailed enough, and the corresponding collaboration diagram is a good base for the refinements that must be made during the system design phase.

The means of defining the static structure of the system in UML are class diagrams and object diagrams. A **class diagram** shows the design classes and the static relations (dependencies, associations, and generalizations) among them without any respect to time. It shows important details about classes, such as their members, fields and functions, and furthermore their types, visibility, and so on. The **object diagram** is similar to the class diagram except that it shows the system frozen at a certain moment of time. Typically, the object diagram will show system objects (class instances) with the characteristic and important values of certain field members.

The means of gathering and refining details about the system behavior are the UML interaction diagrams. Two types of interaction diagrams are found, namely collaboration diagrams (introduced in the previous chapter) and **sequence diagrams**. Collaboration diagrams show the interaction organized by the architecture, meaning that their focus is an architectural view of the system. The architecture is adorned by the flow of events. The sequence of evens is shown by adding sequence numbers as prefix labels to the events.

Alternately, sequence diagrams show system interactions from a time progress perspective. The top of the sequence diagram shows the objects of the system without static relations among them. Each object is represented further by a vertical line rendered from its bottom toward the bottom of the diagram. Time advances in the same direction. The interaction itself is shown by the series of events and messages sent among the objects, which are rendered by horizontal arrows from the source object's line to the destination object's line.

The means of specifying complete system behavior are **activity diagrams** and **statechart diagrams** or, more briefly, **statecharts**. An activity diagram shows the action or activity states starting from the initial and ending in the final state. State transitions can be sequential, branching, or concurrent (through forking and joining). The activity diagram is essentially a flowchart that emphasizes the activity that takes place over time, similar to Pert Charts.

Statecharts are the means of specifying finite state machines in UML. They are a type of advanced state transition graphs. A statechart shows simple and composite states starting from the initial and ending in the final state. The composite states are a means to organize states hierarchically. The state transitions can be guarded by conditions and they can indicate firing events and the corresponding actions.

The main goal of the deployment design is the decomposition of the system in two dimensions. Horizontally, the system is partitioned into parts that are deployed onto different network nodes. The term used for nodes by ISO OSI is open systems. Vertically, the system is partitioned into layers. Typical layers recognized by the USDP are the following:

- Application-specific layer
- Application-general layer (e.g., packages common for a set of applications)
- Middleware layer (e.g., Java VM and Java packages)
- System-software layer (e.g., TCP/IP protocol stack)

Furthermore, the system-software layer is generically partitioned by ISO OSI into the following seven layers:

- Application layer
- Presentation layer
- Session layer
- Transport layer
- Network layer
- Data link layer

- Physical layer

Another way to vertically partition is in accordance with the TCP/IP Internet layers, as follows:

- Application layer
- Transport layer
- Network layer
- Network interface layer

In the context of operating systems, we can think of layers as processes. Logically, each process has its own program and the processor that executes it but, in reality, some of the processes may share the program or the processor. The processes sharing the same program are referred to as threads. The processes sharing the same processor constitute the multiprogramming set.

The layers do not exist for themselves — rather, they are typically created to service the requests issued by the upper layers. When the number of requests increases, the engineers face the scalability problem, which can be solved by deploying the same layer on more processors. If the layers are the instances of the same class, we refer to them as replicas. Alternately, on multiprocessor systems with common memory, it might be possible for these layers to share the same program.

The deployment of horizontal system partitions onto different processors or computers is used rather frequently by system designers. Examples include the client-server architecture, the multi-tier architecture, and others. This convenience is why most engineers think of it in the first place when deployment issues are raised. However, the deployment of vertical system partitions onto various processors is also possible. A typical example is the Bluetooth Host Controller Interface (HCI), which is a demarcation line between the host processor that executes the upper layers and the Bluetooth link controller (a microprocessor, a microcontroller, or a digital signal processor) that executes the lower layers.

Horizontal and vertical system partitioning are typically conducted as two interactive activities. The designer typically partitions the horizontal system by rendering the deployment diagram, which shows the network nodes, links between them, and the subsystems deployed on individual nodes. Alternately, vertical partitioning — sometimes referred to as subsystem modeling — results in a class diagram that shows just the subsystems (packages), hierarchically organized in layers, and the dependencies among the subsystems. These two diagrams can be combined in the overall deployment diagram, which shows both the hierarchy and the deployment.

Another important design goal is identifying and providing generic design mechanisms that handle common requirements. The generic design mechanisms can be provided as design classes, collaborations, or subsystems. The

Design 49

examples of the generic design mechanisms in the communication protocol engineering are:

- Protocol (finite state machine, automata) state transition management
- Buffer management
- Timer management
- Message management

These mechanisms are common for all communication protocols. Typically, they are designed and implemented once as a separate subsystem that comprises the set of classes, which is then used and refined on a series of projects. In this book, we will use one such subsystem, entitled the FSM library (see Chapter 6). The design and the implementation of such a library are rather specific and rests more in the domain of operating systems. Additionally, such a library would most frequently already exist and the designers would just use the mechanisms that it provides. Because of these two reasons, we intentionally postpone presenting the FSM library details for the next chapter.

By accepting this approach, we keep the focus on the activities that are normally conducted during the design phase. We just assume that somebody has written the FSM library that provides all the necessary mechanisms (state transition, buffer, timer, and message management) and we concentrate on the design based on these mechanisms. Therefore, for a moment we should simply think of the FSM library as an infrastructure that facilitates the design and implementation of communication protocols.

Going back to the system design itself, this chapter will cover two additional domain-specific languages that have been in use much before UML and are still rather popular today, namely SDL and MSC. The SDL diagrams are semantically equivalent to the UML activity diagrams and statecharts. In principle, establishing a one-to-one mapping between them should not be a problem. The SDL diagram, like the UML activity diagram and statechart, specifies the complete system behavior.

The SDL diagram shows states and state transitions starting from the initial state and ending in the final state. The state transitions are rendered in a style of flowcharts. Each state transition starts with an input message that causes the transition. Typically, a state transition processes the received message and optionally sends the consequent messages.

The MSC chart is semantically equivalent to the UML interaction diagrams, i.e., to both collaboration and sequence diagrams. In fact, the MSC chart can be one-to-one translated to the UML sequence diagram, but the opposite is not the case. By looking at both of them, they make the same impression. Most engineers have the impression that they are almost the same, with the MSC being a little less expressive. Like the UML sequence diagrams, the MSC chart shows the objects that communicate — together with their corresponding vertical lines — and the messages they exchange, which are rendered as horizontal arrows connecting the source and the destination vertical lines.

Finally, this chapter covers the third domain-specific language, TTCN, which is used for making test models more formal than in UML. In contrast to the UML test model, which is rather descriptive and more like a general framework, TTCN is a well-defined language for defining test suites. As already mentioned, it originates from the ISO and has been traditionally used for the conformance testing of communication protocols.

TTCN, much like the higher-level programming language, has built-in types and allows a user to define new types (simple and structured), variables, constraints, and functions in specialized tables. The essence of the TTCN test case specification is an indented tree of events that is filled in a table, which specifies the behavior of the testers that run the test case and the outcomes of the test case (pass, fail, or inconclusive).

The next sections describe the class diagrams, the object diagrams, the sequence diagrams, the activity diagrams, the statechart diagrams, the deployment diagrams, the SDL diagrams, the MSC charts, and the TTCN tables. The chapter ends with a series of design examples.

3.1 Class Diagrams

A class diagram is a special type of graph that consists of a set of vertices interconnected by arcs. They are so popular and widely used that most of the newcomers to UML equate the UML and the class diagrams. Normally, we use the class diagrams to model the static design view of the system. More precisely, we typically use them to model the vocabulary of the system, collaborations, or database schemas.

A vocabulary of the system is a set of abstractions that are parts of the system. A collaboration is a group of classes, interfaces, and other elements that cooperate to provide a more complex functionality. A schema is a blueprint that is used for the conceptual design of a database. In communication protocol engineering, we rarely deal with real databases, but we frequently need to design at least a couple of persistent objects that hold the system configuration or similar information.

The basic class diagram vertices are classes, interfaces, and collaborations. These are interconnected with three types of arcs, with dependency, generalization, and association relations. To keep the size of the class diagrams manageable, we typically render smaller collaborations that describe certain aspects of the system. If we want to put those collaborations in a larger context, we can render the surrounding packages or subsystems. Both packages and subsystems enable hierarchical organization of class diagrams. For

Design

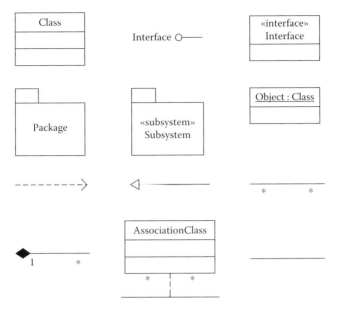

FIGURE 3.1
The basic set of graphical symbols available for rendering class diagrams.

example, we will render the FSM library as a package that is used by the protocols that are the subjects of design and implementation.

We use packages and subsystems to manage complexity. Alternately, we render class instances (objects) in class diagrams to manage ambiguity, especially when we want to explicitly show the dynamic type of an instance or some other hidden details of the system. A special type of class diagrams are object diagrams, which will be described in the next section of this chapter.

Like use case and collaboration diagrams described in the previous chapter, class diagrams are normally also rendered using some of the commercially available graphical tools, e.g., Microsoft Visio. The same is true for other UML diagrams described in this chapter. The basic set of graphical symbols available for rendering class diagrams is shown in Figure 3.1. The design engineer must specify properties for each instance of a symbol in the drawing.

The most frequently used symbol in class diagrams is the class symbol. Eight categories of class properties exist: the general information, the table of attributes, the table of receptions, the table of parameters, the list of components, the table of constraints, and the tagged values. The general information includes the name, the full path, the stereotype (delegate, implementation class, metaclass, structure, type, union, or utility), the visibility (private, protected, or public), and the indicators for the *Root*, *Leaf*, *Abstract*, and *Active* types of classes. The table of attributes comprises columns for the attribute name, the type, the visibility, the multiplicity (1, *, 0..1, 0..*, 1..1, or

1..*), and its initial value. The table of operations comprises columns for the operation name, the return type, the visibility, the scope (classifier or instance), and the indicator for the polymorphic operations. The table of receptions includes columns for the reception name, the corresponding signal name, the visibility, the scope, and the indicator for the polymorphic operations. The table of template parameters stores parameter names and types. The list of components comprises names of the components that implement this class. The table of constraints consists of four columns: the constraint name, the stereotype (precondition, postcondition, or invariant), the language type (OCL, text, pseudocode, or code), and the body of the constraint. The tagged values include the notes for the documentation, the location, the persistence, the responsibility, and the semantics.

Two graphical symbols are available for rendering interfaces. The first shows just the name of the interface, whereas the second also shows the available operations. Being the specialized classifier, the interface properties are a subset of class properties. More precisely, the interface has four categories of properties: the general information, the table of operations, the table of constraints, and the tagged values. Those properties are the same as the corresponding class properties with a single exception. The interface is passive in its nature, hence the general information might not include the indicator of *Active* type.

The package has four categories of properties: the general information, the table of events, the table of constraints, and the tagged values. The general information includes the name, the full path, the stereotype (facade, framework, stub, or system), the visibility (private, protected, or public), and the indicators for the *Root*, *Leaf*, and *Abstract* types of packages. The table of events stores the event names and the types.

The subsystem has four categories of properties: the general information, the table of operations, the table of constraints, and the tagged values. The general information includes the name, the full path, the visibility, and the indicators for the *Root*, *Leaf*, *Abstract*, and *Instantiable* types of subsystems.

The object has four categories of properties: the general information, the table of attributes, the table of constraints, and the tagged values. The general information about the object includes the object name and the corresponding class name. The tagged values are just documentation notes and the tag persistent value.

The dependency relation has three categories of properties: the general information, the table of constraints, and the tagged values. The general information includes the name, the stereotype (becomes, call, copy, derived, friend, import, instance, metaclass, power type, or send), and the description. The tagged values are the notes for the documentation.

The generalization relation has three categories of properties: the general information, the table of constraints, and the tagged values. The general information comprises the name, the full path, the stereotype (extends, inherits, private, protected, subclass, subtype, or uses), and the discriminator. The tagged values are documentation notes.

Design

The association relation has three categories of properties: the general information, the table of constraints, and the tagged values (documentation notes). The general information comprises the name, the full path, the name reading direction (forward or backward), and the information about the association ends, which includes the name, the aggregation (none, composite, or shared), the visibility, the multiplicity, and the indicator *Navigable*. If the end is navigable, it is shown with an arrow symbol, and if not, it is shown without an arrow symbol. Because the composition relation is a specialization of the association relation, it has the same categories of properties (the general information, the table of constraints, and the tagged values), with the exception that the default values for the aggregation and multiplicity (of one of the ends) are composite and 1, respectively.

The association class is a class that models the complex relation; therefore, its set of properties is a union of properties of classes and associations. More precisely, the association class has five categories of properties: the general information, the table of attributes, the table of operations, the table of constraints, and the tagged values. The general information comprises the name, the full path, the information about the association ends (name, aggregation, visibility, multiplicity, and navigability), and the association class details (visibility information and *Root*, *Leaf*, *Abstract*, and *Active* indicators).

The object link has three categories of properties: the general information, the table of constraints, and the tagged values (just documentation notes). The general information includes the name and the information about each of the two link ends. The link end information comprises the name and the stereotype (none, association, global, local, parameter, or self).

This concludes the description of the basic graphical symbols available for rendering class diagrams. The usage of these symbols is illustrated by two examples, shown in Figure 3.2 and Figure 3.3. The first example is a simple model of the TCP/IP protocol stack, and the second example is a simple model of a finite state machine (automata instance).

The TCP/IP protocol stack is modeled by the classes that represent its layers: *Application*, *Transport*, *Network*, and *Interface*. The transport layer has a number of ports, which are modeled by the interface *Port*. The application depends on the transport (this fact is modeled by the dependency relation) and it gets the service it needs through the interface *Port*. Further down, the transport layer depends on the network layer, which in turn is in association with a number of interfaces.

The left side of Figure 3.2 shows the models of the host computers that are connected to the Internet and the routers that interconnect the physical networks that constitute the Internet. The host computer is modeled by the class *Host*. Each host comprises all TCP/IP protocol stack layers. This fact is modeled by the composition relations between the class *Host* and the classes that model the individual layers (*Application*, *Transport*, *Network*, and *Interface*). The router is modeled by the class *Router*. Each router comprises the network and the interface layer. This is modeled by the composition

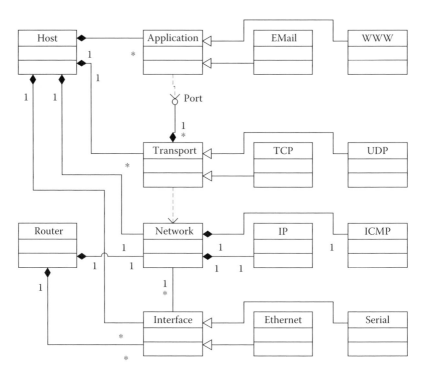

FIGURE 3.2
An example of a simple model of the TCP/IP protocol stack.

relations between the class *Router* and the classes that model the individual layers.

The right side of Figure 3.2 shows some of the applications and protocols available in the TCP/IP family of protocols. The electronic mail and World Wide Web (WWW) applications — and their corresponding protocols — are modeled by the class *Email* and *WWW*, respectively. These two applications are the examples of particular applications, and this fact is modeled by the generalization and specialization relations between the class that models a generic application (*Application*) and the classes that model the particular applications (*Email* and *WWW*). Similarly, TCP and UDP are particular transport protocols (modeled by the classes *TCP* and *UDP*), and this is modeled by the generalization and specialization relations between the class that models a generic transport protocol and the class that model TCP and UDP.

Further down the hierarchy, the Internet network layer comprises the IP and ICMP protocols (modeled as the classes *IP* and *ICMP*). This is modeled by the composition relations between the classes that model the network layer and the IP and ICMP protocols. At the bottom of the hierarchy, we show that various types of interfaces exist, e.g., Ethernet and serial, by generalization and specialization relations between the class *Interface* and the classes *Ethernet* and *Serial*, which model these particular interfaces.

Design

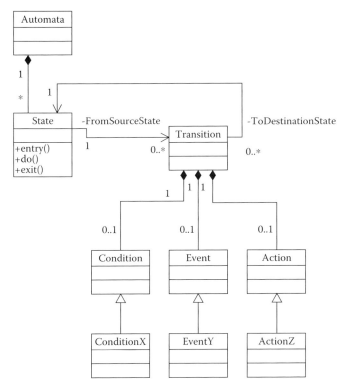

FIGURE 3.3
An example of a simple automata model.

The second example of simple class diagrams is a simple model of a finite state machine (automata instance). The aim of this example is as an easy exercise. We will return to the topic of modeling automata more comprehensively at the beginning of the next chapter. The key abstractions in this example are a finite state machine, a state, and a state transition, which are modeled by the classes *Automata*, *State*, and *Transition*, respectively (Figure 3.3).

The finite state machine comprises a number of states. This fact is modeled by the composition relation between the class *Automata* and the class *State*. The multiplicity from the side of the class *Automata* is 1 and from the side of class *State* is *. (This notation means that a finite state machine must comprise at least one state, which technically sounds like a reasonable requirement.)

The state transition links the source and the destination states, and this is modeled by two association relations between the classes *State* and *Transition*. The ends of these association relations from the side of the class *Transition* are named *FromSourceState* and *ToDestinationState*, respectively. The multiplicity from the side of the class *State* is set to 1 (because each state transition must have exactly one source and one destination state), and from the side of the class *Transition* to 0..* (because a state may have zero or more outgoing and zero or more incoming state transitions). The navigability of these two association relations is set such that the relation *FromSourceState* points from the class *State* to the class *Transition*, whereas the relation *ToDestinationState* points in the opposite direction.

The main problem with this model is ambiguity. The source and the destination states may seem to be always the same (because both *FromSource-State* and *ToDestinationState* association relations are connected to the same class, namely the class *State*). However, source and destination states can be, and most frequently are, different states. We will come back to this point shortly, after introducing additional nodes and relations available for rendering class diagrams, to resolve this problem in a less ambiguous way.

The key abstractions related to the transition are the condition that guards the transition, the event that fires the transition, and the action that is taken by the transition, which are modeled by the classes *Condition*, *Event*, and *Action*. Each transition is characterized by these three optional elements, and that is modeled by the composition relations between the class *Transition* and the classes *Condition*, *Event*, and *Action*. The fact that these elements are optional is modeled by setting the multiplicity to 0..1 from the side of the corresponding classes.

Besides actions that are taken during the transitions, we can define state bound actions, such as the action that is taken at the entrance to a certain state, the action that is performed while the system is in a certain state, and the action that is taken at the exit from a certain state (we will encounter these and more in the UML statecharts later in this chapter). These action types are modeled as the state operations *entry()*, *do()*, and *exit()*, which are defined in the table of operations for the class *State*.

Until now, we were modeling a generic finite state machine. To make this model useful for the implementation of a particular finite state machine, first we need to define the concrete conditions, events, and actions. We do so through the specialization of the base classes *Condition*, *Event*, and *Action*. Figure 3.3 shows the examples of the particular condition, event, and action, which are modeled by the classes *ConditionX*, *EventY*, and *ActionZ*, respectively. Finally, to build the particular finite state machine, we need to instantiate the classes.

This concludes the presentation of two simple examples of class diagrams. To make this graphical language more expressive and to reduce the ambiguity of the class diagrams, the graphical tool provides the additional set of graphical symbols, which are shown in Figure 3.4. The first of them is the metaclass, whose instances are classes that are added to the class diagram. We can resolve the problem of ambiguity in the previous example exactly by using the metaclass instead of the class symbol because it is then clear that the source and the destination state may both be the same state or two completely different states. Again, as for the basic set of symbols, the additional symbols have similar categories of properties. The metaclass has the

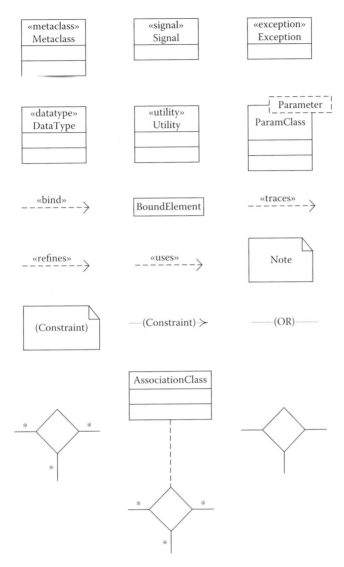

FIGURE 3.4
The additional graphical symbols available for rendering class diagrams.

same properties as the class, with the exception that its stereotype (in general information section) is fixed to metaclass.

Both the signal and the exception symbols have the same four categories of properties, namely, the general information, the table of parameters, the table of constraints, and the tagged values. The general information is the same as for the interfaces (the name, the full path, the visibility, and the indicators *Root*, *Leaf*, and *Abstract*). The table of parameters stores the information about the parameters, which comprise the parameter name, the type, the kind (in, out, or in-out), and the default value.

The data type has five categories of properties. These are the general information, the table of enumeration literals, the table of operations, the table of constraints, and the tagged values. The general information includes the name, the full path, the stereotype (none or enumeration), the visibility, and the indicators *Root*, *Leaf*, and *Abstract*. If the data type is an enumeration, the table of enumeration literals holds the information about the literal names and the corresponding values.

A utility is a special class, therefore it has the same properties as the class with the exception that its stereotype is fixed to utility. Similarly, a parameterized class is a special class that has one or more unbound formal parameters, therefore it has the same categories of properties as the class. Related to the parameterized class is a bind relation, that binds (connects) the designated arguments to the template formal parameters. It has four categories of properties: the general information (just the name and the description), the list of bound arguments, the table of constraints, and the tagged values. The bound element adds the result of binding between the template parameters and their actual values. It has the same categories of properties as the class.

The next three symbols are the traces, refines, and uses relations. We can think of them as specialized dependency relations. The traces relation connects two model elements from two different models. The refines relation connects a more detailed model element to its previous version. The uses relation indicates the dependency relationship between two model elements where one requires another to fully operate. All these relations have the same categories of information as the dependency relation, with the exception that their stereotype is fixed.

The next four symbols are the note, the constraint note, the constraint shown as arrow, and the OR constraint, which we have already encountered in both use case and collaboration diagrams (described in the previous chapter). The last three symbols are used to describe the relations among more than two model elements. The first is the *N*-ary association, which models the association among more than two classifiers. Its properties are the same as for the binary association with the additional properties for each association end (the name, the aggregation, the visibility, the multiplicity, and the navigability indicator).

The second symbol is the *N*-ary association class, which models more complex associations among more than two classifiers. Again, its properties are the same as for the binary association class with additional properties for each association end. The third and the last symbol is the *N*-ary object link, which interconnects more than two objects. Its properties are the same as the binary object link with additional properties for each end (the name and the stereotype).

At the end of this section, we focus on the domain-specific class diagrams. As already mentioned, the reader should assume and accept that somebody has already prepared the infrastructure for the design and implementation of communication protocols. There is no need to start modeling generic

Design

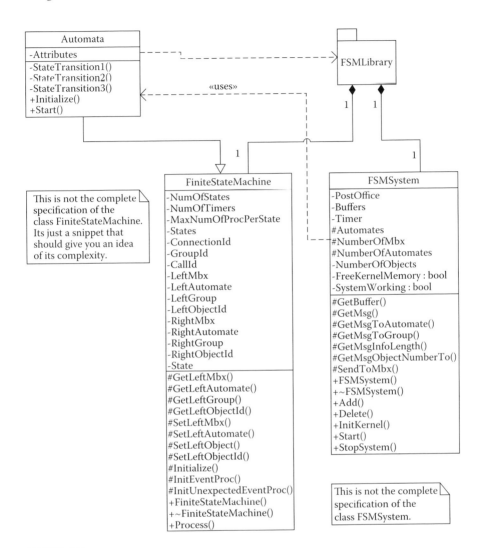

FIGURE 3.5
A typical communication protocol class diagram.

automata every time we start a new project, but rather we do it once and then use it on a number of projects. This practice is what in UML is called providing generic design mechanisms.

In this book, we design and implement communication protocols based on the FSM library. A typical class diagram is shown in Figure 3.5. The FSM library is shown as the package *FSMLibrary* in the diagram and, on most occasions, such representation would be sufficient. It actually comprises a rather ramified hierarchy of C++ classes (we will go into more details in the next chapter). The two most important classes are the *FiniteStateMachine* and *FSMSystem*. The fact that the FSM library contains these classes is modeled by the composition relations between the package *FSMLibrary* and the classes

FiniteStateMachine and *FSMSystem*. The multiplicity is set to 1 on both sides (one library contains one such class).

The communication protocol is modeled by the class *Automata*. The fact that it is a specific type of finite state machine is modeled by the generalization and specialization relation between the class *Automata* and the class *FiniteStateMachine*. The former inherits all the attributes and operations from the latter. The list given in Figure 3.5 is not exhaustive and its purpose is merely to provide the preliminary information about the basic functionality provided by the class *FiniteStateMachine*, and that it is the full set of generic design mechanisms that are needed. Once we have this class, designing a protocol essentially means defining its states and state transitions, and this is basically what we do in this chapter. After the design is finished, implementing the design (in this context) actually means writing the corresponding state transition routines (functions) in C++.

Another important class is the class *FSMSystem*. It actually provides a run-time system for all communication protocols. At the system startup, the main program, here referred to as utility class (not shown in Figure 3.5), registers the given communication protocol by calling the method *Add()* of the class *FSMSystem*, and by giving the reference to the class that models the protocol (*Automata* in this example) as its parameter. Once registered, the protocol can receive, process, and generate events (messages), through the mailboxes provided by the *FSMSystem*.

As we will see in the next chapter, the *FSMSystem* manages all events. It analyzes the event source and destination to locate the destination protocol. Once it is found, the *FSMSystem* looks up its current state, determines the state transition routine based on the event code (type), and calls it. This mechanism is modeled by the uses relation between the class *FSMSystem* and the class *Automata*.

As we can see, the class *Automata* is a specialization of the class *FiniteStateMachine* and is used by the class *FSMSystem* during the system run-time. More briefly stated, the class *Automata* depends on the package *FSMLibrary*. This fact is also modeled in Figure 3.5 by the corresponding dependency relation between the class and the package.

3.2 Object Diagrams

Object diagrams are a special type of class diagrams that typically show a set of objects (instances of classifiers) and their links. Pure object diagrams contain only objects and their links. However, sometimes we may put some classifiers in the object diagram, especially to clarify the relations between the classes and the objects. We may also use packages or subsystems to deal with complexity.

Design 61

Object diagrams, like class diagrams, are used to show the static design view of the system. As already mentioned in the previous chapter, the collaboration diagram is used to model the behavior of the system. It also shows the architecture of the system; hence, we say that the collaboration diagram is organized by the architecture. We can think of the object diagram as one snapshot of the collaboration diagram. Imagine that time is frozen. Whatever we can see in the collaboration diagram at that single moment of time is an object diagram.

Later in this chapter, we will introduce the deployment diagrams, and we will introduce the component diagrams in the next chapter. Both deployment and component diagrams can contain only objects and their links. In such cases, they are actually pure object diagrams.

Clearly, the graphical symbols available for rendering object diagrams are the same as the symbols used for class diagrams (sometimes referred to as a static structure). In practice, we use only a very limited subset of those symbols, most frequently only two of them (object and object link). The properties of these symbols are described in a previous section of this chapter.

Usage of object diagrams can reduce the ambiguity of the static structure twofold. First, by rendering instances of classifiers, we can better understand the relations among them. For example, by rendering just the classes in the TCP/IP protocol stack model, it may not be clear what the network really looks like. Second, by showing the values of the key class attributes, we can recognize reality more easily. For example, by showing the status of the individual protocols, we can also comprehend their expectations from other cooperating protocols.

These ideas are illustrated by the following two examples. The first is the object diagram that shows the snapshot from a simple mail transfer protocol (Figure 3.6). The second is the example of a simple finite state machine object diagram (Figure 3.7).

Figure 3.6 shows the software running on two host computers that are connected to two different local area networks, which are interconnected by the router. The host computers clearly require full protocol stacks whereas the router requires only the two lowest level layers (IP and network interface). One host computer, shown on the left side of the figure, runs the SMTP client on top of the TCP/IP protocol. The other host computer hosts the SMTP server.

The first benefit of this object diagram is that it really makes clear which layers are required by the hosts and which are required by the routers. Graphically, we see the network, which was rather difficult to visualize just by looking at the class diagram shown in Figure 3.3. Enough order is found in this object diagram, too. More symbols are used than in the class diagram, but only five per host and two per router. Of course, if we try to model a large network there would be a flood of objects; therefore, we should always try to restrict our modeling to a certain aspect of a system.

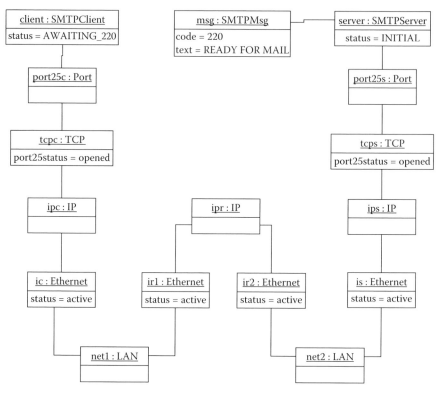

FIGURE 3.6
A snapshot from the simple mail transfer protocol (SMTP).

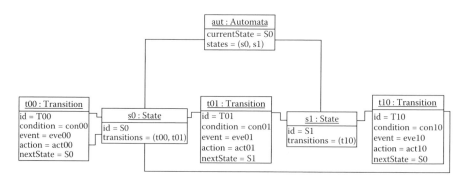

FIGURE 3.7
An example of a simple finite state machine (FSM) object diagram

Design 63

The second benefit is that we can peacefully study all the details of a certain moment in the life of a protocol, in this case SMTP. It is like looking at the photograph of a certain party. This one shows the moment when the SMTP server has prepared the message 220 READY FOR MAIL and its intention was to send it at the moment when the time has been frozen. We can imagine what the sensation of looking at a series of such object diagrams would be, like watching a replica of an important event in a game in slow motion. After receiving the message 220 READY FOR MAIL, the SMTP client would prepare the message HELO, and so fortth.

Besides current messages, other details are also important. For example, we see in Figure 3.6 that the TCP port number 25 is opened from both sides, and from there we can deduce that the SMTP client and server had to establish the TCP connection in the first place before they could proceed any further. Some details may seem obvious, for example that all Ethernet cards and their drivers must be active, but they also help in making the complete picture of the selected moment. In a series of object diagrams, the changes of values of certain attributes, such as status, are the most interesting and most informative part.

The second example of object diagrams is a simple finite state machine object diagram, which is shown in Figure 3.7. A simple finite state machine object, named *aut*, is an instance of the class *Automata* (Figure 3.4). It comprises a set of two state objects, namely *s0* and *s1*, which are the instances of the class *State*. Their identifications are *S0* and *S1*, respectively. The current state of the automata is the state with the identification *S0*.

The state object *s0* contains a set of two transition objects, namely *t00* and *t01*, which are the instances of the class *Transition* (Figure 3.4). Similarly, the state object *s1* contains a set with one transition object, named *t10*. The transition objects *t00*, *t01*, and *t10* model the automata state transitions from the state with the identification *S0* to the state with the identification *S0*, or more briefly from *S0* to *S0*, next from *S0* to *S1*, and last from *S1* to *S0*, respectively.

The attributes of the transition objects are the transition identification, the condition that guards the transition, the event that fires the transition, the action that is taken by the transition, and the next state identification. Their identifiers are *id*, *condition*, *event*, *action*, and *nextSate*, respectively. *id* and *nextState* would typically be strings or integers. *condition*, *event*, and *action* are the instances of the class *Condition*, *Event*, and *Action*.

An important detail is that the values of these attributes are the instances of classes that are specialized from the classes *Condition*, *Event*, and *Action*. For example, the values of the attribute condition, namely *con00*, *con01*, and *con10*, are the instances of the classes, e.g., *Condition00*, *Condition01*, and *Condition10*, which are actually specializations of the class *Condition*. Such modeling allows us to use polymorphism, the most powerful abstraction of object-oriented design and programming.

3.3 Sequence Diagrams

Two types of UML interaction diagrams are used, namely, the sequence diagrams and the collaboration diagrams. We have already introduced the collaboration diagrams in the previous chapter. They can be used in both analysis and design phases of communication protocol engineering. Sequence diagrams are just another type of interaction diagrams and are semantically equivalent to collaboration diagrams. This means that a one-to-one mapping exists between these two formalisms that are used for specifying interactions.

An interaction is basically a set of objects and their relationships, together with the messages that are exchanged among the objects. Both sequence and collaboration diagrams show interactions. The major difference between them is that the sequence diagrams emphasize time ordering of messages whereas the collaboration diagrams emphasize the structural organization of a set of objects. The sequence diagrams are particularly useful for visualizing dynamic behavior in the context of the use case scenario. Generally, they are better suited for modeling sequences of events, simple iterations, and branching. Alternately, collaboration diagrams are more useful for modeling complex iterations and branching and for visualizing multiple concurrent flows of control.

Sequence and collaboration diagrams also differ in appearance. As we have already seen in the previous chapter, a collaboration diagram looks like a graph. It consists of objects that are linked together in a certain arrangement. A sequence diagram appears more like a table whose columns are related to individual objects and whose rows are related to the messages that are exchanged among the objects. We can imagine the horizontal axis x, at the top of the diagram, pointing from left to right, and the vertical axis y that points from top to bottom. The objects that participate in the interaction are arranged across the x-axis, starting on the left with the objects that are initiating the interaction and proceeding to the right with more subordinate objects. The messages that are exchanged among the objects are ordered in increasing time along the y-axis. (Actually, we have already informally encountered sequence diagrams in the previous chapter. See the example of the SIP session setup in Figure 2.13.)

The sequence diagrams have two key features that distinguish them among other diagrams:

- Object lifeline
- Focus of control

An object lifeline is a dashed vertical line that represents the existence of an object over a period of time. The object lifeline starts with the reception of the message stereotyped as <<create>> and ends with the reception of the message stereotyped as <<*destroy*>>. The end of the life of an object is indicated by the mark "X." However, most of the object will exist throughout the interaction. Such objects are normally placed at the top of the diagram and their lifeline typically goes to the end of the diagram.

The focus of control represents the period of time during which the object executes. It is rendered as a long, thin rectangle. We can model recursion, a call to self-operation, or call-back by placing a new focus of control symbol on top of the current focus of control symbol and slightly to the right, so that both of the symbols are visible. We can explicitly show the part of the focus of control where the actual computation takes place by shading the corresponding region.

We can model the mutation of objects in their state, role, or attribute values in sequence diagrams. Two methods to do this exist: The first is by placing a new copy of the object in the sequence diagram and showing the change by connecting the existing and the new object copy with the transition <<*become*>>. This procedure can be repeated if we want to show a sequence of changes. The second method is by placing a new copy of the object directly on the object's lifeline and showing the change of state, role, or attribute values then and there.

The set of graphical symbols available for rendering sequence diagrams is shown in Figure 3.8. Similar to the diagrams that were previously introduced, each of the symbols has its own properties with the exception of the focus of control, which has no properties on its own (it is a symbol that can exist only on top of the object's lifeline). The designer must fill in the properties after adding the symbol to the diagram.

The object and its lifeline have three categories of properties: the general information, the table of constraints, and the tagged values. The general information includes the name, the full path, the classifier, and the multiplicity. Other categories of properties are already explained in the previous sections.

The message has four categories of properties. These are the general information, the table of arguments, the table of constraints, and the tagged values (documentation notes). The general information includes the name, the direction (forward or backward), the operation, and the sequence expression. The table of arguments holds information about the arguments, such as the name, the type, the language, and the value.

The following four types of messages are used:

- Flat
- Call
- Return
- Asynchronous

The flat message models the communication between the objects that convey information, which should result in an action. The call message

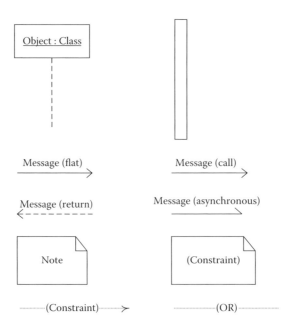

FIGURE 3.8
The set of graphical symbols available for rendering sequence diagrams.

models a synchronous procedure call that should result in some action. The return message models returns from the procedure, which convey the return value that will cause an action. The asynchronous message models the asynchronous communication between two objects, which also carries some information that will trigger an action. The note, the constraint note, the constraint, and the OR constraint are symbols that we have already encountered and explained in previous sections.

Next, we illustrate the use of sequence diagrams by four examples shown in Figure 3.9, Figure 3.10, Figure 3.11, and Figure 3.12, which are semantically equivalent to the collaboration diagrams shown in Figure 2.9, Figure 2.10, Figure 2.11, and Figure 2.12, with one exception. Figure 3.9 and Figure 2.9 do relate to the same interaction, but they are not exactly semantically equivalent because of two reasons. First, the former shows fewer objects than the latter, mainly because of the limited diagram width. Second, the latter shows only a part of the interaction shown by the former. Interestingly enough, this seems to be a general rule. The sequence diagrams typically show fewer objects but more messages than do collaboration diagrams.

The example shown in Figure 3.9 generally illustrates the same use case *Send e-mail* as does the collaboration diagram shown in Figure 2.9. Figure 3.9 shows only the most important subset of objects but, at the same time, it illustrates the interaction long enough to show the moment when the SMTP client sends the SMTP message DATA toward the SMTP server. The

Design

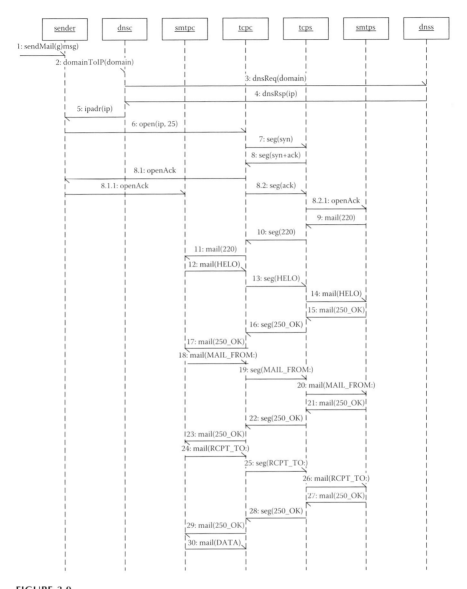

FIGURE 3.9
A sequence diagram showing the interaction between a simple program for sending and receiving e-mails and its environment.

collaboration diagram shown in Figure 2.9 shows the situation only up to the point when the SMTP client receives the message 220 READY FOR MAIL, which is actually the very beginning of the SMTP protocol. The names of the objects, messages (signals), and message arguments used in both figures are explained in the previous chapter. The exact flow of events shown in Figure 3.9 is as follows:

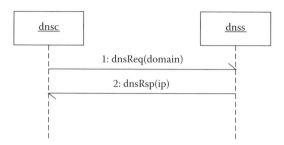

FIGURE 3.10
A sequence diagram showing the interaction between the DNS client and the DNS server.

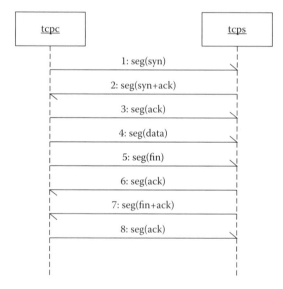

FIGURE 3.11
A sequence diagram showing the interaction between two TCP entities.

1: The object *mailc* (not shown in the diagram) sends the signal *sendMail(msg)* to the object *sender*.
2: The object *sender* sends the signal *domainToIP(domain)* to the object *dnsc*.
3: The object *dnsc* sends the signal *dnsReq(domain)* to the object *dnss*.
4: The object *dnss* sends the signal *dnsRsp(ip)* to the object *dnsc*.
5: The object *dnsc* sends the signal *ipadr(ip)* to the object *sender*.
6: The object *sender* sends the signal *open(ip,25)* to the object *tcpc*.
7: The object *tcpc* sends the signal *seg(syn)* to the object *tcps*.
8: The object *tcps* sends the signal *seg(syn+ack)* to the object *tcpc*. (The event flow now forks into two parallel flows.)

Design

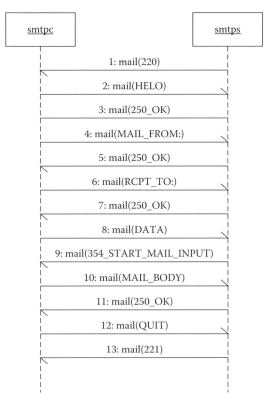

FIGURE 3.12
A sequence diagram showing the interaction between the SMTP client and the SMTP server.

8.1: The object *tcpc* sends the signal *openAck* to the object *sender*. (The first flow begins here.)

8.1.1: The object *sender* sends the signal *openAck* to the object *smtpc*. (The first flow ends here.)

8.2: The object *tcpc* sends the signal *seg(ack)* to the object *tcps*. (The second flow begins here.)

8.2.1: The object *tcps* sends the signal *openAck* to the object *smtps*.

9: The object *smtps* sends the signal *mail(220)* to the object *tcps*. (Note: We have restarted the message numbering here for brevity. We promoted 8.2.2 to 9.)

10: The object *tcps* sends the signal *seg(220)* to the object *tcpc*.

11: The object *tcpc* sends the signal *mail(220)* to the object *smtpc*.

12: The object *smtpc* sends the signal *mail(HELO)* to the object *tcpc*.

13: The object *tcpc* sends the signal *seg(HELO)* to the object *tcps*.

14: The object *tcps* sends the signal *mail(HELO)* to the object *smtps*.

15: The object *smtps* sends the signal *mail(250_OK)* to the object *tcps*.

16: The object *tcps* sends the signal *seg(250_OK)* to the object *tcpc*.
17: The object *tcpc* sends the signal *mail(250_OK)* to the object *smtpc*.
18: The object *smtpc* sends the signal *mail(MAIL_FROM:)* to the object *tcpc*.
19: The object *tcpc* sends the signal *seg(MAIL_FROM:)* to the object *tcps*.
20: The object *tcps* sends the signal *mail(MAIL_FROM:)* to the object *smtps*.
21: The object *smtps* sends the signal *mail(250_OK)* to the object *tcps*.
22: The object *tcps* sends the signal *seg(250_OK)* to the object *tcpc*.
23: The object *tcpc* sends the signal *mail(250_OK)* to the object *smtpc*.
24: The object *smtpc* sends the signal *mail(RCPT_TO:)* to the object *tcpc*.
25: The object *tcpc* sends the signal *seg(RCPT_TO:)* to the object *tcps*.
26: The object *tcps* sends the signal *mail(RCPT_TO:)* to the object *smtps*.
27: The object *smtps* sends the signal *mail(250_OK)* to the object *tcps*.
28: The object *tcps* sends the signal *seg(250_OK)* to the object *tcpc*.
29: The object *tcpc* sends the signal *mail(250_OK)* to the object *smtpc*.
30: The object *smtpc* sends the signal *mail(DATA)* to the object *tcpc*.

Another practical detail about sequence diagrams is that not only their width but also their height is limited. Because of this, we are normally forced to break the flow of events at a certain point. In the previous example, it was after the object *smtpc* has sent the signal *mail(DATA)* to the object *tcpc*. Typically, we would continue that flow on another sequence diagram. Good practice is to pick the breaking points logically, for example, at the beginning or at the end of certain communication phases.

Also important is to emphasize that the sequence diagram in Figure 3.9 shows only main flows of events. It does not show what happens in the case of errors. The error handling is typically shown in separate sequence diagrams. We can use packages to wrap together all the related sequence diagrams.

Figure 3.9 shows also that the real overall interaction can be fairly complex. To deal with complexity, we can focus on the individual virtual interactions instead. For example, the sequence diagram showing the interaction between the DNS client and server is a trivial one (Figure 3.10). The flow of events then reduces to only the following two events:

1: The object *dnsc* sends the signal *dnsReq(domain)* to the object *dnss*.
2: The object *dnss* sends the signal *dnsRsp(ip)* to the object *dnsc*.

Similarly, the virtual interaction between two TCP entities, modeled by the objects *tcpc* and *tcps*, is governed by the TCP protocol. It is slightly more complex and comprises the following flow of events (Figure 3.11):

Design 71

1: The object *tcpc* sends the signal *seg(syn)* to the object *tcps*.
2: The object *tcps* sends the signal *seg(syn+ack)* to the object *tcpc*.
3: The object *tcpc* sends the signal *seg(ack)* to the object *tcps*.
4: The object *tcpc* sends the signal *seg(data)* to the object *tcps*. (This is the data transmission phase.)
5: The object *tcpc* sends the signal *seg(fin)* to the object *tcps*.
6: The object *tcps* sends the signal *seg(ack)* to the object *tcpc*.
7: The object *tcps* sends the signal *seg(fin+ack)* to the object *tcpc*.
8: The object *tcpc* sends the signal *seg(ack)* to the object *tcps*.

Finally, the virtual interaction between the SMTP client and server, modeled by the objects *smtpc* and *smtps*, is of the same order of complexity (Figure 3.12). The interaction is governed by the SMTP protocol. The corresponding flow of events is the following:

1: The object *smtps* sends the signal *mail(220)* to the object *smtpc*.
2: The object *smtpc* sends the signal *mail(HELO)* to the object *smtps*.
3: The object *smtps* sends the signal *mail(250_OK)* to the object *smtpc*.
4: The object *smtpc* sends the signal *mail(MAIL_FROM:)* to the object *smtps*.
5: The object *smtps* sends the signal *mail(250_OK)* to the object *smtpc*.
6: The object *smtpc* sends the signal *mail(RCPT_TO:)* to the object *smtps*.
7: The object *smtps* sends the signal *mail(250_OK)* to the object *smtpc*.
8: The object *smtpc* sends the signal *mail(DATA)* to the object *smtps*.
9: The object *smtps* sends the signal *mail(354_START_MAIL_INPUT)* to the object *smtpc*.
10: The object *smtpc* sends the signal *mail(MAIL_BODY)* to the object *smtps*.
11: The object *smtps* sends the signal *mail(250_OK)* to the object *smtpc*.
12: The object *smtpc* sends the signal *mail(QUIT)* to the object *smtps*.
13: The object *smtps* sends the signal *mail(221)* to the object *smtpc*.

3.4 Activity Diagrams

Up to now, we have introduced three types of diagrams that are used for modeling dynamic aspects of systems. These are the use case, the collaboration, and the sequence diagrams. The use case diagrams are used first for capturing the requirements of the system. They are then translated into collaboration diagrams that model the architecture of the system. Next, at the beginning of the design phase, both collaboration and sequence diagrams are used for building up the storyboards of scenarios.

These scenarios describe the interaction among the most interesting objects; hence, we refer to them as interaction diagrams. The interaction itself is shown by the messages that are dispatched among the objects. Generally, interaction (collaboration and sequence) diagrams are similar to Gantt charts. The main difference between the collaboration and sequence diagrams is that the former emphasize the structural relations whereas the latter emphasize the time ordering of messages.

The storyboards of scenarios are a good place to start the design — therefore, they are a type of design front-end. Although the interaction diagrams make a perfect start of the design, they are seldom used as the final artifacts of the design phase because of the following two problems:

- The interaction diagrams are most frequently incomplete.
- The interaction diagrams specify the external behavior of individual objects, leaving their internal behavior unknown.

As already mentioned, the interaction diagrams typically cover the main flow of events and, because of the limited space in the diagrams, even the main flow must be partitioned into logical communication phases. Other less frequent flows, including error handling, are modeled in additional interaction diagrams. All these diagrams can be sorted into packages for easier manipulation. However, no matter how pedantic the engineer is, the set of interaction diagrams remains incomplete by unwritten rule. Some scenarios are always missing. In the area that is of primary interest for this book, the packages of interaction diagrams are especially vulnerable on specification of timers and complex unforeseen error scenarios.

Another problem we encounter while trying to make the packages of interaction diagrams complete is that they become voluminous and, because of that, hard to comprehend. But this behavior is what we should expect when we try to enumerate and describe the cases instead of trying to create the rules that generate these cases. Even a simple program performing some simple arithmetic calculations can produce enormous numbers of execution cases when we take into account the cardinal numbers of sets of values the common variable types can have. Because of the coverage problems, an implicit engineering rule is that the design based solely on the interaction diagrams is considered as incomplete. This may not be true in the case of simple systems, but generally it is. Therefore, we need the design back-end, the means to end the design.

The secret of how to finish the design is found by turning our attention to the internal behavior of the objects and trying to specify it. This attitude is like turning the interaction diagrams inside out. We want to specify the activities that should take place to provide the desired external behavior and

Design

what should be the order (flow) of the activities in the scope of a single object or in the scope of a set of objects that are involved in the interaction. The means to do this in UML are the activity diagrams, which are similar to Pert network charts. The alternative means to specify the behavior of single objects in UML are statecharts, which will be introduced in the next section.

An activity diagram is essentially a flowchart that shows the flow of control from activity to activity. If we model the behavior of a single object, we render the flow of control within that single object. The activity diagrams are even more powerful and they allow us to model the behavior of a group of objects by rendering the flow of control in that larger scope. Additionally, we can model a single flow of control or more concurrent flows of control within both a single object and a group of objects.

In the context of a single object, we typically partition its behavior into a set of its operations and then model the flow of control of these operations individually. Therefore, the most elementary level of modeling by using activity diagrams is the level of the object's operation. On the opposite side of the scope scale, we can model the workflow of a group of cooperating objects, and we will return to that point shortly.

The most elementary activity is an **action state**. It is defined as an atomic (i.e., uninterruptible) program computation. Examples of action sates are the following:

- Create another object
- Destroy another object
- Call an operation on an object
- Return a value
- Send a signal to an object
- Receive a signal from an object
- Evaluate an expression
- Execute a single statement

The action states can be specified in informal text, pseudocode, or a higher level programming language. Although it is generally assumed that the action state takes a small amount of execution time, that finite amount of time must be taken into account, especially in the models of hard real-time systems.

By combining more action states, we are building more complex activities, which are referred to as **activity states**. We can think of the activity state as a composite state that is made of other activity states and action states. The activity state can also comprise some special actions, such as entry and exit actions. The former is taken at the entrance to the activity state, and the latter is taken at its exit.

The state transitions in activity diagrams normally take place after completion of the last activity in the originating state. A transition without a

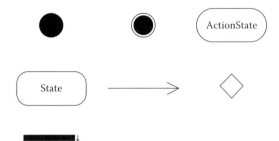

FIGURE 3.13
The basic set of graphical symbols available for rendering activity diagrams.

guard (condition) immediately passes control to the destination state. Such a transition is referred to as a triggerless, or completion, transition. A transition can branch into two or more guarded transitions, or it can fork into more concurrent transitions, and more concurrent transitions can join into a single transition, as we will explain shortly with some simple examples.

An activity diagram is a special type of a graph that comprises a set of vertices that are interconnected by arcs. The basic set of graphical symbols available for rendering activity diagrams is shown in Figure 3.13. Each symbol has a set of properties that must be set by the designer once they add a symbol to the diagram.

The initial state has three categories of properties. These are the general information, the table of constraints, and the tagged values (documentation notes). The general information is just the name and the type (initial). Each activity diagram must start with this symbol.

The final state has the same categories of properties as the initial state symbol, with the exception that its type is final. If the activities specified by the activity diagram go on forever, the diagram will not contain this symbol. Alternately, it can contain one or more such symbols.

The action state has five categories of properties, namely, the general information, the call action, the list of deferred events, the table of constraints, and the tagged values (documentation notes). The general information comprises the name, the stereotype, and the partition. The call action specifies the name of the operation and the table of its arguments, which holds information about the argument name, type, language, and value.

The activity state has six categories of properties. These include the general information, the table of entry actions, the table of exit actions, the table of internal transitions, the table of constraints, and the tagged values. The general information is just the name and the stereotype. Both the table of entry and the table of exit actions store the corresponding action names and their types. The table of internal transitions comprises their properties. Each internal transition is characterized by its name, its stereotype, and the event that triggers the transition.

The control flow transition has four categories of properties, including the general information, the table of actions, the table of constraints, and the

Design 75

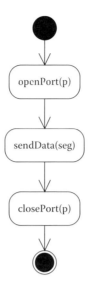

FIGURE 3.14
An example of a simple sequence of activity states.

tagged values (documentation notes). The general information comprises the name and optionally the corresponding event and the guard expression. The table of actions holds action names and their types. The decisions, as well as the fork and join transitions, have three categories of properties, namely, the general information (just the name), the table of constraints, and the tagged values.

We illustrate the usage of these basic symbols by the following four simple examples shown in Figure 3.14, Figure 3.15, Figure 3.16, and Figure 3.17. The example in Figure 3.14 shows a simple sequence of interruptible activities, i.e., activity states, namely, *openPort(p)*, *sendData(seg)*, and *closePort(p)*. Normally, these activity states would be modeled by the activity diagrams themselves on the subordinated level of the hierarchy. The control flow transitions between the individual activity states in this example are triggerless, or completion, transitions, which means that they are not triggered by other events. They also may not be guarded because their sources are not decisions.

The exact semantics of the states in this example are not really important; for example, we can interpret it as open the given port, send the given segment of data, and close the port at the end. Generally, we should think of the activity state as an operation (procedure, function), which consists of executable statements or calls to other operations, including calls to itself (recursion). Thinking about forward engineering helps make useful activity diagrams. Try to imagine how the model would map to the code. It really does not make any difference how the mapping would be made, automatically with a tool or by hand.

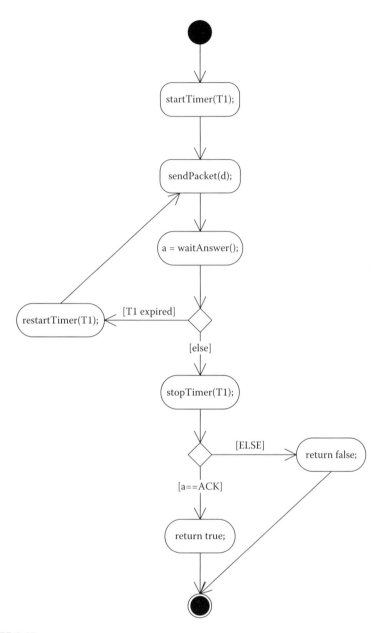

FIGURE 3.15
An example of a simple flow of activities with branching.

The example in Figure 3.15 is an illustration of activity flow with branching. Actually, it is a simplified implementation of the reliable transport mechanism known as Automatic Repeat Question (ARQ). The whole operation begins by starting the retransmission timer *T1*. This beginning is modeled by the activity state *startTimer(T1)*. The operation then sends the datagram

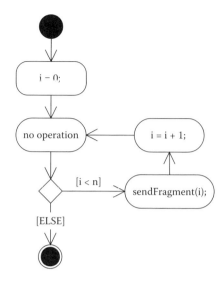

FIGURE 3.16
An example of a loop in an activity diagram.

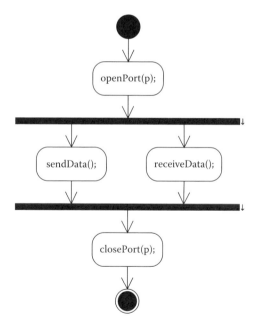

FIGURE 3.17
An example of a simple set of concurrent flows.

and waits for the answer. These two activities are modeled by the activity state *sendPacket(d)* and *a=waitAnswer()*, respectively.

If the retransmission timer expires, the packet is retransmitted. This mechanism is modeled by the transition guarded by the expression *[T1 expired]*, the activity state *restartTimer(T1)*, and the completion transition back to the activity state *sendPacket(d)*. The reception of the answer is modeled by the transition that covers all the other cases (guard expression *[ELSE]*). The operation proceeds by stopping the retransmission timer, and this action is modeled by the activity state *stopTimer(T1)*. If the answer is the acknowledgment (*ACK*), the operation returns the *value true*; otherwise, it returns the *value false*.

The previous example uses two branches. Each branch has one incoming and two or more outgoing transitions. The outgoing transitions are guarded by the Boolean expressions that are evaluated at the entrance to the branch. The set of guards has two important features:

- The guards must not overlap — this makes the flow of control unambiguous.
- The guards must cover all possibilities — this ensures that the flow of control is not going to freeze.

Precisely these two features force us to make complete models and specifications of activities that describe the behavior of the system. When we render interaction (collaboration and sequence) diagrams, no such enforcements are present and, mainly because of that, they remain unfinished. Of course, at the time when we render interaction diagrams, we really do not want to make them final; rather, we want to check the most important aspects and scenarios and to make our analysis more comprehensive and useful for the finalization later. Therefore, when we start rendering the activity diagram, we already have a good overall vision, but non-overlapping and complete coverage features are the driving forces of the design finalization.

One safe way to provide both of these features is to use only the decisions with two outgoing transitions and to guard one of them by the keyword *ELSE*, as in the example in Figure 3.15. Special attention should be paid to the decisions with more outgoing transitions, which are guarded by explicit expressions (i.e., without the keyword *ELSE*). However, the price that we may pay for safety is ambiguity. For example, if the operation in the previous example returns the *value false*, it might do so because the correct not acknowledge answer (*NAK*) has been received. However, the operation will return the same value if any other message (including corrupted *ACK* or *NAK*) has been received.

The example in Figure 3.16 illustrates the usage of loops in activity diagrams. Imagine that the IP protocol must route a datagram over a physical network, which has the Maximal Transfer Unit (MTU) smaller than the datagram size. Normally, the IP protocol partitions the datagram into fragments (that fit MTU) and routes the resulting fragments individually in such cases. The standard means to model repetitive activities in activity diagrams are loops.

The example in Figure 3.16 starts by setting the control variable i to the value 0. It continues with no operation activity state, followed by the decision that checks the loop continuation condition ($i < n$). If the condition is satisfied, the flow enters the loop body (*sendFragment(i)*). The loop body is followed by the activity state that updates the control variable ($i = i + 1$). The example terminates when the loop continuation condition becomes false.

The example in Figure 3.17 shows the usage of concurrent control flows. Imagine that we want to model a simple communication over the TCP connection. First, we must establish the TCP connection by opening a particular TCP port. We model this by the activity state *openPort(p)*. Once the connection is established, the TCP protocol provides simultaneous transfer of data in both directions (full-duplex). To model that, we need to fork a single flow of control into two parallel (concurrent) flows of control. One of them enters the activity state *sendData*, which models the activity of sending the data to the remote site. The other control flow enters the activity state *receiveData*, which models the activity of receiving the data from the remote site.

These two activities logically evolve in parallel over time. On a multiprocessor system, they can be deployed on two different processors to maximize the system throughput. In such a case, these two activities would be parallel in reality, also. Alternately, single-processor systems create quasi-parallelism using the time-sharing operating system. The activities are then not parallel in reality, but they are still concurrent because they can compete for the same resources. Additionally, the activities can communicate using signals. Traditionally, such communicating sequential processes are referred to as **coroutines**.

Although the model shown in Figure 3.17 is fairly simple, it may reflect a realistic communication, such as a Telnet session. Imagine that the activity state *sendData* is a composite state that reads the user keystrokes and sends them to the Telnet server over the TCP connection, in a loop, until the end-of-file key combination is detected. The activity state *receiveData* in this scenario would be also a composite activity state, which receives the responses from the Telnet server and displays them on the monitor, in a loop, until the end-of-communication signal is detected (typically, it would be sent when the end-of-file key combination is detected).

Once one of the parallel activities finishes, it proceeds to the control flow join synchronization point where it waits for the other parallel activity to finish. When both of the activities are finished, the corresponding parallel control flow joins into a single control flow, which enters the activity state *closePort(p)* and, after finishing that activity, it terminates.

As we have seen from the previous example, fork and join synchronization points are rendered as either thick horizontal or vertical lines. It is important to remember that they must be balanced. Similar to the subexpression — which must begin with the opening parenthesis and end with the closing one — each nesting level of the concurrent control flows must begin with the fork symbol and end with the corresponding join symbol. Apart from

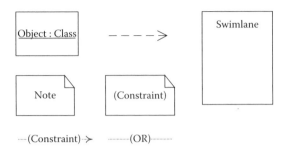

FIGURE 3.18
The additional graphical symbols available for rendering activity diagrams.

that, no restrictions are placed on the number of nesting levels, at least not in theory. Of course, in practice we should not go beyond a manageable number.

The set of additional symbols that are available for rendering activity diagrams is shown in Figure 3.18. These are the object in state, the object flow, and the swim lane symbols, as well as the symbols common for all diagrams, namely, the note, the constraint note, the two-element constraint, and the OR constraint.

The object flow transition enables us to show how the object state changes in the activity diagrams. Typically, we render the objects showing the current and the new states and we connect them by the object flow transition. The objects themselves may be results of activity states and can be used by other activity states. The object flow symbol has the same four categories of properties as the control flow symbol (described previously in this section).

The swim lane has no strict semantics. It is normally used to show individual parties in the workflows. The swim lane is typically implemented as a class or a set of classes. It is better suited for modeling business processes, but it can also be used for modeling communication protocols. The swim lane has three categories of properties: the general information (essentially, its name), the table of constraints, and the tagged values.

The example in Figure 3.19 illustrates the usage of objects, data flow transitions, and swim lanes, with the example of activities initiated by the Domain Name System (DNS) client request for mapping a given domain name onto the corresponding IP address. Figure 3.19 is a type of a workflow conducted by the DNS client and server in their cooperative work of translating a domain name into the IP address. The DNS client is represented by the first swim lane and the DNS server is represented by the second. This activity diagram shows both the control flow among individual activity states and data flow, which are created by a series of objects that are consumed and produced by the activity states of both DNS client and server.

The given domain name is the input parameter of the DNS client operation that translates the domain name into the corresponding IP address. This operation starts by the activity state *createDNSmsg()*, which creates an empty

Design

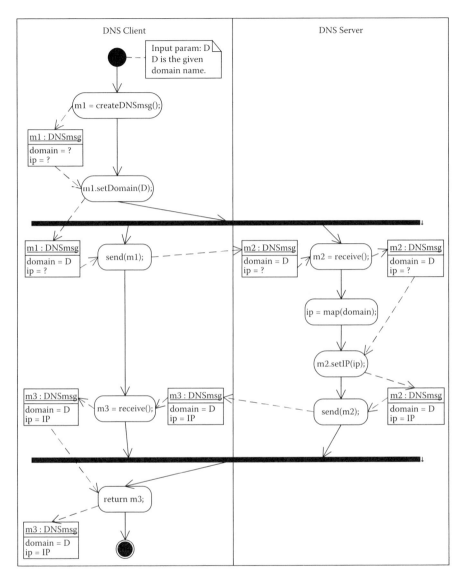

FIGURE 3.19
The workflow between the DNS client and server with the message flow.

DNS message. This action is modeled by placing the object *m1* that represents the DNS message in the activity diagram and by connecting it to the activity state *createDNSmsg()* with the arrow pointing toward the object *m1*. This means that the object *m1* is produced by the activity state *createDNSmsg()*. The fact that the message is empty is indicated by showing that the values of both attributes *domain* and *ip* are unknown (the unknown value is denoted by the question mark character, "?").

Next, the activity state sets the attribute domain to the value of the input parameter *D*, thus creating a new state of the object *m1*. This action is modeled by placing a new copy of the object *m1* in the activity diagram and by adding two object flow arcs. The first connects the previous object copy and the activity state *m1.setDomain(D)*. The arrow points toward the activity state, which means that the state consumes the object. The second object flow arc connects the activity state and the new copy of the object *m1*, thus implying that the activity state produces it.

The control flow then forks into two independent flows. One is conducted by the DNS client and the other is conducted by the DNS server. The DNS client continues by sending the DNS message, as a DNS request, to the DNS server. The corresponding activity state creates a new object, named *m2*, and places it in the second swim lane, because we assume that the DNS server runs on a different machine, or at least in a different address space. The DNS server, in its turn, receives the DNS message. A common mechanism for copying the message from an internal operating system buffer to the buffer that is located within the address space of the DNS server is modeled by placing two different copies of the object *m2*.

The DNS server continues by translating the given domain name into the corresponding IP address and by setting the attribute *ip* to the value IP, which denotes the result of that translation. This fact is shown in the third copy of the object *m2*. The DNS server proceeds by sending the completed DNS message, which models the DNS response message, to the DNS client, which in its turn receives it and creates the copy of the object *m3* in its address space. Finally, two independent control flows join together and the DNS client returns the completed DNS message to its user, thus creating the final copy of the object *m3*.

As this example shows, the models of the workflows are useful because they show and specify the external behavior, i.e., the interface and protocol between the objects in the form of the corresponding sequence of messages exchanged by the objects, as well as the internal behavior of objects in the form of the series of activity states visited by them. The first is created by modeling the data and object flow, and the second is created by modeling the control flow across the objects. Again, by taking care of the complete coverage of possibilities, without any overlaps, we ensure that the model is complete. (This was not the main goal of the last example, at least not to the extent as in the previous one, but we should keep that in mind.)

Figure 3.20 shows the activity diagram for one real protocol, TCP, and it follows the conventions introduced by the corresponding IETF RFC 793. The user requests are written in capital letters. The user requests are *OPEN*, *SEND*, and *CLOSE*. Two types of *OPEN* requests are used, namely active *OPEN* and passive *OPEN*. The difference between the two is who is taking the initiative in the connection establishment procedure.

The next convention is that the names of the events and actions are written in lowercase letters, with the following abbreviations:

Design

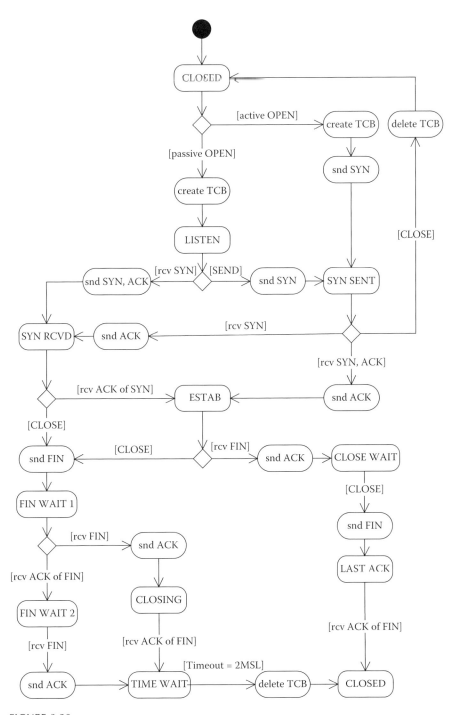

FIGURE 3.20
A TCP activity diagram.

- TCB (Transmission Control Block)
- snd (send)
- rcv (receive)
- SYN (indicates that the synchronization bit of the TCP segment is set)
- ACK (indicates that the acknowledgment bit of the TCP segment is set)
- SYN, ACK (both SYN and ACK bits are set)
- FIN (indicates that the final bit of the TCP segment is set)
- ACK of SYN (denotes the acknowledgement of the SYN segment)
- ACK of FIN (denotes the acknowledgement of the FIN segment)
- MSL (Maximum Segment Lifetime)

The TCP events are actually modeled as guard expressions whereas the TCP activities are modeled as UML action states (relatively short and uninterruptible series of executable statements). Notice that we could model the TCP activities either by action or by activity states because these activities are essentially interruptible. However, because they can be implemented as rather short routines — which do not involve reception of any signals — modeling them as action states makes more sense than as activity states.

The TCP protocol spends most of the time in one of its stable states waiting for a certain event to occur. The TCP stable states are modeled by the UML activity states. While being in one of its stable states, the TCP protocol just waits for an event (it does not execute any statements). The process that executes the TCP protocol is blocked and it does not compete for the processor execution time. Therefore, the activity corresponding to the stable state is more than interruptible — it is blocked. Because such an abstraction is missing in the UML activity diagrams, we are forced to model it with an abstraction that is the most close to it, and that is the activity state. The model of the TCP protocol shown in Figure 3.20 comprises the following activity states (the names of the states are taken from the RFC 793):

- CLOSED (no connection exists)
- LISTEN (wait for a connection request from any remote TCP and port)
- SYN SENT (wait for a matching connection request after having sent a connection request)
- SYN RCVD (wait for a confirming connection request acknowledgement after having both received and sent a connection request)
- ESTAB (the connection is established, i.e., open)
- FIN WAIT 1 (wait for a connection termination request from the remote TCP, or an acknowledgment of the connection termination request that was previously sent)

Design 85

- CLOSING (wait for a connection termination request acknowledgment from the remote TCP)
- FIN WAIT 2 (wait for a connection termination request from the remote TCP)
- TIME WAIT (wait for enough time to pass to be sure that the remote TCP has received the acknowledgment of its connection termination request)
- CLOSE WAIT (wait for a connection termination request from the local user)
- LAST ACK (wait for an acknowledgment of the connection termination request previously sent to the remote TCP, which includes an acknowledgment of its connection termination request)

The activity diagram shown in Figure 3.20 is fully compliant with the original TCP standard. Interested readers can refer to IETF RFC 793 for more details.

The last example in this section shows a model of a simplified send e-mail operation. The corresponding activity diagram (Figure 3.21) is a straightforward implementation of a typical SMTP scenario (client side), which has already been introduced in this chapter (Figure 3.12) and in the previous chapter (Figure 2.12). Although simplified, in a sense that it just follows the successful path of the SMTP scenario, it is a complete specification of a desired behavior because it covers all possibilities in a non-overlapping manner.

Again, as in the previous example, the events associated with the reception of the corresponding messages are modeled as guard expressions, while the actions taken by the SMTP client are modeled by the corresponding action states. Additional similarity with the previous example is that the SMTP client, like the TCP protocol, spends most of the time in its stable states, waiting for a message from the SMTP server. If the received message is the one expected, the SMTP client sends the next message, prescribed by the ideal SMTP scenario, and proceeds to the next stable state. If the received message is not the one expected, the SMTP client returns the *value false* and the operation terminates.

The e-mail is successfully sent if all of the prescribed messages between the SMTP client and server are successfully exchanged. In this case, the send e-mail operation returns the *value true* and terminates.

3.5 Statechart Diagrams

In contrast to activity diagrams — which can be used for modeling activities both inside the individual objects and across the workflow of objects — the

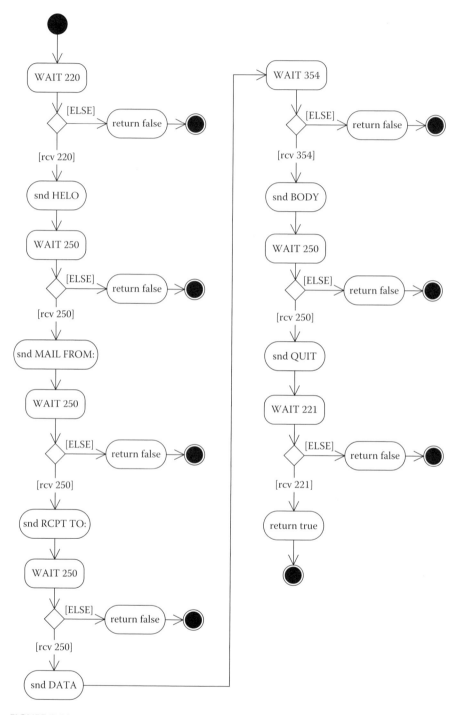

FIGURE 3.21
A simple send e-mail operation activity diagram (SMTP client side).

statechart diagrams are normally used for modeling the lifetime of a single object (typically, an instance of a class) or a use case. The activity diagrams emphasize the flow of the action and the activity states, whereas the statecharts emphasize the event-ordered behavior of an object, which is especially suitable for modeling reactive systems.

The common feature of both activity diagrams and statechart diagrams is that they aim at making complete models of behavior, i.e., for use in the design back-end. The driving forces for providing complete behavior specifications are the same, namely, the complete coverage of possibilities without overlaps. The styles differ a bit. By unwritten rule, the decision symbols are extensively used in activity diagrams and seldom in statechart diagrams. Therefore, the coverage of possibilities is shown explicitly in activity diagrams and more implicitly in statechart diagrams.

That the activity diagrams and statechart diagrams are semantically equivalent is also important to emphasize, i.e., we can use both of them for modeling the same behavior on the comparable level of details. They merely provide two different views of the same behavior. The activity diagrams are better suited for modeling individual operations whereas the statechart diagrams are better for modeling the behavior of entire stateful objects, especially if the behavior is driven by events (messages).

Statecharts were originally invented for modeling state machines, which makes them a perfect tool for modeling communication protocols because the protocols are essentially state machines. According to the UML terminology, a **state machine** is a sequence of states an object goes through in its lifetime. A **state** is a situation during which an object satisfies a certain condition, performs an activity, or waits for an event. An **event** is an occurrence of a stimulus that triggers the state transition. An **action** is an atomic executable statement (computation). An **activity** is a non-atomic execution composed of actions and other activities. A **transition** is a relation between the source and the target states (these can be different states or the same state) that specifies the actions to be taken when the given event occurs and the given guard condition is satisfied.

The key abstractions in the context of state machines are the object state and the state transition. We can think of the object state as a period of an object's lifetime (it can be just a moment characterized by a certain condition, a period of a certain activity, or an interval of time in which the object waits for a certain event). Alternately, we can think of the state transition as a rather short interval of object's lifetime, which is related to actions caused by a certain event and which is defined by the following five attributes:

- The source state
- The event trigger
- The guard condition
- The actions
- The target state

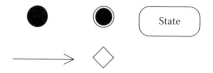

FIGURE 3.22
The basic set of symbols available for rendering statecharts.

A statechart diagram is a special type of graph that comprises a set of vertices that are interconnected by arcs. The basic set of graphical symbols available for rendering statechart diagrams is shown in Figure 3.22. Each symbol has a set of properties that must be set by the designer once they add the symbol to a diagram.

The initial state has three categories of properties. These are the general information, the table of constraints, and the tagged values (documentation notes). The general information is just the name and the type (initial). Each statechart diagram must start with this symbol.

The final state has the same categories of properties as the initial state symbol, with the exception that its type is final. If the lifetime specified by the statechart diagram is infinite the diagram will not contain this symbol. Alternately, it can contain one or more such symbols.

The state has six categories of properties. These include the general information, the table of entry actions, the table of exit actions, the table of internal transitions, the table of constraints, and the tagged values. The general information is just the name and the stereotype. Both the table of entry and the table of exit actions store the corresponding action names and their types. The table of internal transitions comprises their properties. Each internal transition is characterized by its name, its stereotype, and the event that triggers the transition.

The following eight common types of actions are used:

- Create an object
- Destroy an object
- Call an operation on another object
- Call an operation on this object (local invocation)
- Send a signal (message) to another (or this) object
- Return a value
- Terminate execution
- Uninterrupted action (other unclassified types of actions)

Four common types of events are also used:

- Signal event. This object has caught (received) the signal (message) that was thrown (sent) by another (or this) object. In UML, we model the signal by the class stereotyped as <<*signal*>>. We can also use a dependency relation, stereotyped as <<*send*>>, between the operation of the class that sends the signal and the class that defines the signal, to explicitly show the source of the signal. A signal is an asynchronous event.

- Call event. The object's operation is called by another (or this) object. A call event is a synchronous event. The event name and the parameters are the names and the parameters of the corresponding operation, respectively.
- Change event. The given condition is satisfied. Generally, the condition is related to the state of this object (value of its attributes) or to absolute time. We use the keyword *when* to specify the condition, e.g., *when((time == 17:00)*, or *when(key == pressed)*. A change event is an asynchronous event.
- Time event. The given interval of time has expired. We use the keyword *after* to specify the expression that evaluates to a period of time, e.g., *after(10s)*, or more symbolically *after(T1)*, which means that the timer *T1* has expired. By default, the starting time of such an expression is the time since entering the current state. If we want the starting time to be other than that, we must specify it explicitly. We should note that time events enable implicit timer management, as will be illustrated shortly.

The transition has four categories of properties. These are the general information, the table of actions, the table of constraints, and the tagged values (documentation notes). The general information comprises the name and optionally the corresponding event and the guard expression. The table of actions holds action names and their types. The decision has three categories of properties, namely, the general information (just the name), the table of constraints, and the tagged values (same as the decision in activity diagrams).

Simple examples that illustrate the usage of the basic set of graphical symbols for rendering statechart diagrams seem to be appropriate at this point. The following two examples, shown in Figure 3.23 and Figure 3.24, are semantically equivalent to the simple examples of activity diagrams shown in Figure 3.14 and Figure 3.15, respectively. The activity diagram shown in Figure 3.14 illustrates a sequence of three activity states, namely, *openPort(p)*, *sendData(seg)*, and *closePort(p)*. Figure 3.23 shows three versions of statechart diagrams that model the same behavior. These are the versions A, B, and C.

Version A models the behavior by a sequence of three transient states, namely, *Opening*, *Sending*, and *Closing*. By selecting appropriate names, we can indicate what type of activity is taking place in each of the states. The original activities *openPort(p)*, *sendData(seg)*, and *closePort(p)* are modeled as internal transitions of the states *Opening*, *Sending*, and *Closing*, respectively.

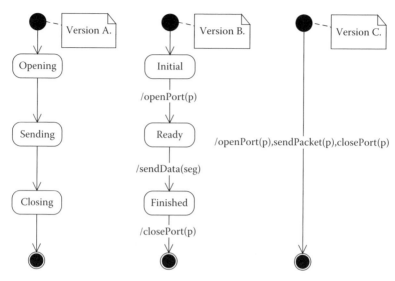

FIGURE 3.23
An example of a simple state machine with a single path of evolution.

We could also use entry or exit actions instead of internal transitions. Alternately, we could model this simple behavior by only one transient state with three internal transitions. Generally, by compressing models we decrease their clarity, and we should seek the compromise appropriate for the project at hand. Of course, defining clarity is tricky because it is essentially a matter of taste.

Version B is the model of the same behavior that employs another way of modeling activities in the statechart diagrams, and that is by actions taken by state transitions. This version of the model comprises three transient states, namely, *Initial*, *Ready*, and *Finished*, which are connected by triggerless transitions. Such transitions take place immediately after their source state is left (finished). The original activities *openPort(p)*, *sendData(seg)*, and *closePort(p)* are modeled here by the actions of the corresponding state transitions.

Finally, version C is the most compressed form of the model with the equivalent semantics. It comprises only one state transition, from the initial to the final state, which conducts a series of actions, namely, *openPort(p)*, *sendData(seg)*, and *closePort(p)*. This extreme shows the power of statechart diagrams. Generally, statecharts are more expressive than activity diagrams when it comes to modeling state machines, therefore we can model the same behavior in less space.

The activity diagram shown in Figure 3.15 is a model of a reliable packet delivery operation, which starts the timer T1, sends a packet, and waits for the answer from the remote site. If the timer T1 expires before the answer is received, the packet is sent again. If the answer is ACK, the operation returns the *value true*. Otherwise, it returns the *value false*. Figure 3.24 shows

Design

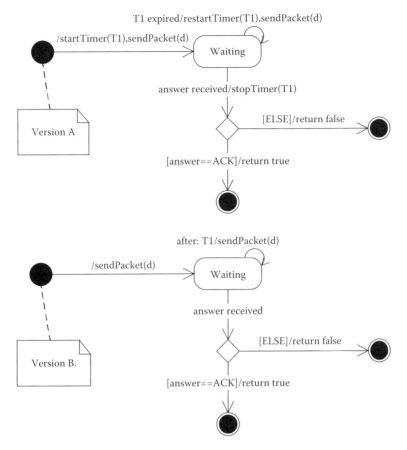

FIGURE 3.24
An example of a simple state machine with alternative paths and loops of evolution.

two versions of statechart diagrams that are models of the same behavior, namely, versions A and B.

Version A models the given behavior by explicit, rather than implicit, timer management. The triggerless transition from the initial state to the state *Waiting* starts the retransmission timer T1 and sends the packet by conducting the actions *startTimer(T1)* and *sendPacket(d)*. The expiration of the timer T1 is modeled here by the signal event *T1 expired*. The corresponding transition restarts the timer T1 and sends the packet again. The reception of the answer from the remote site is modeled by the signal *answer received*. The corresponding transition stops the timer T1 and leads to the decision with two outgoing transitions. The first is taken if the answer is ACK; otherwise, the second is taken. Those who prefer not to use decision symbols in their statechart diagrams should delete it, as well as the previous transition, and add the event answer received to both transitions that are leading to the final state.

Version B, in contrast to version A, models the given behavior by implicit timer management. Here the triggerless state transition from the initial state to the state *Waiting* just sends the packet by conducting the action *sendPacket(d)*. The existence of the state transition triggered by the time event *after: T1* implicitly implies that the timer T1 is started at the entrance to the state *Waiting*. If the timer T1 expires, the packet is sent again by the action *sendPacket(d)* and the timer T1 is restarted at the new entrance to the state *Waiting*. The event *answer received* occurs when this object receives the answer from the remote side. This event triggers the transition that leads to the decision and, later, to the final state. The timer T1 is implicitly stopped at the exit from the state *Waiting*. The result is a more compressed form of a model with more implicit details, which may not be seen at first glance. We can use either one of these two styles, but we should be consistent and stick to one on a certain project.

Now that we have covered the basics of statechart diagrams, we proceed to their more advanced abstractions. First, besides entry and exit actions and internal transitions, a state can perform an ongoing activity that we can specify by using the keyword *do*. Most of the states are stable states, which means that the object is blocked while waiting for an event. Some of the states are transient, which means that they perform certain computation and finish. Sometimes we need to also model active states, which perform some activity all the time while waiting for an event to occur, and we do this by using the keyword *do*. Generally, the special *do* transition can name another state machine or a sequence of actions.

Deferred events are the next important abstraction in the context of states. Until now, we were not interested in the events that occur during the state that does not react to them. What happens to these events? They are simply lost. If we want to save them so that they can be processed later in some other states, we must specify that they are to be deferred by using the special action named *defer*. Each event that is associated with this special action will be saved for further processing by the states that explicitly name that event in one of their transitions.

We have already shown how to manage complexity by using hierarchical organization. Statechart diagrams allow us to use that powerful concept in the context of states. Until now, we have dealt with simple states. Actually, a state in UML can also be a composite state, which means that it can comprise simple states and other composite states. This nesting of states can go to an unlimited depth, at least in theory.

A composite state can contain either sequential or concurrent substates. The sequential substates are disjoint, i.e., an object can be in only one of them at a certain point in time. The concurrent substates are orthogonal, which means that an object at a certain point in time is in all of the concurrent substates that are active at that point. We can think of a concurrent state as one aspect (orthogonal axis) of the object's lifetime.

The state transitions until now were transitions between simple states. After the introduction of composite states, the situation becomes more complex in this respect. Besides the transitions between simple states, there exist the transitions between simple states and composite states, as well as the transitions from substates to external states. The transitions from external states to substates of a composite state are not allowed. This asymmetrical relation raises the following question: What happens with the flow of states inside a composite state if a transition from that composite state to another state is triggered?

The answer is that the information about the point of interruption inside the composite state is lost by default. This means that the processing will be restarted from the very beginning when that composite state is re-entered once again later. This means that the composite state operates without context saving, which is referred to as a **history** in the UML.

If we want the composite state to operate with the history — which means it is able to restart from the point of interruption at its re-entrance — we can use the special history state. The history state is a special type of an initial state that is the target for the transitions from the external states. Once activated, it restarts the operation at the point of interruption. The following two types of history states are used:

- The shallow history state (marked with the symbol H)
- The deep history state (marked with the symbol H*)

The shallow history state ensures context-saving only on the first level of nesting of composite states. Alternately, the deep history state provides context-saving on the innermost state at any depth. If there are more nesting levels, the shallow history remembers the outermost nested state and the deep history remembers the innermost nested state.

Like activity diagrams, statechart diagrams also support modeling concurrency. We model concurrent activities in statechart diagrams by using concurrent sequences of substates inside a certain composite state. Typically, each such sequence begins with the initial state and ends with the final state. The transition, from the external state to this composite state, forks to concurrent substates, which at the end join in the transition from this composite state to the external state. The usage of concurrent substates is advisable only if the behavior of one of these concurrent flows is affected by the state of another. Alternately, if the behavior of the concurrent flows is driven by the signals (messages) they exchange, partitioning the object into more active objects is preferable.

The set of additional symbols that are available for rendering statechart diagrams is shown in Figure 3.25. These are the composite state, the shallow history state, the deep history state, the fork or join synchronization point, the note, the constraint note, the constraint and the OR constraint. These symbols, like others, have their properties. The composite state has the same categories of properties as a simple state and two additional indicators, namely, *Concurrent* and *Region*, which determine whether the composite state

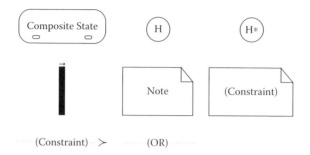

FIGURE 3.25
The additional graphical symbols available for rendering statecharts.

is concurrent or not and if it is a region or not. Both shallow and deep history states have the same three categories of properties. These are the name, the table of constraints, and the tagged values. The rest of the symbols have already been introduced.

Figure 3.26 shows the simple example of a statechart diagram that uses the shallow history state. Imagine a simple state machine that starts from the state *Idle*. The event *sendCharacter(ch)* triggers its transition to the composite state *Sending Segment*, which starts with the shallow history state to ensure context saving. Because this state comprises only simple states, the application of the deep history state, instead of the shallow history state, would have the same effect because only one level of nesting of composite states is found.

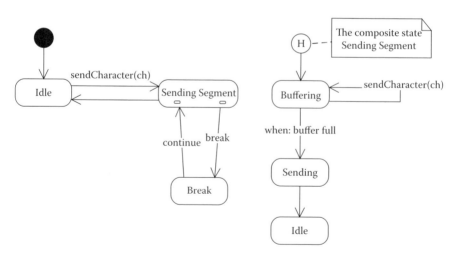

FIGURE 3.26
An example of a composite state that uses the shallow history state.

Design 95

The state machine remains in the substate *Buffering* while it is filling the corresponding buffer with new incoming characters. This status means that the state machine will wait for the additional event *sendCharacter(ch)* until the buffer becomes full, when the state machine will proceed to the state *Sending*. After it sends all the characters from the buffer, the state machine leaves the compound state *Sending Segment* and triggerlessly transits to the state *Idle*.

If the event *break* occurs while the state machine is in the compound state *Sending Segment*, its context will be saved and the state machine will leave it and move to the state *Break*. It will remain in this state until the event *continue* occurs. Then the state machine will re-enter the compound state *Sending Segment*, the context will be restored, and the state machine will resume the processing from the point of interruption.

The example in Figure 3.27 shows a simplified DNS client and server statechart diagrams. Both of them have just a single state. Being simple enough, these diagrams make very clear how statechart diagrams are used to make complete designs of communication protocols. Typically, a job performed by the communication protocol is to receive a message, process it, and send one or more messages as the result of this processing. Both DNS client and server go along this simple scheme.

The DNS client starts from the initial state by receiving a call to map the given domain name into the corresponding IP address. This action is modeled by the call event *map(d)* in Figure 3.27. This event triggers the transition of the DNS client from the initial state to the state *Wait DNS Response*. During the course of this transition, the DNS client sends the signal (message) *DNSrequest(d)*, which causes the signal event *receive DNSrequest(d)* at the DNS server side.

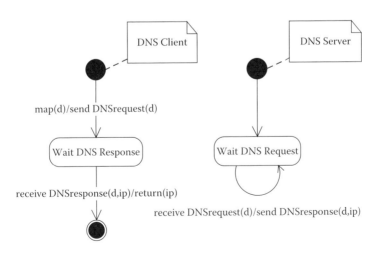

FIGURE 3.27
A DNS client and server statechart diagrams.

The DNS client is simply blocked in the state *Wait DNS Response* while waiting for the signal *DNSresponse(d,ip)*. The signal event *receive DNSresponse(d,ip)* triggers the DNS client transition to its final state. During this transition, the DNS client extracts the IP address from the received signal and returns it as its return value. This is modeled by the return action *return(ip)*.

The DNS server starts with the triggerless transition from its initial state to the state *Wait DNS Request*, where it is blocked while waiting for the signal *DNSrequest(d)*. The signal event *receive DNSrequest(d)* causes the DNS server to map the given domain name to the corresponding IP address, to create the signal (message) *DNSresponse(d,ip)*, and to send it to the DNS client. The DNS server performs all these actions during the transition to the same state, i.e., *Wait DNS Request*. This ensures that after servicing the current request, the DNS server remains available for servicing the next DNS request.

The example in Figure 3.28 shows the statechart diagram for one real protocol, namely TCP. It starts with the triggerless transition from the initial state to the state *CLOSED* in which it awaits one of the two possible call events. The call event *passive OPEN* causes TCP to create TCB (modeled with the action *create TCB*) and to move to the state *LISTEN*. Alternately, the call event *active OPEN* causes TCP to additionally send the signal *SYN* (TCP segment with the bit SYN set in the header) to the remote TCP entity. This is modeled with the actions *create TCB* and *snd SYN*.

TCP is blocked in the state *LISTEN* while waiting for one of the two possible events. The signal event *rcv SYN* triggers it to send the signal *SYN, ACK* (TCP segment with both bits SYN and ACK set) to the remote TCP entity and to move to the state *SYN RCVD*. The call signal *SEND* causes TCP to send the signal *SYN* to the remote TCP entity and to move to the state *SYN SENT*.

While being blocked in the state *SYN SENT*, TCP can be triggered by one of the three possible events. If the call event *CLOSE* occurs, TCP deletes TCB (modeled with the action *delete TCB*) and returns to the initial state. If the signal event *rcv SYN* occurs, TCP sends the signal *ACK* and moves to the state *SYN RCVD*. If the signal event *rcv SYN, ACK* occurs, TCP sends the signal *ACK* to the remote TCP entity and moves to the state *ESTAB*.

After reaching the state *SYN RCVD*, TCP can react to one of the two possible events. If the call event *CLOSE* occurs, TCP sends the signal *FIN* to the remote TCP entity and moves to the state *FIN WAIT 1*. If the signal event *rcv ACK of SYN*, occurs, TCP moves to the state *ESTAB*.

Two events are recognizable in the state *ESTAB*. If the call event *CLOSE* occurs, TCP sends the signal *FIN* to the remote TCP entity and moves to the state *FIN WAIT 1*. If the signal event *rcv FIN* occurs, TCP sends the signal *ACK* and moves to the state *CLOSE WAIT*.

In the state *FIN WAIT 1*, TCP may receive either *FIN* or *ACK of FIN* signals. In the former case, it sends the signal *ACK* and moves to the state *CLOSING*, whereas in the latter case it just moves to the state *FIN WAIT 2*, where it waits for the signal *FIN* to send the signal *ACK* and move to the state *TIME*

Design 97

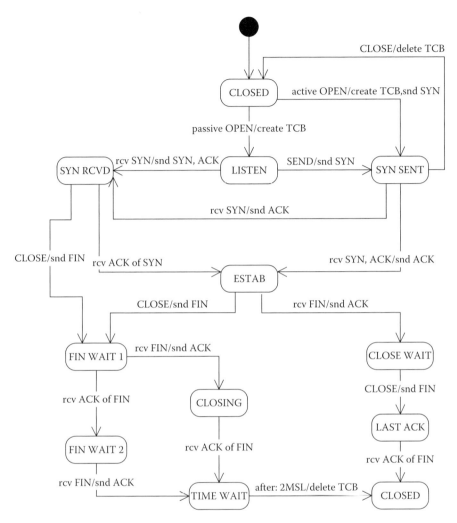

FIGURE 3.28
A TCP statechart diagram.

WAIT. On the alternative path, TCP moves from the state *CLOSING* to the state *TIME WAIT* after it receives the signal *ACK of FIN*.

Upon the entrance to the state *TIME WAIT*, a timer with the period 2MSL is started. When this period expires, TCP deletes TCB and moves back to its initial state *CLOSED*. After reaching the state *CLOSE WAIT*, TCP waits for the call event *CLOSE* to send the signal *FIN* and move to the state *LAST ACK*, and from there to the initial state *CLOSED* after it receives the signal *ACK of FIN*.

The example in Figure 3.29 shows the statechart diagram of a simple send e-mail operation (SMTP client side). It starts with the triggerless transition from its initial state to the state *WAIT 220* where it waits for the signal

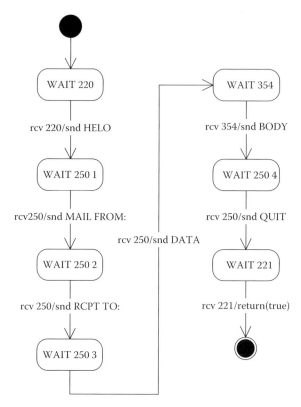

FIGURE 3.29
A simple send e-mail operation statechart diagram (SMTP client side).

(message) *220* from the SMTP server. When the signal event *rcv 220* occurs, the SMTP client sends the signal *HELO* to the SMTP server and moves to the state *WAIT 250 1*. After receiving the signal *250*, the SMTP client sends the message *MAIL FROM:* to the SMTP server and moves to the state *WAIT 250 2*.

Next, two signals of *250* in succession cause the SMTP client first to send the signal *RCPT TO:*, then to send the signal *DATA* to the SMTP server, and finally to reach the state *WAIT 354*. Upon reception of the signal *354*, the SMTP client sends the body of the e-mail message and moves to the state *WAIT 250 4*. After receiving the signal *250*, it sends the signal *QUIT* to the SMTP server and finally, after receiving the signal *250* again, it returns the value *true* and moves to its final state.

The main problem in this oversimplified version of the SMTP client is that it can block indefinitely while waiting for a signal from the SMTP server. The first thing that would be added in a more realistic design is a time limit on waiting for signals, which would be modeled with timers (keyword *after:*). The reaction to the expiration of a timer could be as simple as returning the

Design

value *false* and moving to the final state, or it can include some type of a recovery mechanism.

3.6 Deployment Diagrams

Deployment diagrams are used to model the deployment of the components, the component instances, objects, and packages, on nodes and node instances. A **component** is a part of the system that implements a set of interfaces. It typically models a physical package of logical elements, such as classes, interfaces, and collaborations. The common forms of packages are the following:

- Executables
- Libraries
- Tables
- Files
- Documents

A **node** is a physical element that models a computational platform, which comprises a set of resources, such as memory banks, buses, I/O channels, controllers, processors, and so on. The examples of nodes are the following:

- Personal computers
- Mainframes
- Embedded controllers
- Mobile or cellular phones
- Network routers

We use deployment diagrams in the design phase of communication protocol engineering for the following two main purposes:

- To identify network nodes and configurations
- To identify design subsystems and interfaces

The software architecture is closely related to the structure of the physical network. Sometimes the latter can be fixed and, in such a case, it governs the distribution of functionality across the network nodes as well as the selection of active classes. Alternately, both software architecture and network structure can be subjects of design and, in that case, some particular network structure can yield a more appropriate software architecture and system solution.

While trying to identify network nodes and configurations, we typically render network nodes as cubes, interconnect them with association relations, and think how to deploy individual components on these nodes. We show the deployment in the deployment diagrams by adding the component symbols (rectangles with tabs) and by connecting the related nodes and components with the dependency relations. Another way to do this is to adorn the node instances by the names of the components that are deployed on them.

Similarly, while trying to identify the subsystems and interfaces, we typically render the packages with their corresponding interfaces. We try to organize them into hierarchical layers (e.g., application-specific, application general, middleware, and system-software). Finally, we show which interfaces (services) are provided by which packages or components and also which packages or components are users of the services provided through those interfaces.

Deployment diagrams are a special type of graph that comprise the set of vertices that are interconnected with the corresponding arcs. Figure 3.30 shows the basic set of graphical symbols available for rendering deployment diagrams. These are the node, the node instance, the component, the component instance, the object, the package, the interface, the association relation, the aggregation relation, the dependency relation, the note, the constraint note, the two-element constraint, and the OR constraint. Each symbol has a set of properties, which must be set by the designer once they add the symbol to the diagram. The new symbols are the symbols representing the nodes, the components, and their instances. The rest of the symbols are already introduced in the previous sections about class and object diagrams (called together a static structure).

The node has six categories of properties. These are the general information, the table of attributes, the table of operations, the list of components, the table of constraints, and the tagged values. The general information includes the name, the full path, the stereotype, the visibility, and the indicators *Root*, *Leaf*, and *Abstract*. The list of the components comprises the names of the components that are deployed by this node.

The component has seven categories of properties, including the general information, the table of attributes, the table of operations, the list of nodes, the list of classes, the table of constraints, and the tagged values. The general information comprises the name, the full path, the stereotype, the visibility, and the indicators *Root*, *Leaf*, and *Abstract*. The list of nodes holds the names of the nodes that deploy this component. The list of classes stores the names of the classes that are implemented in this component.

The node instance has four categories of properties. These are the general information, the table of attribute values, the table of constraints, and the tagged values (documentation and persistent). The general information comprises the node instance name and the node name. The table of attribute

Design

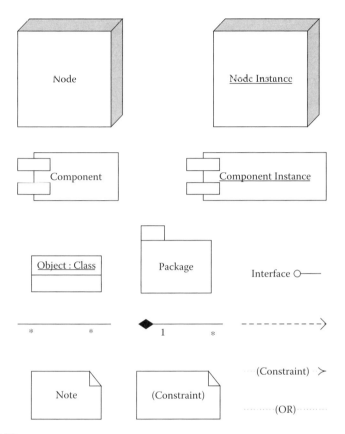

FIGURE 3.30
The basic set of symbols available for deployment diagrams.

values stores the name, the stereotype, the type, and the value for each attribute. The component instance has the same categories of properties as the node instance, with the exception that its general information differs and it comprises the name of the component instance and the component name.

The deployment diagram in Figure 3.31 shows an example of a network configuration comprised of three personal computers that are connected to the Internet. A personal computer is modeled as the node *PC*. Individual PCs are modeled as node instances, namely *Machine1*, *Machine2*, and *Machine3*. Internet is modeled as the node instance, named *Network*, of the node type named *Internet*. The real links that connect PCs to the Internet are modeled with the association relations between the node instances *Machine1*, *Machine2*, and *Machine3*, and the node instance *Internet*. The one-to-one nature of these links is modeled by setting the multiplicities on both sides of associations to 1.

This diagram is what the physical infrastructure of this example looks like. The software components are deployed as follows: The e-mail client executable is deployed to the first PC, the DNS server executable is deployed to

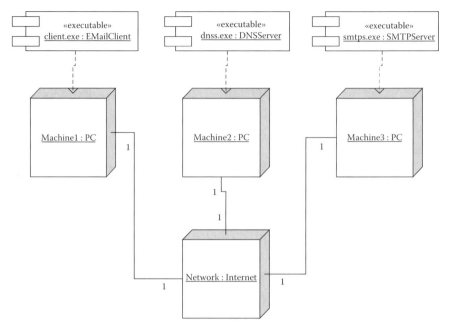

FIGURE 3.31
An example of a network configuration.

the second PC, and the SMTP server is deployed to the third PC. We model the e-mail client executable with the component *EMailClient*, which is stereotyped as the *<<executable>>*, and its particular instance deployed to the first PC with the component instance *client.exe*. Similarly, the DNS server executable is modeled with the component *DNSServer* and its particular instance deployed to the second PC with the component instance *dnss.exe*. Finally, the SMTP server is modeled with the component *SMTPServer* and its particular instance deployed to the third PC with the component instance *smtps.exe*.

The deployment diagram in Figure 3.32 shows the example of subsystems and interfaces. While thinking about the system shown in the previous example (Figure 3.31), we can identify three application layer packages, two system-software layer packages, and three interfaces. The application layer packages are the packages *EMailClient*, *SMTPServer*, and *DNSServer*, whereas the system-software packages are the packages *TCP/IP* and *OS*.

The package *TCP/IP* provides two service types through the interface *TCPport* and *IPint*, respectively. The services provided through the former interface are used by the package *EMailClient* and *SMTPServer*, whereas the services provided through the latter interface are used by the package *EMailClient* and *DNSServer*. Similarly, the package *OS* provides services through its interface *OSapi*. These services are used by the package *TCP/IP*.

Interested readers can find more information about the UML diagrams in the original books by Booch, Rumbaugh, and Jacobson (Booch et al., 1998).

Design

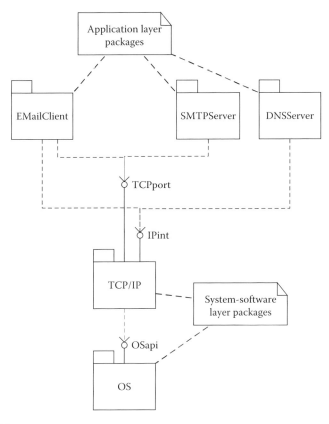

FIGURE 3.32
An example of subsystems and interfaces.

This section concludes the part of this chapter based on UML. The second part of the chapter is based on domain-specific languages.

3.7 Specification and Description Language

Software for real communication systems and devices (concentrators, packet switches, gateways, routers, and so on) is very complex and because of that, hard to understand. Proving that it is correct is very difficult, thus special attention is paid to software design. Software of this type can be modeled in the form of an individual or a group of finite state machines. Japanese designers were the first to apply this method of specification and description of communication protocols in the 1970s. Not long after its initiation, the CCITT (predecessor of ITU-T) has standardized it in the form of the so-called Specification and Description Language (SDL).

SDL creators have been facing the following dilemma. Traditionally, a finite state machine (FSM) has been modeled by a state transition graph. Typically, a state transition graph is graphically illustrated by circular symbols representing states and arrows representing state transitions. State labels are state names whereas state transition labels indicate FSM input that causes the corresponding state transition and FSM output produced by the same transition. An advantage of this type of FSM representation is that all the stable FSM states are clearly indicated and can be easily noticed. Alternately, a disadvantage of this type of FSM representation is that message-processing procedures are not defined formally. Informally written state transition labels, placed close to the corresponding arrows, indicate only the FSM input causing the transition and the output that the FSM must produce. This information is far from being sufficient for writing the software that implements the given FSM — it only provides some hints to programmers.

Another approach would be to use a flow chart, a traditional way of specifying data-processing algorithms. An advantage of this type of FSM representation is that message-processing procedures are clearly and precisely defined. A disadvantage is that stable FSM states are not clearly indicated, therefore they can hardly be noticed. The FSM states can be marked as certain points in a flow chart by using informal annotations, and that is simply not comprehensible enough.

The creators of the SDL language have found a solution to this dilemma by combining the above mentioned approaches, namely, the state transition graph-based approach and the flowchart approach. This combination has been cleverly made by simple extension of the set of graphical symbols available for drawing flowcharts. The key new graphical symbols introduced are the symbol corresponding to an FSM stable state and the symbols that represent FSM inputs and outputs (input and output messages). We will fully describe all the SDL graphical symbols later in this chapter.

The protocol designer uses SDL language to specify and describe the corresponding automata instance by listing all its states and state transitions. Although the number of states can be very large, this task is simplified by the fact that in a given state, only a limited number of events can occur, and this means that the automata instance can evolve from a given state only to a limited number of new states. For example, consider a telephone call automata instance waiting for the first digit to be dialed (the automata instance enters this state immediately after the user has initiated an outgoing call, i.e., after the so-called "hook-off" event). The telephone call automata instance cannot evolve from this state to any other arbitrary state. More precisely, in this state only the following three events are possible:

- The user ends the call (hook-on event), which causes the automata instance to evolve to its initial idle state.
- The user dials a digit (digit event). This event triggers the state transition from the current state to the state of waiting for the second digit.

Design 105

- The user does nothing during a certain interval of time. This will cause the expiration of the corresponding timer and a state transition to the state in which the telephone line is blocked.

Communication protocol is by nature a reactive system. Normally, it is blocked in its current state while waiting for one of a few recognizable events to occur. Statistically, it is inactive most of the time. A recognizable event triggers the corresponding state transition to a new state, where the protocol is again blocked while waiting for further events. The state transitions comprise a finite number of primitive operations that are statistically rather short.

An important characteristic of program implementations of the protocols is that they are not trying to monopolize the CPU. This implies that the execution of this type of a program should be organized as a **process with stable states**. In contrast to the conventional time-slicing system, where the task switching is driven by timer interrupts, switching of processes with stable states is performed at the moment at which the running process reaches its new stable state. Whereas conventional tasks can be interrupted in an arbitrary point of time (determined by the asynchronous occurrence of timer interrupt signal), a process with stable states is normally not a subject to preemption because, unlike conventional tasks, they are not monopolizing the processor. Of course, a process with stable states can be interruptible so that the whole system can react to the urgent events handled by the higher priority tasks.

Enumeration of the possible states and state transitions, as described above, is a logical process that seems to be straightforward for the experts. However, graphical language, such as SDL, is needed to make it possible for the design engineers to easily make complete formal specifications of the protocols. The main advantages of graphically-oriented languages are the following:

- Graphical language is easy to read and, because of that, it is easy to check specification completeness and correctness.
- The specification can be easily extended.
- The specification can be directly implemented in software. This means that if the specification is correct, a high probability exists that the software implementation is also correct.

According to ITU-T, the complete software (system) is decomposed into a set of **functional blocks**. Each functional block consists of a set of processes and each process comprises a number of **tasks** (Figure 3.33).

A **process** is essentially an execution of a logical function, which consists of a series of operations applied to message information elements (referred to as tasks) in discrete points of time. Either it is in some of its stable states or it makes its transition from the current to the next state. (In Chapter 4, we refer to the state transition as **unstable states**).

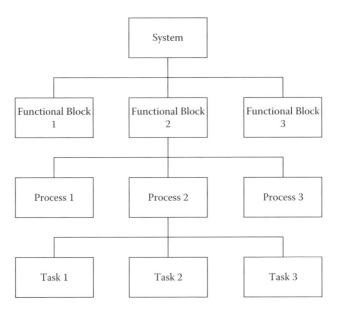

FIGURE 3.33
The structure of the communication software according to ITU-T.

A **signal** is defined as a data stream that delivers information to the receiving process. A data stream among the processes inside the same functional block represents the **internal signal**, whereas a data stream between the processes that are parts of different functional blocks represents the **external signal** to the receiving process. Therefore, from the receiving process point of view, the signal can be classified as internal or external, depending on whether it originates from the same or from a different functional block.

Today, SDL is a standard design language that can be used to specify and describe any system implemented in hardware or software, particularly real-time systems. In this book, we are especially interested in one type of the real-time systems — communications systems.

The basic set of SDL rules is given in ITU-T recommendation Z.100e. Additional explanations are given in a series of subsequent ITU-T recommendations, namely Z.100d1e, Z.100nce, Z.100nfe, Z.100p1e, and Z.100s1e. The main characteristics of the SDL language are the following:

- It is easy to learn.
- It is easy to extend the specification in case of the new requirements.
- In principle, it can support various methodologies for making the system specifications.

Two forms of SDL language exist, graphical (SDL-GR) and program (SDL-PR). The graphical form has been widely accepted for two reasons. First, it is closer to human understanding because it is easier to understand and follow. Second, in principle, it does not require the support by special, and frequently very expensive, software tools. Of course, a piece of paper and a pencil is hardly sufficient for a professional work. At least a modern graphical editor that supports the SDL set of graphical symbols is needed to enable the making of decent specifications. In this book, we use Microsoft Visio for that purpose.

The second SDL form, SDL-PR, is practically a higher-level programming language of textual type (similar to C/C++ and Java programming languages). Clearly, this programming language is less synoptic and is harder to follow than the graphical form. It is intended to be used mainly by the accompanying software tools, such as Telelogic® Software Development Tools (SDT). The goal of using such software tools is not just to make isolated specification and description documents, but rather to make electronic specifications, essentially models of protocols. The software tools can then be used to interpret the models and generate the corresponding program code.

In addition to the tools provided by Telelogic, other tools exist based on this philosophy that is, as already mentioned, referred to as model integrated computing (MIC). One of them is also already mentioned, GME.

The main SDL applications are the following:

- Call processing in switching systems
- Error supervision and management in telecommunication systems
- Supervision, control, and data acquisition systems
- Telecommunication services
- Data transfer protocols
- Protocols in computer communications

The SDL language basics are as follows: SDL is based on a set of special symbols and the rules for their application. The graphical form (SDL-GR) is based on special graphical symbols whereas the program form (SDL-PR) is based on a set of special keywords. Both SDL forms use the same set of keywords specialized for data representation.

Later, we assume that a system consists of a number of protocols. Also, we refer to a set of hierarchically organized protocols as a **family of protocols** or a **protocol stack**. Typically, each protocol that is a part of the family performs its well-defined task. The family of protocols conducts rather complex tasks by cooperation of its members.

A system is described as a set of interconnected functional blocks. **Channels** are defined as communication links that are used for the interblock communication and for the communication between the blocks and the environment. Each block comprises a number of processes that communicate by exchanging signals. A channel is typically implemented as a FIFO (First-In-First-Out) queue that stores the signals (i.e., messages) to be transferred through the channel. A process is defined as a finite state machine (automata instance) that is described by the given set of states and state transitions.

The next simple example illustrates the notions and terms introduced above. Both graphical and program SDL forms are presented. The only goal of presenting the program form is to provide the intuition for the reader that will help them understand the main differences between the graphical and program forms of the SDL language. The aim of this book is not to fully cover the program form of the SDL language.

The example is a simple game called *Daemongame*. The core of the game is a simple FSM that has only two states, *even* and *odd*. Timing is controlled with a single timer. The expiration of the timer (this event is labeled *none*) causes the FSM to switch from *even* state to *odd* state. The player presses a button when they wish (this event is labeled *Probe*), i.e., at arbitrary points of time. If the FSM is in *even* state, the player gets one negative point (*Lose*). If the FSM is in *odd* state, the player gets one positive point (*Win*). If the player scores more *Win* than *Lose* points, they win the game.

The first step in describing this simple system is to define input and output signals. Input signals are the following:

- *Newgame*: the player wants to start the game
- *Probe*: the player has pressed a button
- *Result*: the player wants to see the current score
- *Endgame*: the player wants to quit the game

Output signals are the following:

- *Gameid*: current game identification
- *Win*: positive point
- *Lose*: negative point
- *Score*: total amount of points (number of *Win* points minus number of *Lose* points)

The specification of the game *Daemongame* in the graphical form of SDL is shown in Figure 3.34. It contains a single functional block labeled *Game*. Input signals are *Newgame*, *Probe*, *Result*, and *Endgame*. Output signals are *Gameid*, *Win*, *Lose*, and *Score*. Signal declarations are shown in the upper left corner of the figure.

The *Daemongame* system specification in the program form of SDL is the following:

```
system Daemongame
   signal Newgame, Probe, Result, Endgame, Gameid, Win, Lose, Score(Integer);
   channel Gameserver.in
      from env to Game
```

Design

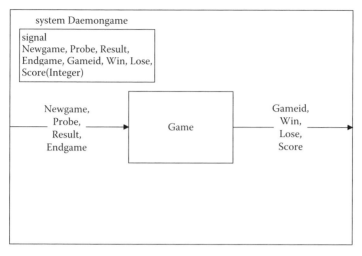

FIGURE 3.34
The structure of the system *Daemongame*.

```
    with Newgame, Probe, Result, Endgame;
  endchannel Gameserver.in;
  channel Gameserver.out
    from Game to env
    with Gameid, Win, Lose, Score;
  endchannel Gameserver.out;
block Game referenced;
endsystem Daemongame;
```

Generally, any system SDL program specification starts with the keyword *system* and ends with the keyword *endsystem*. This particular program defines all the required signals (*Newgame, Probe, Result, Endgame, Gameid, Win, Lose,* and *Score*), the input channel *Gameserver.in*, and the output channel *Gameserver.out*.

In contrast with the graphical form, which is easy to understand, the program form represents a lower-level specification, closer to the machine and with more details. For example, in the graphical form a channel is simply represented by an arrow pointing to or from the functional block. The channel declaration in the program form is much more detailed: It comprises the channel name (e.g., *Gameserver.in*), its direction (e.g., from environment toward the functional block *Game*), and a list of signals that must be transferred over the channel (e.g., *Newgame, Probe, Result,* and *Endgame*).

The next lower hierarchical level of detail describes a single functional block of this simple system, namely, the block *Game*. Its specification is given in both forms of SDL. The graphical form of the specification is given in Figure 3.35. The program form of the specification is given immediately after a short explanation of Figure 3.35.

Figure 3.35 shows that the block *Game* consists of two processes, namely *Monitor* and *Game*. The processes are connected to the environment, and to

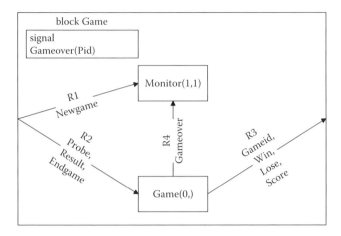

FIGURE 3.35
The structure of the functional block *Game*.

each other, by signaling paths. It also shows that the input channel *Gameserver.in* consists of two signaling paths, the signaling path R1 (which is used to carry *Newgame* signal) and the signaling path R2 (which is used to carry the signals *Probe*, *Result*, and *Endgame*). The output channel *Gameserver.out* comprises the single signaling path R3. A single internal signaling path exists inside the block *Game*, the path R4, which is used to carry the internal signal *Gameover* from the process *Game* to the process *Monitor*. This new signal is declared in the upper left corner of the graphical specification.

The specification of the block *Game* in SDL-PR is the following:

```
block Game;
  signal Gameover(Pid);
  connect Gameserver.in and R1, R2;
  connect Gameserver.out and R3;
  signalroute R1 from env to Monitor with Newgame;
  signalroute R2 from env to Game with Probe,Result,Endgame;
  signalroute R3 from Game to env with  Gameid,Win,Lose,Score;
  signalroute R4 from Game to Monitor with Gameover;
  process Monitor(1,1) referenced;
  process Game(0,) referenced;
endblock Game;
```

The specification given above starts with the keyword *block* and ends with *endblock*. Inside the body of the definition of the block *Game*, we start with the declaration of the internal signal *Gameover* by declaring its name, followed by the list of its parameters enclosed in parenthesis. The signal *Gameover* has a single parameter, the identification of a process (*Pid*) that is sending this signal.

Design 111

Further on, we connect the channel *Gameserver.in* with the signaling paths R1 and R2. We also connect the channel *Gameserver.out* with the signaling path R3. We proceed with the declarations of signaling paths (keyword *signalroute*). Each declaration indicates the signaling path name, its direction (by using the keywords *from* and *to*), the names of the processes it connects (note that *env* is the special process which represents the environment), and a list of signals it carries (by using the keyword *with*). For example, the first signal path declaration shown in SDL-PR above declares the signaling path R1, which carries the signal *Newgame* from the process *env* (environment) to the process *Monitor*.

We end the definition of the functional block *Game* by declaring the processes it contains. A process in general is declared by the keyword *process*. A process declaration indicates the name of the process followed by the initial and maximal number of process instances that can appear in the system. The maximal number of process instances is an optional parameter, i.e., it can be omitted.

The process *Monitor* is declared as *Monitor(1,1)*, which means that the block *Game* should initially create one instance of this process and, at the same time, it is also the maximal number of *Monitor* instances that can be created in this block. Alternately, the process *Game* is declared as *Game(0,)*, which means that initially there are no *Game* instances, but also that the maximal number of Game instances is not limited, i.e., in theory it is allowed to create an infinite number of process *Game* instances inside the functional block *Game*. Of course, in reality this number is always limited to the available hardware resources.

In this particular example, we have declared two processes, *Monitor* and *Game*, that operate inside the functional block *Game*. The process *Monitor* handles the interaction with a player. It is a mediator between the player and the process *Game*, which is essentially a model of the win-lose game. Due to the fact that the process *Monitor* is trivial and actually insignificant for this example, we will define only the process *Game* on the next hierarchically lower level of abstraction. On this level of detail, the process *Game* is modeled as a finite state machine (automata instance).

As already mentioned, the creators of SDL-GR (graphical form of SDL) have extended the basic set of traditional flow chart symbols with a set of graphical symbols specialized for modeling finite state machines. The complete set of graphical symbols available for describing processes in SDL-GR is shown in Figure 3.36.

The meaning of the individual graphical symbols shown in Figure 3.36 is as follows:

- *state*: specifies a stable state in which a process is blocked while waiting for one of the recognizable signals (referred to as *input*)
- *input*: specifies the reception of a given input signal (i.e., the occurrence of a certain event)

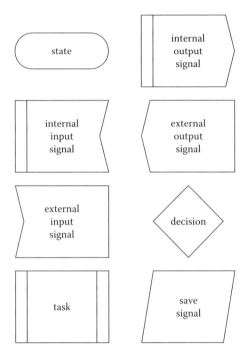

FIGURE 3.36
The set of graphical symbols available in SDL-GR.

- *output*: specifies the transmission of a given output signal (normally the output signal generated by a certain process represents an input signal for a process that receives it)
- *decision*: specifies an operation that checks if a given condition is true or false and, based on the outcome, selects one of the two possible paths in the current state transition
- *task*: specifies an action in the course of current state transition that is neither *decision* nor *output*
- *save signal*: specifies that recognition (processing) of a given signal should be postponed until it reaches a state where it is recognizable This symbol is used in specifications of signaling systems (e.g., SS7). It is seldom used in other applications, such as call processing.

The specification of a process in SDL-GR is generally made as a combination of the instances of the graphical symbols shown and explained above. An example of this type of specification is shown in Figure 3.37. It specifies and describes the process *Game*, the core of the win-lose game.

The evolution of the process starts from an unnamed state in the upper right corner of the graphical presentation (Figure 3.37). Starting from this state, the process unconditionally transits to its next stable state *even*. During this transition, the process *Game* sends the signal *Gameid* to the player.

Design

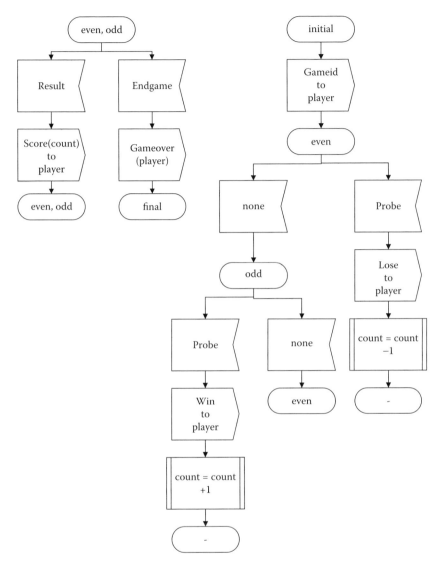

FIGURE 3.37
The process *Game* specification in SDL-GR.

While the process *Game* is in its stable state *even*, it awaits one of two possible events, the reception of the signal *Probe* or the expiration of the timer labeled *none*. If the timer expires, the process *Game* receives the corresponding signal *none*, and this causes the process to evolve into the next stable state *odd*. If the process receives the signal *Probe*, it sends the signal *Lose* to the player and updates the player's score, which is stored in the variable *count*, by adding one negative point. The process does not change its stable state, i.e., it remains in its current state (which is denoted with the character "–"), and that is the state *even*.

In its stable state *odd*, the process *Game* recognizes two same possible events, the reception of the signal *Probe* or the expiration of the timer labeled *none*. Actually, the timer *none* determines the time interval the process will spend in either the *even* or *odd* state before switching to the other one. Hence, if the timer *none* expires, the process evolves into the stable state *even*. Alternatively, if the process receives the signal *Probe*, it sends the signal *Win* to the player and updates the player's score (value of the variable *count*) by adding one positive point. The process remains in its current state (i.e., the state *odd*).

The upper left corner of the graphical representation of the process *Game* (Figure 3.37) shows one important example of simplifying SDL-GR diagrams. Because the reception of the input signals *Result* and *Endgame* is possible in both *even* and *odd* states, a straightforward solution would be to mechanically add these inputs and their processing to both states. The result would be a diagram that is much more complex and harder to understand and follow. A more elegant solution is to draw the description of the processing of the inputs *Result* and *Endgame* in both states as a separate drawing in the diagram, as shown in Figure 3.37.

Generally, it is always useful to try to find identical processing of input signals (state transitions) that repeat in a number of stable states and to simplify the specification by drawing these parts separately in the diagram. This type of a model reduction is really easy. We just draw an oval state symbol and write a list of the states (the list comprises the state names separated by commas) that share the common inputs inside the state symbol. Then we can copy and paste common state transitions. At the end, we can just remove the redundant state transitions. Of course, in the simple diagrams such as in the example at hand, we can see this in advance and draw accordingly, as we did for the processing of the inputs *Result* and *Endgame* in the states *even* and *odd*.

If the process *Game* receives the signal *Result*, which comes from the environment, i.e., from the player, the process sends the signal *Score(count)* to the environment (actually to the player) and it remains in its current state (*even* or *odd*). Alternately, if the process *Game* receives the signal *Endgame*, it sends the signal *Gameover* to the process *Monitor* and the game ends, i.e., the functional block deletes the process *Game*.

The specification of the process *Game* in SDL-PR (SDL program form) is the following:

```
process Game(0,); fpar player Pid;
  dcl count Integer := 0; /* the counter that contains the  result */
  start;
    output Gameid to player;
      nextstate even;
  state even;
```

Design

```
      input none;
        nextstate odd;
      input Probe;
        output Lose to player;
        task count:=count-1;
        nextstate -;
    state odd;
      input Probe;
        output Win to player;
        task count:=count+1;
        nextstate -;
      input none;
        nextstate even;
    state even,odd;
      input Result;
        output Score(count) to player;
        nextstate -;
      input Endgame;
        output Gameover(player);
        stop;
endprocess Game;
```

The definition of the process starts with the keyword *process* and it ends with the keyword *endprocess*. As already mentioned, initially no instances of the process *Game* are used, and the maximal number of its instances is unlimited. The process declaration is followed by the construct *fpar player Pid*, which defines the formal process parameter *player* that is assigned the value *Pid*. At the beginning of the game, the run-time environment creates an instance of the process, and assigns a unique *Pid* number to it.

Next, we declare the integer variable *count* (using the keyword *Integer*), which contains the current total value of points that the player has scored so far. After the label *start*, we define a series of statements that are executed by the process at its startup. In this example, the process *Game* at its startup sends the signal *Gameid* to the player and enters its initial stable state *even* (next state of the process is defined by the keyword *nextstate*).

For each stable state (keyword *state*) of the process, we define all the recognizable input signals (using the keyword *input*) and on the next level of indentation, we define the corresponding state transition as a series of statements that ends with the *nextstate* statement. For example, the recognizable input signals in the stable state *even* are the signal *none*, which relates to the expiration of the corresponding timer, and the signal *Probe* generated by the player's stroke of the pushbutton. In the case the timer *none* expires, the process evolves to its next stable state *odd*. Alternatively, if the process receives the signal *Probe*, it sends the signal *Lose* to the player (using the keyword *output*), performs the task of decrementing the score by 1 (using the keyword *task*), and remains in its current state (the statement *nextstate -;*).

The stable state *odd* is defined in a similar manner. The input signals recognized by the process in its stable state *odd* are the signal *Probe* and the expiration of the timer *none*. If the process receives the signal *Probe*, it sends the signal *Win* to the player, increments the score by 1, and remains in its current stable state *odd*. Alternatively, if the timer *none* expires, the process evolves into its stable state *even*. Finally, we define the state transitions initiated by the reception of the input signals *Result* and *Endgame* in either the state *even* or *odd*.

Understanding the principals of SDL-PR helps in more easily understanding the communications protocol software implementation in the state-of-the-art, higher-level programming languages such as C/C++ or Java. Although SDL-PR can resemble a pseudolanguage when compared to these programming languages, in reality it is a specialized language of higher level abstraction and it is feasible to construct a compiler for it. However, the study of the compilers is out of the scope of this book. The primary goal of this book in this respect is to provide an insight into the manual coding of SDL graphical diagrams in some of the above mentioned programming languages (C/C++ or Java).

The example under study can help in this respect. Obviously, two levels of nesting are included in it. The first level of nesting corresponds to the current stable state, in which the process is blocked while waiting for the next input signal, i.e., *start*, *even*, and *odd*. The second level of nesting corresponds to the type of input signal, i.e., *Probe*, *none*, *Result*, or *Endgame*.

The simplest method to implement this selection construction with two levels of nesting in the C/C++ or Java programming language is to use nested *switch-case* statements. The first *switch-case* statement is used to locate the current state. Then in each *case* clause of the first *switch-case* statement, another *switch-case* statement is used to locate the state transition statements that correspond to the given input signal. This type of protocol implementation will be covered in detail in the next chapter.

3.7.1 Telephone Call Processing Example

The second example of the system specification made in SDL-GR is the specification of the telephone call processing system. The description of this system is given in the separate ITU-T recommendation Q.71. The Q.71 compliant program system consists of six mutually interconnected functional entities (referred to as functional blocks), namely FE1, FE2, FE3, FE4, FE5, and FE6 (Figure 3.38). The aim of this example is just to illustrate SDL-GR applicability and the details of the recommendation Q.71 (such as the concrete names of the entities, their types and links, i.e., relations) are not really significant for the comprehension of the usage of SDL-GR. The reader that is more interested in Q.71 details can refer to the corresponding ITU-T recommendation.

We use the hypothetical telephone call processing system *CallProcessor* to make further illustrations more concrete, but without diving into the bulk of details of Q.71 recommendation. Comparing it to the real Q.71-compliant system, the *CallProcessor* is a very simplified academic example that consists

Design

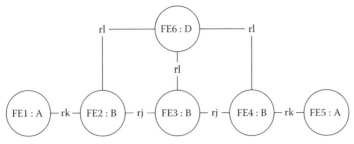

Where:
A, B, and D are the types of functional entities
FE1, FE2, FE3, FE4, FE5, and FE6 are the names of the functional entities
rk, rj, and rl are the relations between the functional entity types

FIGURE 3.38
A functional model of the telephone call processing system.

of a single functional block, namely *TelephoneLine* (Figure 3.39). This functional block is linked with the environment by one input channel, named *input*, and one output channel, named *output*. So far, this example is very similar to the previous example *Daemongame*, which also comprises the single functional block *Game* that is interconnected with the environment with one input and one output channel.

The functional block *TelephoneLine* is shown in Figure 3.40. This simple functional block consists of the single process *FE1*. Two lists of signals are declared (using the keyword *signallist*) in the upper left corner of Figure 3.39, namely, *input* and *output*. The process *FE1* is connected both to the telephone user (shown by the arrows placed at the right of *FE1*) and to the telephone exchange (indicated by the arrows placed at the bottom of *FE1*). It can receive

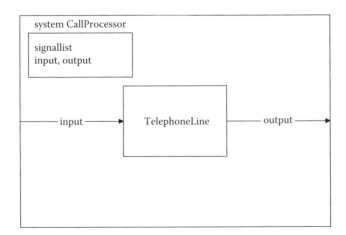

FIGURE 3.39
The hypothetical system *CallProcessor*.

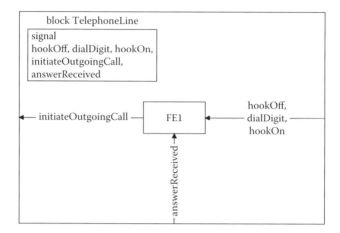

FIGURE 3.40
The structure of the functional block *TelephoneLine*.

one of the three possible input signals (*hookOff*, *dialDigit*, and *hookOn*) from the telephone user's side. Alternately, it can send the output signal *initiateOutgoingCall* to the telephone exchange or it can receive the input signal *asnwerReceived* from the exchange.

The process *FE1* is specified in the graphical form of SDL, SDL-GR, in Figure 3.41. This process resides in the telephone exchange and it communicates with the human that uses the telephone to establish a call, talk to the called party, and release the call at the end of the conversation. In reality, such a process must handle many scenarios, e.g., the user picks up the receiver but does not dial the number, or stops after dialing an insufficient number of digits.

The process specified in Figure 3.41 is rather simplified but it still captures the most significant part of the telephone line functionality on the calling party side. The telephone line in this context is a processor that hosts *FE1*, together with the interfacing hardware that connects it to both the calling party user's telephone and switching unit of the telephone exchange. For brevity, we refer to the former simply as a user and to the latter as a telephone exchange, or just an exchange.

The process *FE1* has four stable states, namely, *IDLE*, *WAIT_DIGIT*, *WAIT_ANSWER*, and *CONVERSATION*. The evolution of the process starts from the state *IDLE*. The single recognizable input signal in this state is the signal *hookOff*. If the process *FE1* receives the signal *hookOff*, it performs the task *prepareForDialing* and moves to its next stable state *WAIT_DIGIT*. While performing the task *prepareForDialing*, the process connects the free-to-dial tone to the calling party user. This tone serves as the indication to the user that they can start dialing the number of another user to which they wish to talk.

Two recognizable input signals are used in the stable state *WAIT_DIGIT*, i.e., the process can either receive the input signal *hookOn* or the input signal

Design

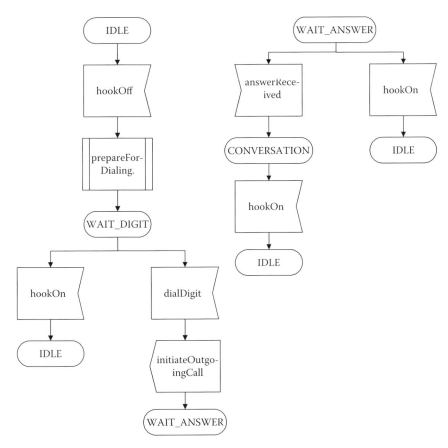

FIGURE 3.41
A simplified model of the Q.71 FE1 in SDL-GR.

dialDigit. In this simplified example, we assume that the telephone number of the called party consists of a single digit. However, in real ISDN telephone networks, a so-called *enblock* dialing mode exists in which the ISDN terminal sends the complete telephone number to the telephone exchange in a single *SETUP* message. Therefore, this simplified example is not so far from reality. If the process *FE1* receives the input signal *hookOn*, it evolves into its initial state *IDLE*. If it receives the input signal *dialDigit*, it sends the output signal *initiateOutgoingCall* to the telephone exchange and it moves to the stable state *WAIT_ANSWER*.

In the stable state *WAIT_ANSWER*, two events are again possible — the reception of the input signal *hookOn* or the reception of the input signal *asnwerReceived*. In the former case, the process goes back to its initial state *IDLE*, whereas in the latter it evolves into its next stable state *CONVERSATION*. The input signal *asnwerReceived* is actually the result of the series of events that start with the input signal *hookOff* at the called party side. The telephone line entity at the called party translates it to the signal *answerIncomingCall* and sends it to the exchange at the called party side, which in turn sends it to the exchange at the calling party side. Finally, the exchange at the calling party side translates it to *asnwerReceived* and sends it to *FE1*.

In the final stable state *CONVERSATION*, only a single event is possible. The process *FE1* can receive the input signal *hookOff*, and if it does, that is the end of the conversation phase of the call and the process will return to its initial stable state *IDLE*. This closes the circle and the process is ready to process a new call originating from the same telephone line. Clearly, an instance of the process *FE1* is assigned to each telephone line in the telephone exchange.

In this example, we described the process *FE1* that is assigned to the calling party telephone line without going into a detailed specification of the operations performed by the telephone exchanges and the called party telephone line involved in the call. Obvious from this example should be that SDL diagrams are self-documented formal specifications and that no need really exists for any additional textual descriptions.

The SDL diagram shows the possible evolution paths of a process (a call processing in the example above). It defines unambiguously all telephone stable states, as well as all possible input signals for each state. The functional specification is based on the logical advance of a call, expressed in terms of telephony events. This makes it completely independent of both hardware structure of the hosting system and selected programming language and framework.

The SDL diagram is drawn based on the observations of a single telephone call without thinking about other calls, which are processed simultaneously (quasi-parallel by a single CPU or genuinely parallel by a multi-CPU system). This approach greatly simplifies software design. Finally, the existing SDL diagram can be easily extended by adding new states and input signals without the need to start drawing a new diagram from the very beginning. This possibility also enables the easy removal of revealed design errors.

3.8 Message Sequence Charts

An alternative method of specifying communication systems is by drawing message sequence charts that show the sequences of messages (signals) exchanged by the communicating entities. The ITU-T has developed a special language for this purpose, briefly referred to as MSC (Message Sequence Charts), and it has standardized it in Z.120 series of ITU-T recommendations.

MSC is based on the idea of following a single evolution path of a process. We start from a certain, most frequently initial, state of the process (e.g., the state *IDLE* in the previous example). After that, we select one of the possible input signals and follow the evolution path to which it points. In the previous example, a single input signal can be received in the state *IDLE*: signal *hookOff*, which causes the transition to the state *WAIT_DIGIT*.

In the newly reached stable state, we select again one of the recognizable events (the input signals that may be received in the stable state *WAIT_DIGIT* are *hookOff* or *dialDigit*; let us assume that we have selected *dialDigit*) and we follow the process evolution along the corresponding path (in the case of the input signal *dialDigit*, the process moves to the state *WAIT_ANSWER*). At the same time, as we mentally follow the evolution path of the process, we draw on the paper, or even better in the corresponding graphical editor, the messages that are exchanged between the process and its environment. The messages are represented by the graphical arrow symbols that are labeled by the message names. This is how we get the MSC charts.

Clearly, an MSC chart represents a single trace over the corresponding path, through the SDL diagram, or some other form of specifying finite state machines. We can see intuitively that for the real automata that we come across in practical applications, a finite number of paths exist that cover the SDL diagram. The set of the MSC charts that are obtained by visiting these paths represents the specification that is in a logical sense equivalent to the SDL diagram.

However, an obvious disadvantage of this type of a specification, in a form of a set of the MSC charts, is that it is much less evident than the SDL diagram. Therefore, when communication protocol designers refer to the formal specification, they really assume the SDL diagram. This disadvantage becomes obvious if instead of dealing with a single automaton, we try to follow the evolution of a group of automata, which communicate between themselves, as well as with the environment, e.g., the group of automata defined in the above mentioned recommendation Q.71. The number of evolution traces of such systems can be extraordinarily large.

Not only must we select the initial state of a single automata, we must do it for all the automata from the group we want to analyze. Furthermore, in the case of simple and loosely coupled automata, an increase in the number of possible path combinations is not so high, but in the case of complex or tightly coupled automata, it is clear that the number of evolutions of the system can be huge.

The discussion above naturally raises the following questions: For what purpose are the MSC charts useful? Do we need them at all? Practical experience shows that making the MSC charts can be useful at the beginning of the design process, when the designers talk rather freely about possible communications scenarios. These scenarios of message exchange most frequently represent the so-called main branches, i.e., main paths, through the protocol. Typically, they go from the beginning (the initial state) to the end (logically, the last state in the chain of states), e.g., from the state *IDLE* to the state *CONVERSATION* in the previous example, without any errors or other exceptional events. Later, after finishing the analysis of the main paths, the paths of minor importance are analyzed. These are related to various less frequent cases, such as handling timer expirations, error recovery procedures, and so on.

All these scenarios, in the form of MSC charts, would be very useful in the later stages. Actually, these charts will be used as individual test cases during the implementation phase to partially check the functionality of the individual software modules (this is the so-called unit testing). They are also used during the final phase of the software verification as test cases for the compliance testing. The goal of **compliance testing** is to check if the software is compliant with the specification.

In most cases, the number of manually written MSC charts is finite and not too large (on the order of a few hundred at most). Later, during the testing and verification phase, automatically generating a much larger number of test cases would be normal (logically equivalent to MSC charts) to check the system much more thoroughly. This testing most frequently takes the form of statistical usage testing, and it enables quality engineers to estimate the software reliability without any previous knowledge about the system under examination.

As already mentioned, the MSC language — similar to the SDL language — has both the graphical and program form. The graphical form of the MSC language is more interesting than the program form for developing communications software. The next example illustrates the message exchange among the functional entities *FE1*, *FE2*, *FE3*, *FE4*, and *FE5*, in the case of the successful establishment and successful release of the ISDN connection between two subscribers. From this example, MSC is obviously useful for tracing the message exchange between more processes, which is not so easy and clear by looking at the set of corresponding SDL diagrams.

We start drawing the MSC chart by placing the rectangle graphical symbols that represent the communicating entities (i.e., processes) at the top of the chart sheet. The names of the entities are used to label these rectangle symbols. Next, we draw a vertical line from each rectangle symbol to the bottom of the sheet. After that, we enter a series of messages exchanged by the processes shown on the top of the chart. Each message (i.e., signal) is represented by the arrow symbol labeled with the message name. The arrow starts from the vertical line that represents the process sending the message and ends at the vertical line that represents the process receiving the message. The time advances in the direction from top to bottom of the sheet, i.e., the messages that appear on the top of the chart are exchanged before the messages that appear at the bottom of the chart.

An example of the MSC chart is shown in Figure 3.42. This example illustrates the scenario of successful establishment and release of the ISDN connection. The functional entities *FE1* and *FE5* are assigned to the calling and called party user, respectively. Initially, the functional entity *FE1* receives the signal *SETUP_req* from the environment (in reality, this signal is generated by the signaling system DSS1). After receiving the signal *SETUP_req*, *FE1* translates it to the signal *SETUP_req_ind* and sends this new signal to *FE2*. *FE2* forwards this signal to *FE3*, *FE3* forwards it to *FE4*, and finally *FE4* forwards it to *FE5*.

Design 123

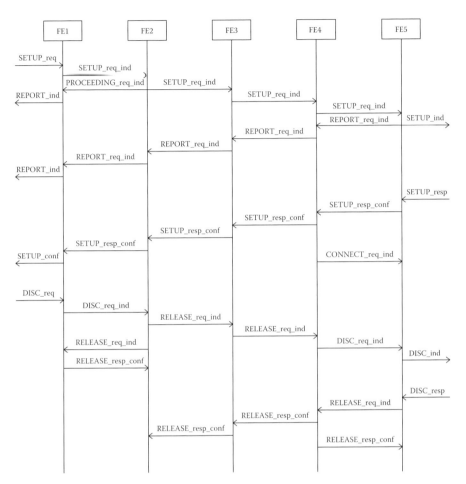

FIGURE 3.42
An example of the MSC chart: Successful ISDN call establishment and release.

After receiving the signal *SETUP_req_ind*, the functional entity *FE5* immediately sends two signals, the signal *SETUP_ind* to its environment and the signal *REPORT_req_ind* back to *FE4*. The latter signal is forwarded from *FE4* to *FE3*, then from *FE3* to *FE2*, and finally from *FE2* to *FE1*. *FE1* translates this signal to *REPORT_ind* and sends the latter to its environment.

The acceptance of the call by the calling party is signaled to *FE5* by the signal *SETUP_resp*. *FE5* translates this signal to the signal *SETUP_resp_conf* and sends the latter over the chain of FEs back to *FE1*. *FE1* in its turn translates it to *SETUP_conf* and sends the latter to its environment. This is the final step of the connection establishment procedure. The next communication phase is a conversation.

At the end of the conversation, the calling party user initiates the call release procedure by sending the signal *DISC_req* to the functional entity *FE1*, which in turn translates it to *DISC_req_ind* and sends the latter to *FE2*.

The functional entity *FE2* translates this signal to the signal *RELEASE_req_ind* and sends the latter to both *FE1* and *FE3*. From there, we have two parallel flows of messages. *FE1* replies to the signal *RELEASE_req_ind* by the signal *RELEASE_req_conf*. Alternately, *FE3* forwards the signal *RELEASE_req_ind* to *FE4*, which translates it to *DISC_req_ind* and sends the latter to *FE5*. *FE5* indicates the reception of that signal by sending the signal *DISC_ind* to its environment.

The environment answers with the signal *DISC_resp*, which is then translated to *RELEASE_req_ind* and sent to *FE4*. The functional entity *FE4* translates that to the signal *RELEASE_resp_conf* and sends the latter to both *FE3* and *FE5*. Finally, *FE3* forwards that final signal to *FE2*. This is the final step of the call release procedure.

This real-world example shows the main advantage of using MSC charts — instead of speculatively analyzing the parallel work of five finite state machines (*FE1*, *FE2*, *FE3*, *FE4*, and *FE5*) by looking at their SDL diagrams, here on a single chart we see how the system evolves through the procedures of call establishment and release. At this level of abstraction, we are not interested in the individual work of the individual automata. We just follow the interaction based on the message exchange between the automata in a given group.

3.9 Tree and Tabular Combined Notation

Tree and tabular combined notation (TTCN) is a language originally standardized by the International Standardization Organization (ISO). A group of designers can employ TTCN to make a formal specification of test procedures that are used to check if the implementation behaves in conformance with the system's formal specification. The type of testing that is conducted in accordance with such test procedures is referred to as **conformance testing**. The implementation that is the object of the testing is called an **implementation under test** (IUT). A primitive test procedure is specified in a form of a **test case**. A **test suite**, as defined by TTCN, is a collection of various test cases along with all requisite declarations and components.

The philosophy of treating software as art has caused the "big software crisis," which clearly indicated the need to treat the programs as products of a well-defined production process to ensure their quality. Software quality assurance required development of new methods and supporting tools for verification and validation of both software standard specification and its implementation. One of the results of research and development efforts in this area is the ISO/IEC 9646 (X.290) standard, which defines TTCN language for writing test suites — collections of test cases to be used for verification and validation of software products. Soon after its emergence, TTCN was

Design

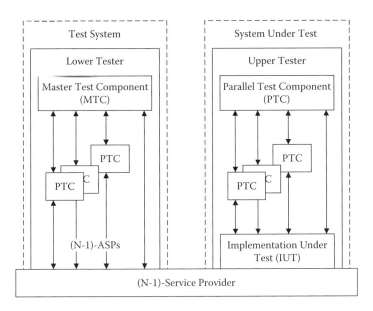

FIGURE 3.43
An illustration of the test configuration.

widely accepted by the organizations involved in defining standards and making test procedures for testing implementations of standards.

The main goal of TTCN is to enable conformance testing of software products. Conformance testing is used to check if the software product is compliant with its specification. Conformance test cases deal only with the external behavior of a software product. Actually, conformance testing is based on the application of the "black box" principle. The product must interact with its environment as specified. Its internal structure and behavior are not significant.

One of the most important concepts behind the TTCN standard is the concept of the so-called **test configuration** (Figure 3.43). The *upper tester* (UP) and the *lower tester* are driving the implementation under test (IUT) to check the correctness of its behavior. This concept addresses conformance testing of standardized protocols on the level N of the communication protocol hierarchy, i.e., in the layer N of the communication protocol stack.

Note that the test configuration concept (Figure 3.43) is based on the assumption that a reliable lower-level service already exists, provided by the group of $(N-1)$ service providers. Typically, this lower-level layer of the protocol stack would be developed by the same group of engineers, but it can also be ordered as a commercial off-the-self (COTS) component. In any case, the test configuration concept implies the bottom-up implementation paradigm. The engineering team starts from the very bottom of the protocol stack by implementing the lowest layer and proceeds toward the top of the stack by building a new layer on top of the previous layer. Each layer is checked for the conformance with the corresponding standard or specification before proceeding to the next one.

The lower tester is a program object on the same level, i.e., the level N of the hierarchy, and is actually a peer object of the object under test (IUT). The upper tester is a program object on the upper level, i.e., the level $(N + 1)$ of the protocol stack hierarchy. It imitates the future users of the IUT.

During the elaboration of the test configuration concept, the TTCN creators have concluded that it is of great importance to provide the means for abstract representation of certain details that are related to test configuration and to make them available during the startup procedures. The **protocol implementation conformance statements** (PICS) and **protocol implementation extra information** (PIXIT) have been conceived as the informal questionary. The answers to the questions of this questionary are mapped to the corresponding TTCN parameters.

The protocol implementation conformance statements (PICS) contain the information related to the protocol, such as optional parts, specific constraints, annexes, and so on. These data are the base for making the decision about the applicability of the test cases under the given circumstances. The protocol implementation extra information (PIXIT) contains the information related to the physical architecture of the system, such as the physical interconnection and startup procedure, which are not the constitutive parts of the protocol itself. By making a list of details of this type, the test suite becomes more flexible and makes it possible for the test personnel to stay out of the TTCN specification. Test personnel are people responsible for running the test suite on the test configuration.

In the text that follows, we briefly describe the TTCN language. As already mentioned, the TTCN test suite is a collection of various test cases along with all the necessary declarations and components. The test case descriptions are based on the application of the black box model. Each test case is a sequence, in a form of a tree of events, which describes the external behavior of the implementation under test, IUT.

The TTCN test suite specification consists of the following parts:

- Test suite overview part
- Test suite declaration part
- Test suite constraints part
- Test suite dynamic part

The test suite overview part is a table that contains an overall description of the test suite. The purpose of this part of the TTCN specification is to provide high-level information about the test suite to make it clear and more readable.

The test suite declaration part contains declarations of the following specification elements:

- User defined types
- Variables
- Timers
- Points of control and observation (PCO)

The declarations of types in TTCN are similar to the declarations of types in any conventional higher-level programming language, with the exception that in TTCN we use tables for that purpose rather than writing textual statements. The user can use the so-called built-in data types, i.e., the types defined by the TTCN language itself (e.g., *INTEGER*), and other previously defined user types, to define new user-defined types. The main constraint concerning TTCN types is that no equivalence exists for the pointers, i.e., the TTCN types may not be recursive.

Two structural types that are specific for the normal TTCN test suite applications are available. These are **protocol data unit** (PDU) and **abstract service primitive** (ASP). The protocol data unit (PDU) is a packet of data exchanged by the communication entities on the same level of the protocol stack hierarchy. These are referred to as **peer entities**, or just **peers**.

The abstract service primitive (ASP) is the type that encapsulates the PDU for the purpose of the PDU transfer between the peers, which is conducted by the lower-level layers of the protocol stack. Actually, ASP describes the message that the layer N entity uses to send the PDU to its peer by sending the ASP to the layer (N-1) service provider. Both ASPs and PDUs are related to a given point of control and observation (PCO).

The variables used in TTCN test suites are declared by the usage of the basic (built-in) and supplementary user-defined types. The TTCN language also includes the notions of the test suite constants and parameters. The value of a parameter is provided at the beginning of the system testing.

The TTCN language allows the user to extend the basic set of TTCN operations by defining the so-called supplementary introduced operations that are specific to a test case at hand. The design engineer defines the supplementary introduced operations according to the needs that they face during their work on the TTCN test suite specification.

The concept of a point of control and observation (PCO) defines the logical point for sending or receiving messages to or from the implementation under test (IUT). The PCO is identified by its name (address). The mapping of an abstract PCO to a real connection is outside of the domain of TTCN language, i.e., it is not defined by the TTCN test suite. The timers must be defined in the TTCN test suite declaration part to be available later in the TTCN dynamic part. The timer declaration contains its name (i.e., identification) and its duration.

A few examples seem to be appropriate at this point to make it easier for the reader to comprehend the concepts and notions introduced previously. We start with the example of the declaration of types in the TTCN language shown in Table 3.1. The table that is used for the type declarations starts

TABLE 3.1

An Example of Type Declarations in TTCN Language

Simple Type Definitions			
Type Name	Type Definition	Type Encoding	Comments
B_1	BITSTRING[1]		
O_1	OCTETSTRING[1]		
O_2	OCTETSTRING[2]		
Detailed Comments:			

with the row that names the table (in the example shown in Table 3.1, the name of the table is *Simple Type Definitions*). Each next row of the type declaration table is used to declare an individual data type additionally introduced by the user. The last row of the type declaration table is reserved for the detailed textual comments that the design engineer can provide to improve the quality of the TTCN test suite specification.

Each individual data type is declared by writing the type name, type definition, type encoding, and a textual comment in separate columns of the type declaration table. The type encoding information and a textual comment are optional. Table 3.1 declares three types, namely *B_1* (a bit string comprising a single bit), *O_1* (an octet string comprising a single octet), and *O_2* (an octet string comprising two elements, i.e., octets).

The next example illustrates the ASP type definition. Table 3.2 defines ASP type *dl_data_ind*. The table includes the information about its point of control and observation type, namely *DSAP*, and about two abstract service primitive parameters, *v5dl_address* and *user_data*. The type of the former parameter is *O_2* whereas the type of the latter is *PDU*. A single textual sentence is included to improve the readability of this definition.

We proceed with an example of a PDU type efinition. Table 3.3 shows the example of PDU type *bcc_allocation_cpl* definition. The point of control and observation type for this PDU is *DSAP*. The PDU has three fields, namely,

TABLE 3.2

An Example of an ASP Type Definition

ASP Type Definition		
ASP Name: *dl_data_ind*		
PCO Type: *DSAP*		
Comments:		
Parameter Name	Parameter Type	Comments
v5dl_address_value	O_2	
user_data	PDU	
Detailed Comments: ASP definition of the primitive received from *DSAP* PCO		

Design

TABLE 3.3

An Example of a PDU Type Definition

PDU Type Definition			
PDU Name: *bcc_allocation_cpl*			
PCO Type: *DSAP*			
Encoding Rule Name:			
Encoding Variation:			
Comments:			
Field Name	Field Type	Field Encoding	Comments
protocol_discriminator	O_1		m
bcc_reference_number	O_2		m
message_type	O_1		m
Detailed Comments: m = mandatory			

TABLE 3.4

An Example of Test Case Variables Declarations

Test Case Variables Declarations			
Variable Name	Type	Value	Comments
TestCaseV	INTEGER	10	
TestCaseV2	IA5String	"DefaultValue"	
Detailed Comments:			

protocol_discriminator, *bcc_reference_name*, and *message_type* (look at the column "Field Name"). The types of the fields are *O_1*, *O_2*, and *O_1*, respectively (see the column "Field Type"). All the fields are mandatory (see the column "Comments" and the row "Detailed comments").

The next example illustrates the declarations of test case variables in the TTCN language. Table 3.4 defines two variables, *TestCaseV* and *TestCaseV2*. The type of the former is *INTEGER* and the type of the latter is *IA5String*. Their default values are decimal number *10* and IA5 string of characters, "DefaultValue."

As already mentioned, the design engineer can define a supplementary test suite operation, which corresponds to a function or a macro in a higher-level programming language. The example of such a definition is shown in Table 3.5. It defines the test suite operation *TSO_GET_BCC_REF_NUM*. This operation has the single parameter named *bcc_message*. The type of the parameter is *dl_data_ind* (the parameter name and the parameter type name are separated by the character ":"). The type of the result of this operation is *O_2*.

Normally, in the TTCN test suite specifications, the test suite constraints part follows the test suite declaration part. The constraints part contains the constraint definitions that describe the messages that are sent or received by the IUT. The structural TTCN types, PDUs, and ASPs are used as the models

TABLE 3.5

An Example of a Test Suite Operation Definition

Test Suite Operation Definition
Operation Name: *TSO_GET_BCC_REF_NUM(bcc_message:dl_data_ind)*   Result Type: *O_2*   Comments: Description: The operation *TSO_GET_BCC_REF_NUM* returns the value of the field *reference_number*, which is a part of the received message (*bcc_message*).

to describe the messages exchanged over the PCOs. The constraints can be parameterized to make them usable in various contexts. In each of the contexts, real values are passed as constraint arguments.

The next example illustrates a constraint declaration. Table 3.6 declares the PDU constraint *Bcc_allocation_cpl*. This constraint relates to the packet data unit type *bcc_allocation_cpl*. This constraint concretely requires that three PDU fields that are listed in the table (see the column "Field Name") have particular values (see the column "Field Value"), as specified in the table. The field *protocol_discriminator* must contain the value *TSC_V5_PD*, the field *Bcc_reference_number* must contain the value *TSPX_BCC_REF_NUM*, and the field *message_type* must contain the value *TSC_METY_ALLOCATION_CPL*. Of course, the constants *TSC_V5_PD*, *TSPX_BCC_REF_NUM*, and *TSC_METY_ALLOCATION_CPL* must be declared earlier in the test suite declaration part.

The last part of the TTCN test suite specification is the dynamic part, which describes the individual test cases. The dynamic part comprises test cases, test steps, and default behavior tables, with all the **test events** and **test verdicts**. A test verdict describes the current IUT behavior. The dynamic part

TABLE 3.6

An Example of a PDU Constraint Declaration

PDU Constraint Declaration			
Constraint Name: *Bcc_allocation_cpl* PDU Type: *bcc_allocation_cpl* Derivation Path: Encoding Rule Name: Encoding Variation: Comments:			
Field Name	**Field Value**	**Field Encoding**	**Comments**
protocol_discriminator	TSC_V5_PD		
Bcc_reference_number	TSPX_BCC_REF_NUM		
message_type	TSC_METY_ALLOCATION_CPL		
Detailed Comments:			

is created in a hierarchical and nested manner. The building blocks are test groups, test cases, test steps, and test events.

Starting from the basic toward more complex constructions, the following hierarchical levels of constructions exist:

- Test event: the basic element that is used to build all other more complex constructions.
- Test step: a sequence of test events.
- Test case: a sequence of test steps.
- Test group: a set of test cases.
- Test suite: a collection of individual test cases and groups of test cases. Normally, the test suite is stored in a file system hierarchy. The test suite itself corresponds to the top-level directory, the test group corresponds to the subdirectory (so it is possible to have groups of the groups of test cases), and the individual test case corresponds to the file that contains the series of the test steps.

Test constructions are built as behavior trees that are placed into behavior tables. This is where the name of the language comes from, tree and table combined notation (TTCN). The behavior table contains the behavior tree defined by writing test events in lines with different nesting (indentation) levels. The lines on the same nesting level (i.e., with the same depth of indentation) represent alternative test events. The line on the next nesting level (i.e., with the next deeper indentation) executes after the line on the previous nesting level. The line is successfully executed when the event assigned to it has been processed.

This process means that the evolution of the behavior tree (table) begins from the lines that start at the beginning of the corresponding rows of the table, i.e., with the indentation of zero spaces. Of course, generally there may be more such lines. One of them is arbitrarily selected and the execution continues on the next level of nesting (indentation) where again more alternative lines may exist. One of them is selected randomly and this procedure is repeated until the end of the tree, i.e., until its leaf is reached. The leaf of the tree contains the test verdict.

An example of the successfully processed event is the reception of the expected message. Once a line has been executed, the tester goes to the next level of indentation. The tester is a program that executes the test case script corresponding to the test case definition table. No return to the previous level of indentation is allowed, except by using the *GOTO* construct. The lines on the same level of indentation have equal probabilities.

The line in the behavior table can contain the following:

- The send message statement, e.g., *L!CONNECTrequest* (where *CONNECTrequest* is the message name and *L* is the name of the corresponding point of control and observation, PCO)

- The receive message statement, e.g., *L?DISCONNECTindication*
- The assignment statement, which is used to assign a value to the specified variable, e.g., $V\_R := V\_R + 1$
- The timer operation statement that is used to start or cancel the specified timer, e.g., *START T01_max* or *CANCEL T01_max*
- The timer expiration statement, which is the reception of an internal message that signals the timer expiration, e.g., *?TIMEOUT T01_max*
- The Boolean operation that qualifies the execution, e.g., *[TSPC_PSTN OR TSPC_ISDN]*

The send message statement always executes successfully. The receive message statement executes successfully if the message of the correct type has been received and if its contents satisfy the given constraints. The value assignment statement always executes successfully, as well as the timer operation statements. The statement that is guarded with the Boolean expression executes if the expression evaluates to a true value.

Each leaf of the tree of events is assigned a single test verdict, which can be *pass*, *fail*, or *inconclusive*. The verdict *pass* means that the test case has been completed without errors, whereas the verdict *fail* indicates that the IUT is not compliant with the specification. The verdict *inconclusive* means that not enough evidence exists to proclaim that the IUT is conformant to the specification.

In practice, the user can face the need to construct new behavior trees by concatenating the already existing trees to the newly made constructions. This is frequently the case when writing the test cases for the certain phases of communication that come after some initial phases. The naive approach would be to write again, or copy, the tree lines that correspond to the initial phase in every test case that targets the subsequent phases of communication. A more profound approach would be to construct a common behavior tree by writing these common lines once and then to use the common behavior tree as a building element for other test cases. A tree can be concatenated with another tree by the operator "+". Thus, we make the so-called attach construct:

```
+STEP_CHECK_STATE
```

Another problem that can appear in practice is the need to repeat certain test cases. A single test case can be repeated the given number of times by using the **repeat construct**. For example, the next iteration ends when the expression *FLAG* evaluates as true:

```
REPEAT STEP1 UNTIL (FLAG)
```

The example of a behavior tree is given in Figure 3.44. This example illustrates the IUT behavior for the case of the outgoing telephone call. This

Design 133

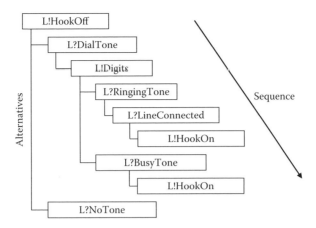

FIGURE 3.44
An example of the behavior tree for the outgoing telephone call.

TABLE 3.7

An Example of a Test Case Dynamic Behavior Specification

Test Case Dynamic Behavior					

Test Case Name: Basic outgoing call to the conversation phase.
Group:
Purpose: To check whether the base outgoing call can be established.
Configuration:
Default:
Comments:

No.	Label	Behavior Description	Constraint Ref	Verdict	Comments
1		L!HookOff			
2		L?DialTone			
3		L!Digits	CallSubscr2		
4		L?RingTone			
5	L1	L?LineConnected	ConnSubscr2	PASS	
6		L!HookOn			
7	L2	L?BusyTone		INCONC	
8		L!HookOn			
9	L3	L?NoTone		FAIL	
Detailed Comments:					

behavior tree is then placed in the behavior table to construct the test case shown in Table 3.7.

The behavior tree starts with sending the hook-off signal to the IUT (the line *L!HookOff*) because that line is a single line on that level of nesting (indentation). On the next level of nesting, there are two equal native lines (the line *L?DialTone* and the line *L?NoTone*). If the upper tester receives a dial tone (the line *L?DialTone*), the execution of the test case may continue.

Alternatively, if the upper tester receives some other tone or no tone at all (the line L?NoTone), that run of the test case is definitely not successful and we reach the test verdict FAIL. If the upper tester receives a dial tone, it sends the address of the called party (digits) to the IUT (the line L!Digits) and after that, it waits for the IUT answer. On this level of indentation, there are again two equal lines (the line L?RingTone and the line L?BusyTone).

If the upper tester receives a busy tone (the line L?BusyTone), that execution of the test case is inconclusive (the test verdict is INCON). Alternately, if the upper tester receives a ringing tone (the line L?RingTone), the upper tester checks if the connection is successfully established (the line L?LineConnected), e.g., by sending DTMF tones from the lower to the upper tester. If the upper tester receives the LineConnected signal, that run of the test case is definitely successful and we reach the test verdict PASS. In both cases, the upper tester ends the test case by sending the hook-on signal to IUT (the line L!HookOn).

3.10 Examples

This section contains some examples that are related to the communication protocol design. These should help the reader to consolidate their understanding of the concepts and techniques introduced so far.

3.10.1 Example 1

This example demonstrates the procedures for connection establishment and release that are performed by two communicating processes, namely TE1 and TE2. The processes TE1 and TE2 are specified by their statechart diagrams shown in Figure 3.45 and Figure 3.46, respectively. The semantically equivalent SDL diagrams are shown in Figure 3.47 and Figure 3.48, respectively.

The process TE1 has four stable states, labeled TE1_IDLE, TE1_CONNECTING, TE1_CONNECTED, and TE1_DISCONNECTING. While the process TE1 is in the state TE1_IDLE, it can receive only the message CONNECT_req from the user and after receiving that message, the process TE1 sends the message CONNECT_ind to the process TE2, and evolves to its next stable state TE1_CONNECTING. In that state, the process may receive one of two possible input messages, namely CONNECT_conf or CONNECT_reject. In the former case, the process moves to the stable state TE1_CONNECTED, whereas in the latter case, it evolves to its initial stable state TE1_IDLE.

In its stable state TE1_CONNECTED, the process TE1 may receive the message DISCONNECT_req from the user. In that case, it sends the message DISCONNECT_ind to the process TE2 and evolves to the stable state

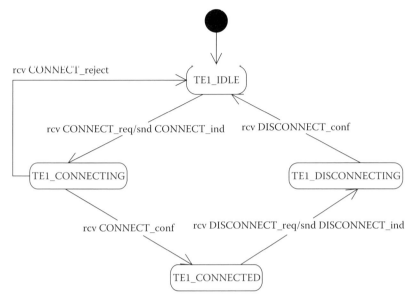

FIGURE 3.45
The statechart diagram of the process *TE1*.

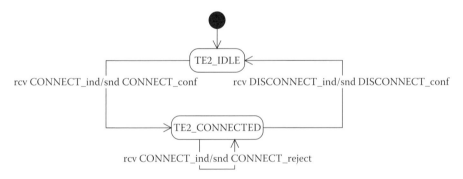

FIGURE 3.46
The statechart diagram of the process *TE2*.

TE1_DISCONNECTING. From that stable state, it returns to its initial stable state *TE1_IDLE* after receiving the message *DISCONNECT_conf* from its peer process *TE2*.

The SDL diagram specification of the process *TE2* is much simpler because it comprises only two stable states, namely, *TE2_IDLE* and *TE2_CONNECTED*. In the former state, the process *TE2* may receive only the message *CONNECT_ind*, to which it replies by the message *CONNECT_conf* and after that, it evolves to the state *TE2_CONNECTED*. In the latter state, the process may receive one of two possible messages, *CONNECT_ind* or *DISCONNECT_ind*. In the former case, the process *TE2*

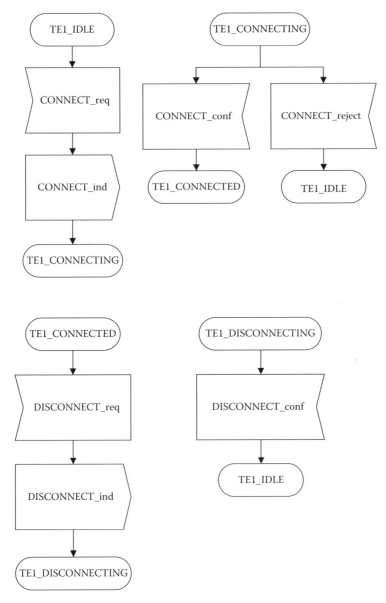

FIGURE 3.47
The SDL diagram of the process *TE1*.

replies with the message *CONNECT_reject* and remains in its current state. In the latter case, it replies with the message *DISCONNECT_conf* and goes back to its initial state *TE2_IDLE*.

The scenario of a successful connection establishment and release is illustrated by the MSC chart shown in Figure 3.49. The top of the chart shows the communicating entities, the human user, and the program processes *TE1* and

Design

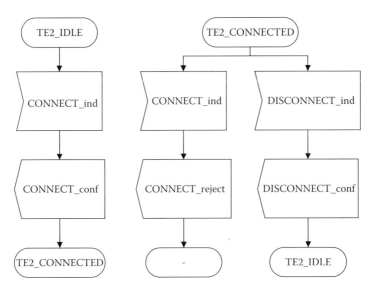

FIGURE 3.48
The SDL diagram of the process *TE2*.

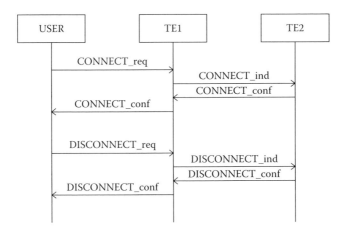

FIGURE 3.49
A successful connection establishment and release MSC.

TE2. The vertical lines are drawn from the rectangular graphical symbols down to the bottom of the sheet. The time advances in the same direction.

The connection establishment procedure starts when the user sends the message *CONNECT_req* to the process *TE1* (this event is noted by the arrow drawn from the vertical line labeled *USER* to the vertical line labeled *TE1*), which in turn sends the message *CONNECT_ind* to the process *TE2*. The process *TE2*, in its turn, replies with the message *CONNECT_conf*. Upon recept of the message *CONNECT_conf*, the process *TE1* forwards it to the user. This completes the connection establishment procedure. The next communication phase is normally used for the desired data transfer. Because of that, it is most frequently referred to as a data transfer phase.

The connection release procedure starts when the user sends the message *DISCONNECT_req* to the process *TE1*, which translates it to the message *DISCONNECT_ind* and sends it to the process *TE2*, which in turn replies by the message *DISCONNECT_conf*. Upon reception of the message *DISCONNECT_conf*, the process *TE1* forwards it to the user. This completes the connection release procedure.

The tables shown represent a simple TTCN test suite specification for this example. This simple test suite comprises Table 3.8 with simple type declarations, Table 3.9 with the PDU type declarations, Table 3.10 with PDU constraint declarations, and two tables (Table 3.11 and Table 3.12) with dynamic behavior description, i.e., test cases.

The reader is encouraged to play more with this simple example. For example, we can change the previous example so that before the existing connection is established, the process *User* checks if the process *TE1* is ready for the communication. The MSC chart that specifies a new connection establishment procedure is shown in Figure 3.50.

TABLE 3.8

The Example 1 Simple Type Declarations

Simple Type Declarations			
Type Name	Type Definition	Type Encoding	Comments
O_1	OCTETSTRING[1]		
O_2	OCTETSTRING[2]		
Detailed Comments:			

TABLE 3.9

The Example 1 PDU Type Declarations

PDU Type Declaration			
PDU Name: CONNECT_ind			
PCO Type:			
Encoding Rule Name:			
Encoding Variation:			
Comments: Example of PDU declaration			
Field Name	Field Type	Field Encoding	Comments
Source_address	O_1		
Destination_address	O_1		
User_data	O_2		
Detailed Comments:			

Design

TABLE 3.10

The Example 1 PDU Constraint Declarations

PDU Constraint Declaration			
Constraint Name: CONNECT_ind			
PDU Type: CONNECT_ind			
Derivation Path:			
Encoding Rule Name:			
Encoding Variation:			
Comments:			
Field Name	**Field Value**	**Field Encoding**	**Comments**
Source_address	–		
Destination_address	–		
User_data	–		
Detailed Comments:			

TABLE 3.11

The Example 1 Test Case 1

Test Case Dynamic Behavior					
Test Case Name: Basic Connect *TE2*					
Group:					
Purpose: Check if a normal connection can be established					
Configuration:					
Default:					
Comments:					
No.	**Label**	**Behavior Description**	**Constraint Ref**	**Verdict**	**Comments**
1		L?CONNECT_req			
2		L!CONNECT_ind			
3		L?CONNECT_conf	CallEstablished	PASS	
4		L?CONNECT_reject	CallNotEstablished	INCONC	
Detailed Comments:					

3.10.2 Example 2

Figure 3.51 shows a hypothetical computer network with a star topology. Three terminal nodes, N1, N2, and N3, are connected to one transit node TN. The routing table residing in TN is shown in Figure 3.51 to the right of TN. Terminal nodes generate messages for other terminal nodes in the network. Depending on the value of the message parameter (1, 2, or 3), a transit node delivers the message to its destination by sending it to the corresponding port (A, B, or C).

The communication process that resides in the terminal node of the network is specified by the statechart diagram shown in Figure 3.52. The process that executes in the transit node is described by the statechart diagram shown

TABLE 3.12

The Example 1 Test Case 2

Test Case Dynamic Behavior				
Test Case Name: Basic Disconnect TE2				
Group:				
Purpose: Check call disconnect				
Configuration:				
Default:				
Comments:				

No.	Label	Behavior Description	Constraint Ref	Verdict	Comments
1		L?CONNECT_req			
2		L!CONNECT_ind			
3		L?CONNECT_conf	CallEstablished		
4		L?DISCONNECT_ind			
5		L!DISCONNECT_ind	Disconnect		
6		L?DISCONNECT_conf		PASS	
7		L?CONNECT_conf		FAIL	
8		L?CONNECT_reject		FAIL	
9		L?CONNECT_reject			
Detailed Comments:					

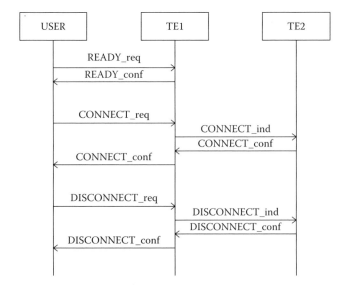

FIGURE 3.50
A new connection establishment procedure MSC.

in Figure 3.53. The semantically equivalent SDL diagrams are shown in Figure 3.54 and Figure 3.55, respectively.

The process that runs in the terminal node of the network has two stable states, *N123_IDLE* and *N123_MSG_SENT*. The state transition is initiated by

Design

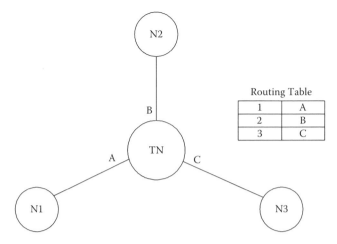

FIGURE 3.51
A hypothetical star network with one transit and three terminal nodes.

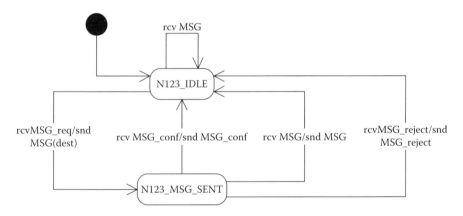

FIGURE 3.52
The statechart diagram of the process that runs in a terminal node of the network.

the user message *MSG_req*. The process returns to its initial state after the reception of one of three possible messages, namely, *MSG_conf*, *MSG*, or *MSG_reject*. The process that resides in the transit node of the network has a single state, *TN_IDLE*. This process routes the input message toward its destination.

Figure 3.56 shows the scenario of a successful message delivery. The node N1 sends the correct message to the node N3 over the node TN. The user is informed about the successful delivery by the message *MSG_conf*. Figure 3.57 shows the scenario of an unsuccessful message delivery. The node N1 has sent the message to the unknown destination, which has been rejected from the node TN by the message *MSG_reject*.

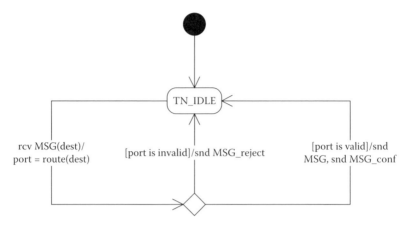

FIGURE 3.53
The statechart diagram of the process that resides in the transit node of the network.

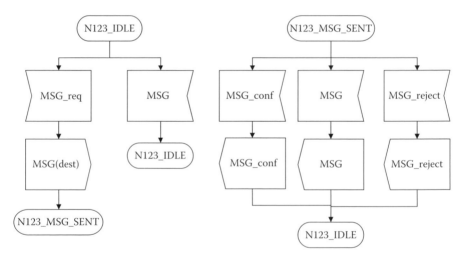

FIGURE 3.54
The SDL diagram of the process that runs in a terminal node of the network.

The next five tables constitute a simple TTCN test suite specification for this example, as follows:

- Table 3.13 contains the PDU type declaration for the message *MSG_req*.
- Table 3.14 contains the PDU constraint declaration *MSG_req_destination_addr_ok*.
- Table 3.15 contains the PDU constraint declaration *MSG_req_destination_addr_not_ok*.

Design

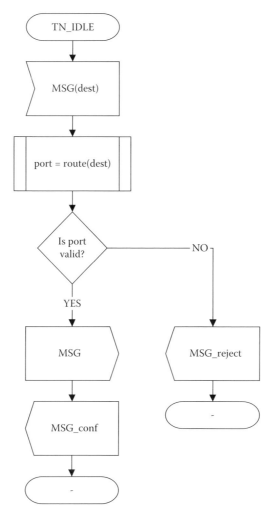

FIGURE 3.55
The SDL diagram of the process that resides in the transit node of the network.

- Table 3.16 contains the test case that corresponds to the scenario of sending a message to a known terminal node.
- Table 3.17 contains the test case that corresponds to the scenario of sending a message to an unknown terminal node.

The reader is encouraged to play more with this example. One interesting direction of generalization would be to consider a more complex network, such as the one shown in Figure 3.58.

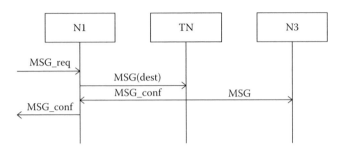

FIGURE 3.56
A successful message delivery MSC.

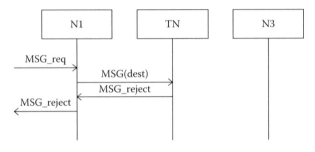

FIGURE 3.57
An unsuccessful message delivery MSC.

TABLE 3.13

The Example 2 PDU Type Declaration *MSG_req*

PDU Type Declaration

PDU Name: *MSG_req*
PCO Type:
Encoding Rule Name:
Encoding Variation:
Comments: Example of PDU declaration

Field Name	Field Type	Field Encoding	Comments
Destination_address	INTEGER		m
User_data	OCTETSTRING[2]		
Detailed Comments:			

3.10.3 Example 3

This example illustrates reliable packet delivery based on the message acknowledgment. Each communicating process expects the acknowledgment of the message that it has previously sent. If the acknowledgment is not received within the limited period of time, the corresponding timer will

TABLE 3.14

The Example 2 PDU Constraint Declaration
MSG_req_destination_addr_ok

PDU Constraint Declaration			
Constraint Name: *MSG_req_destination_addr_ok*			
PDU Type: *MSG_req*			
Derivation Path:			
Encoding Rule Name:			
Encoding Variation:			
Comments:			
Field Name	**Field Value**	**Field Encoding**	**Comments**
Destination_address	1		
User_data	*		
Detailed Comments:			

TABLE 3.15

The Example 2 PDU Constraint Declaration
MSG_req_destination_addr_not_ok

PDU Constraint Declaration			
Constraint Name: *MSG_req_ destination_addr_not_ok*			
PDU Type: *MSG_req*			
Derivation Path:			
Encoding Rule Name:			
Encoding Variation:			
Comments:			
Field Name	**Field Value**	**Field Encoding**	**Comments**
Destination_address	4		
User_data	*		
Detailed Comments:			

expire and the process will assume that the message or its acknowledgment have been lost and will retransmit the message once again.

The statechart diagram and the SDL diagram of the process are shown in Figure 3.59 and Figure 3.60, respectively. The process has two stable states, *FSM_IDLE* and *FSM_MSG_SENT*. In its initial state, the process starts the timer *T1*, sends the message with the sequence number *SN*, and evolves into its next stable state *FSM_MSG_SENT*. In that state, the process either receives the acknowledgment, stops the timer *T1*, and returns to its initial state, or the timer *T1* expires and in its turn the process retransmits the message.

In any state (*FSM_IDLE* or *FSM_MSG_SENT*), the process can receive a message from its peer process. The process acknowledges the message if the sequence number of the message is valid (in communication protocols, the

TABLE 3.16

The Example 2 Test Case 1

Test Case Dynamic Behavior				
Test Case Name: Sending a message to a known terminal node Group: Purpose: Configuration: Default: Comments:				
No.	Label	Behavior Description	Constraint Ref	Verdict	Comments
1		L?MSG_req			
2		L!MSG_req_destination_addr_ok			
3		L?MSG_conf	Message sent	PASS	
4		L?MSG_reject		FAIL	
Detailed Comments:				

TABLE 3.17

The Example 2 Test Case 2

Test Case Dynamic Behavior				
Test Case Name: Sending a message to an unknown terminal node Group: Purpose: Configuration: Default: Comments:				
No.	Label	Behavior Description	Constraint Ref	Verdict	Comments
1		L?MSG_req			
2		L!MSG_req_ destination_addr_not_ok			
3		L?MSG_conf	Message sent	FAIL	
4		L?MSG_reject		PASS	
Detailed Comments:				

process would normally maintain the counter of the next expected message in a sequence by incrementing its contents for each received message — a validity check in this context would be to compare the sequence number in the received message with the contents of this counter). If the sequence number, RN, of the message is invalid, the process throws the message away.

Figure 3.61 illustrates two scenarios of the communication between two peer processes. The MSC on the left in Figure 3.61 shows a successful message delivery. The process $FSM1$ sends the message $M1$ to the process $FSM2$, which in turn sends the acknowledgment ACK to the process $FSM1$.

The MSC on the right in Figure 3.61 shows a more complex scenario of successful message retransmission after the unsuccessful first message

Design

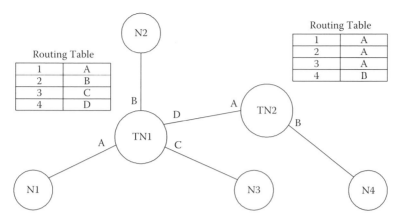

FIGURE 3.58
The topology of a more complex hypothetical network.

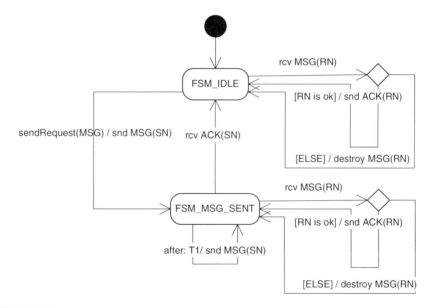

FIGURE 3.59
The statechart diagram of the communicating process that provides the reliable message delivery based on the re-transmission scheme.

delivery attempt. The process *FSM1* sends the message *M1*, the process *FSM2* receives it and sends its acknowledgment *ACK*, but it gets lost. The timer *T1* expires and the process *FSM1* retransmits the message *M1*. The process *FSM2* receives it and sends its acknowledgment *ACK*, which is successfully received by *FSM1*.

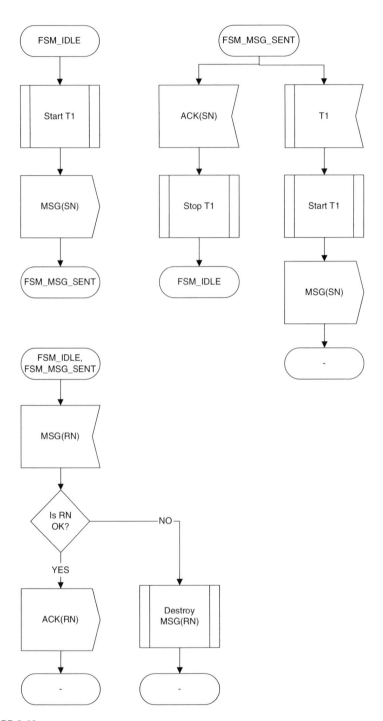

FIGURE 3.60
The SDL diagram of the communicating process that provides the reliable message delivery based on the re-transmission scheme.

Design 149

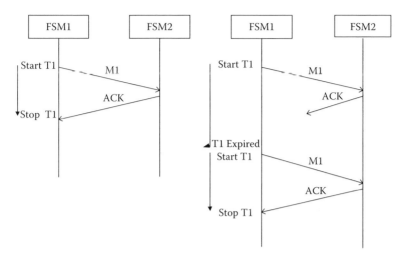

FIGURE 3.61
An example with two scenarios (with and without message retransmission).

3.10.4 Example 4

This example illustrates the sliding window concept, which provides a reliable and efficient transport service. Voluminous literature can be found that addresses this topic (Halsall, 1988). The design shown here is based on *Go-back-N* retransmission mechanism. It also supports the robust frame acknowledgement procedure (one ACK may acknowledge more than one frame).

The collaboration diagram in Figure 3.62 shows two distributed applications that communicate with the help of two communication objects, which are deployed at the local and remote side. The application *a1* sends the data packed into messages (*M*) to the object *p* (primary), which in its turn encapsulates the messages into *I* (information) frames, together with its sequence number V(*s*), and sends them to the object *s* (secondary). The object *s* checks the frame *I* sequence number against the number it expects V(*r*), and if they match, it accepts the frame *I* and acknowledges it by sending the message *ACK* to the object *p*. If these numbers do not match, the object *s* rejects the received *I* frame and sends the corresponding message *NAK*. We assume that the numbers V(*s*) and V(*r*) are maintained in the variables *vs* and *vr*, respectively. The object *s* delivers all the correctly received messages to the remote application *a2*.

In this example, we are mainly interested in the communication protocol between the primary and the secondary side of the communication link, which is established by the corresponding communication processes, *p* and *s*. The process *p* is modeled with the activity diagram shown in Figure 3.63 and Figure 3.64, whereas the process *s* is modeled with the activity diagram shown in Figure 3.65.

FIGURE 3.62
The Example 4 collaboration diagram.

Assume that the variable *rc* holds the number of the *I* frames that were sent by the process *p* but still not acknowledged by the process *s*. The activity diagram in Figure 3.63 starts with the transition from the initial state to the state *IDLE*. During this transition the variables *vs* and *rc* are reset. After receiving a message *M* from the application *a1*, *p* checks if the send window is full. If the send window is not full, *p* calls the procedure *send(M)* to encapsulate *M* into *I* and sends it toward *s*. If the send window is full, *p* adds *M* to the input queue (*inputQueue*). In both cases, it returns to the state *IDLE*.

The procedure *send(M)* first creates the frame *I* and encapsulates the current value of the variable *vs* and the message *M* in it by supplying them as arguments of the corresponding constructor. It then adds the frame to the retransmission queue (*retransmissionQueue*), allocates and starts a new timer (*T*), adds the pair (*T,I*) to the map *mapTtoI*, adds the pair (*I,T*) to the map *mapItoT*, increments *vs* and *rc*, and sends the frame *I* toward *s*. The map *mapTtoI* is used to search for the frame *I* that corresponds to the given timer *T*, whereas the map *mapItoT* is used to search for the timer *T* that corresponds to the given frame *I*. Notice that the procedure *send(M)* assigns a timer to each frame it sends. When the timer expires, *p* restarts the timer (*restartTimer(T)*), finds the corresponding frame by using the map *mapTtoI*, and retransmits the frame toward *s*.

When *p* receives the message *ACK* from *s*, it provides the iterator on the list *retransmissionQueue* and starts iterating through this list. For all the frames whose sequence number is smaller than the sequence number in the received *ACK* message, *p* finds the corresponding timer (by using the map *mapItoT*), stops it, and removes both the pair (*T,I*) from the map *mapTtoI* and the pair (*I,T*) from the map *mapItoT*.

Because some of the slots (or at least one of them) should be free after the previous iteration, *p* provides the iterator on the list *inputQueue* and starts iterating through it. It iterates while empty slots exist in the send window, and while iterating, it removes the messages from the input queue and sends them by calling the procedure *send(M)* as explained previously.

If the process *p* receives the message *NAK*, it performs the *Go-back-N* retransmission procedure. Essentially, *p* scans the whole retransmission queue and for each frame whose sequence number is greater than or equal to the sequence number in the receive message *ACK*, it finds the corresponding timer, restarts it, and retransmits the frame toward *s*.

The activity diagram shown in Figure 3.65 models the process *s*. It starts with the triggerless transition from the initial state to the state *IDLE*. During this transition, the variable *vr* is reset. After receiving the frame *I*, *s* checks

Design

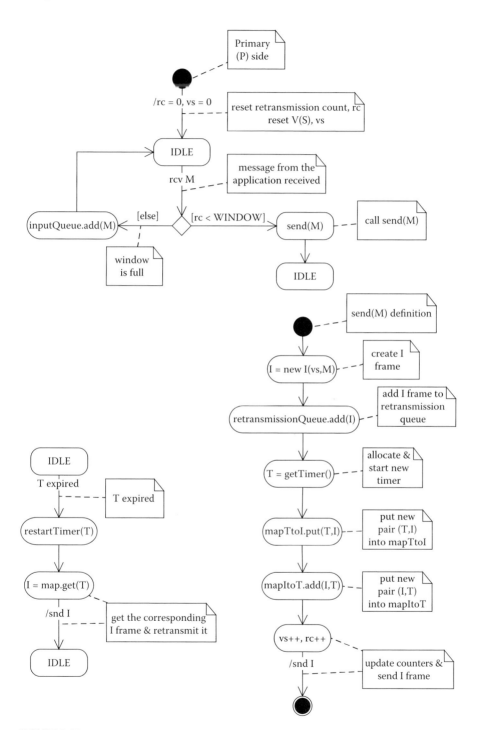

FIGURE 3.63
The Example 4 activity diagram, part I.

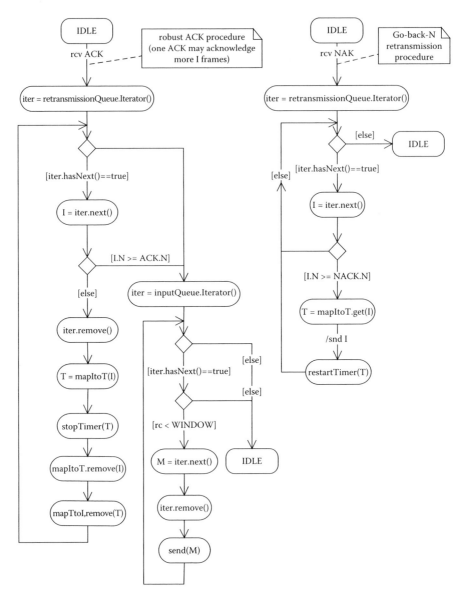

FIGURE 3.64
The Example 4 activity diagram, part II.

its sequence number equal to the value of the variable *vr*. If the values are the same, *s* accepts the frame by incrementing *vs*, creating the message *ACK*, and sending it to *p*. If the values are different, *s* rejects the frame by sending the message *NAK* to *p*.

The next three figures show three typical scenarios. The sequence diagram shown in Figure 3.66 illustrates a successful frame delivery scenario. The frames *I(0)* and *I(1)* are sent through the window and are acknowledged

Design

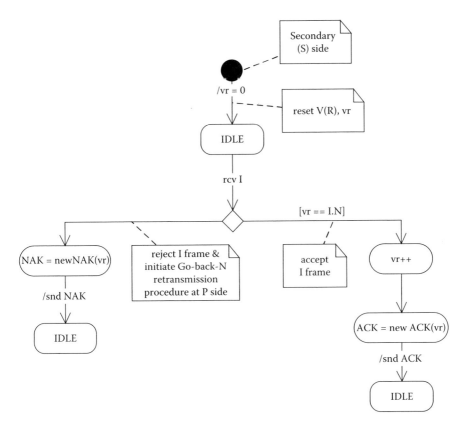

FIGURE 3.65
The Example 4 activity diagram, part III.

with *ACK(1)* and *ACK(2)*, respectively. After some delay, *I(2)* is sent and it is also successfully acknowledged with *ACK(3)*.

The sequence diagram shown in Figure 3.67 illustrates the *Go-back-N* procedure. The process *p* starts by sending the frames *I(0)* and *I(1)*. The frame arrives at *s* side regularly but *I(1)* gets lost. This causes the mismatch of sequence numbers at the secondary side when it successfully receives *I(2)*, because the value of the variable *vr* is 1 (which indicates that *s* is awaiting *I(1)* instead of *I(2)*). Because the sequence number of the frame and the value of the variable are not the same, *s* rejects the frame by sending the message *NAK(1)*. The process *p* in its turn retransmits both *I(1)* and *I(2)*.

The sequence diagram shown in Figure 3.68 illustrates the frame retransmission triggered by the retransmission timer. The process *p* starts again by sending *I(0)* and *I(1)* in succession. The process *s* in its turn acknowledges them by *ACK(1)* and *ACK(2)*, respectively. The message *ACK(1)* arrives successfully at the primary side, but the message *ACK(2)* gets lost. This causes the corresponding timer to expire after a while. Triggered by that event, *p* restarts the timer and retransmits the frame *I(1)*. The second time both *I(1)*

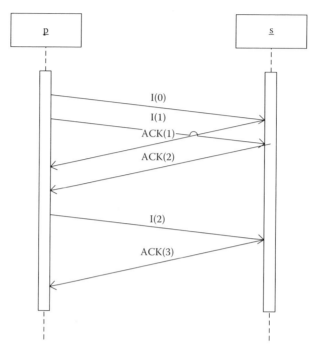

FIGURE 3.66
The Example 4 MSC diagram: Successful frame delivery.

and the corresponding *ACK(2)* are successfully transferred over the communication link. After receiving *ACK(2)*, p stops the timer and removes *I(1)* from the retransmission queue.

3.10.5 Example 5

In this example, we design the SIP INVITE client transaction in accordance with RFC 3261, Section 17.11. First, let us return to the requirements and analysis of a SIP Softphone, introduced as an example at the end of the previous chapter. Briefly, in that example we have constructed the use case diagram and have transformed it into the corresponding general collaboration diagram. At the very end of that example, we have shown the one particular collaboration related to the successful session establishment.

Now let us zoom in on the general collaboration diagram of a SIP Softphone with the focus on the SIP INVITE client transaction and the surrounding objects with which it directly communicates. The resulting general collaboration diagram is shown in Figure 3.69. The SIP INVITE client transaction is modeled as an unnamed object of the class *InClientT* because this object is dynamically created upon user request. It collaborates with the following three objects:

Design

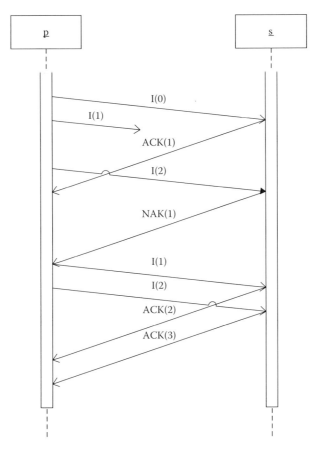

FIGURE 3.67
The Example 4 MSC diagram: *Go-back-N* retransmission.

- *tud*, which represents the transaction user dispatcher
- *tald*, which represents the transaction layer dispatcher
- *tlid*, which represents the transport layer dispatcher

Similarly, we can zoom in on the particular collaboration diagram that illustrates a successful session establishment scenario (Figure 2.17) to provide the corresponding particular collaboration of the SIP INVITE client transaction with its surrounding objects (Figure 3.70). As already mentioned in the previous chapter, *req()* and *rsp()* designate requests and responses, respectively. More precisely, *req(INVITE)* is the SIP invite request, *rsp(1xx)* is the SIP provisional response, and *rsp(200)* is the SIP final response.

Another particular collaboration that corresponds to an unsuccessful session establishment scenario is shown in Figure 3.71. This scenario is the same as the previous one up to the step number 6, when instead of the successful final response *rsp(200)*, the unsuccessful final response *rsp(300-699)*

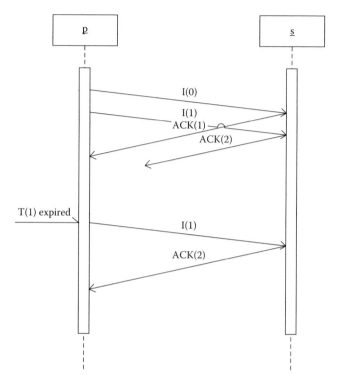

FIGURE 3.68
The Example 4 MSC diagram: *I* frame retransmission triggered by the retransmission timer.

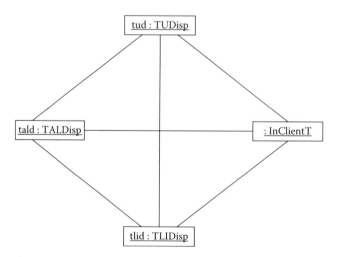

FIGURE 3.69
The SIP INVITE client transaction collaboration diagram.

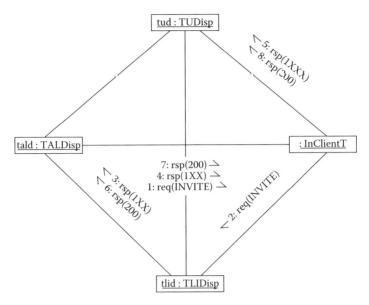

FIGURE 3.70
A successful session establishment collaboration diagram.

is received. In step 7, *tald* forwards *rsp(300-699)* to SIP INVITE client transaction, which in accordance with RFC 3261 forwards it toward the upper layer and sends the message *ACK* to the remote site. These two actions are performed in steps 8 and 9, respectively. Semantically equivalent sequence diagrams are shown in Figure 3.72 and Figure 3.73. Figure 3.72 illustrates a successful session establishment, whereas Figure 3.73 shows an unsuccessful session establishment scenario.

Based on the SIP INVITE client transaction state transition graph (RFC 3261, page 128) we can construct the corresponding statechart diagram (Figure 3.74). This statechart diagram starts with the transition from the initial state to the state *Calling*, which is triggered by the reception of the signal (message) *req(INVITE)* from the transaction user (TU). The signal *req(INVITE)* models the original request SIP INVITE. During this transition, the SIP INVITE client transaction forwards the message *req(INVITE)* to the transport layer.

At the entrance to the state *Calling*, two timers are started, timer A (TA) and timer B (TB). The former corresponds to the time interval that must elapse before the response to the request INVITE can be received, whereas the latter limits the time interval during which the SIP INVITE client transaction waits for the response to the request INVITE. Initially, TA is set to the value T1 (estimated round-trip time, RTT, which is by default 500 ms) and TB is set to $64 \times T1$.

If the timer TA expires, the SIP INVITE client transaction restarts it by doubling its current value (TA = TA × 2) and retransmits the signal *req(INVITE)*. Initial values of TA and TB (T1 and 64 × T1, respectively) allow

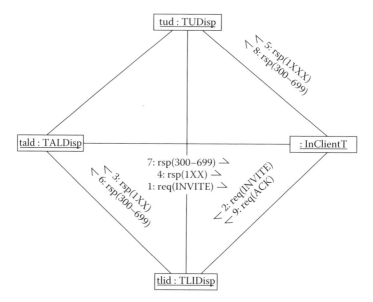

FIGURE 3.71
An unsuccessful session establishment collaboration diagram.

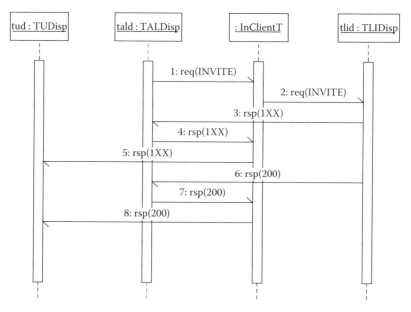

FIGURE 3.72
A successful session establishment sequence diagram.

Design

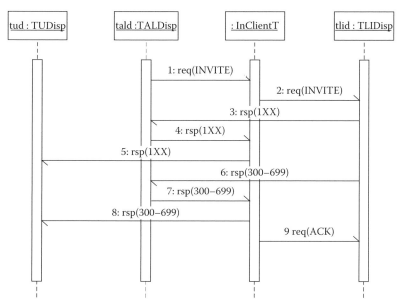

FIGURE 3.73
An unsuccessful session establishment sequence diagram.

this procedure to repeat the maximum of seven times before the timer TB expires. If the timer TB expires (or if a transport error is detected), the SIP INVITE client transaction informs TU accordingly and moves to the state *Terminated*, and from there to its final state.

Most frequently, a response to the request INVITE will be received before the timer B expires. In such a case, the SIP INVITE client transaction stops both timers and moves to the next state, which depends on the type of response. If the provisional response *rsp(1xx)* is received, the SIP INVITE client transaction forwards it to TU and moves to the state *Proceeding*. If the successful final response *rsp(2xx)* is received, the SIP INVITE client transaction forwards it to TU and moves to the state *Terminated*. If the unsuccessful final response *rsp(300-699)* is received, the SIP INVITE client transaction forwards it to TU and sends the signal (message) *ACK* to the remote site.

While being in the state *Proceeding*, the SIP INVITE client transaction simply forwards all the preliminary responses *rsp(1xx)* to TU. Once it receives the successful final response *rsp(2xx)*, it forwards it also to TU and moves to the state *Terminated*. If the SIP INVITE client transaction receives the unsuccessful final response *rsp(300-699)* in the state *Proceeding*, it forwards that response to TU, sends the signal *req(ACK)* to the remote site, and moves to the state *Completed*.

At the entrance to the state *Completed*, the third timer, namely the timer D (TD) is started. While being in the state *Completed*, the SIP INVITE client transaction just confirms any unsuccessful final responses *rsp(300-699)* by sending the SIP message *ACK* to the remote site. If the SIP INVITE client

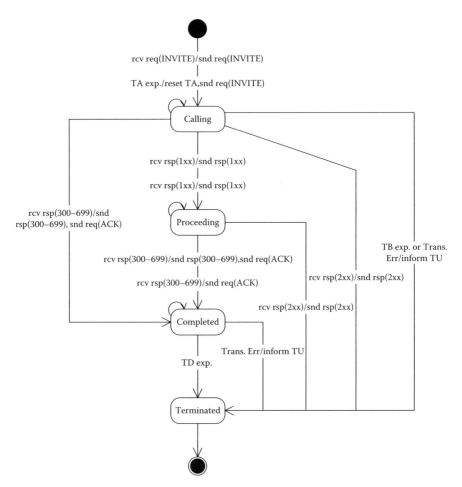

FIGURE 3.74
The statechart diagram of the SIP INVITE client transaction.

transaction detects a transport error, it informs TU accordingly and moves to the state *Terminated*. Finally, when the timer D expires, the SIP INVITE client transaction finishes simply by moving to the state *Terminated*.

We finalize this example with the semantically equivalent SDL diagram, which due to its size is shown in the next four figures (in these figures, TPL stands for the transport layer and TU stands for the transaction user). Figure 3.75, Figure 3.76, Figure 3.77, and Figure 3.78 illustrate the processing of events in the state *Calling*, *Proceeding*, *Completed*, and *Terminated*, respectively.

Design

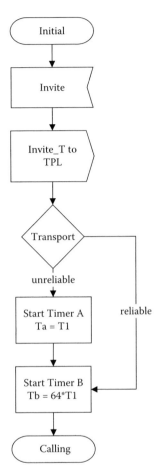

FIGURE 3.75
The SDL diagram of the SIP INVITE client transaction, part I.

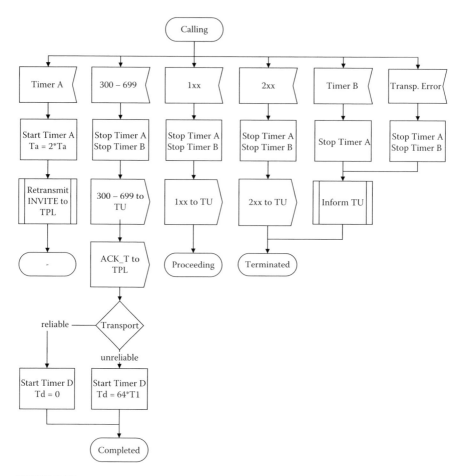

FIGURE 3.76
The SDL diagram of the SIP INVITE client transaction, part II.

Design

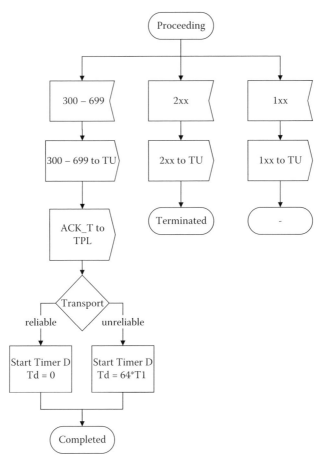

FIGURE 3.77
The SDL diagram of the SIP INVITE client transaction, part III.

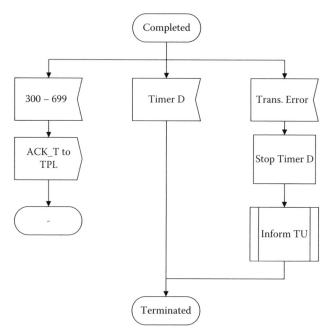

FIGURE 3.78
The SDL diagram of the SIP INVITE client transaction, part IV.

References

Booch, G., Rumbaugh, J., and Jacobson, I., *The Unified Modeling Language User Guide*, Addison-Wesley, Reading, MA, 1998.

Booch, G., Rumbaugh, J., and Jacobson, I., *The Unified Software Development Process*, Addison-Wesley, Reading, MA, 1998.

Halsall, F., *Data Communications, Computer Networks and OSI*, Addison-Wesley, Reading, MA, 1988.

4
Implementation

The system **implementation** is a phase in engineering work that follows the system design phase. This phase consists of the following two steps:

- Transform a design model into the implementation model
- Transform the implementation model into a higher-level programming language code

A design model is given in the form of the corresponding UML (Booch et al., 1998) or SDL diagrams, which are the results of the previous phases of communication protocol engineering, i.e., requirements, analysis, and design. The implementation model takes the form of the corresponding UML component diagram. The output of the implementation phase is a set of source code modules, today most frequently in C/C++ or Java, which is also referred to as the implementation. This may sound confusing, but in reality the correct meaning of the term is easily deduced from its context.

Logically, *implementation as a phase* of the production process is a well-defined mapping of a design model into a higher-level programming language source code. *Implementation as a product* is a result of this mapping. The attribute *well-defined* reflects the assumption that both detailed procedures and adequate tools are provided for transforming models into program source code. This well-defined mapping of a model into the program source code is referred to as **forward engineering** in UML terminology. Likewise, the reverse mapping of a program source code into the model is referred to as **backward engineering**.

In a mathematical sense, both the mapping of a program into the program source code and the result of that mapping (i.e., the implementation in both of its meanings) are not unique. Therefore, logically more than one *correct implementation* exists for a given model of the communication protocol. Under the correct implementation, we assume an implementation that for given inputs produces expected outputs within the expected time frame, which is defined with the corresponding timers. We say for such implementation that it is compliant (conformant) with (to) the given model. The terms *compliant* and *conformant* are synonyms in this context. If the model has been standardized (e.g., by IETF or ITU-T), we say that the implementation is compliant with the standard.

The concept of forward and backward engineering is an intriguing one. Proponents of the model-based software development and various initiatives in Model-Driven Architecture (MDA) strongly believe that forward and backward engineering is possible, and they are putting forth tremendous efforts to make it real. Quite a number of commercially available tools are made with this goal in mind. The agile programming community is strongly opposed to it because their members believe that only the program source code is complete specification of the system. From their point of view, only the set of test cases that successfully pass are the proof that the implementation is correct.

Other groups also exist between these two extremes that are trying to close the gap between software modeling and programming (also called coding). For example, the creators of the StateWORKS® tool and the corresponding approach claim that although UML tools vendors made serious attempts to generate code from models, they are facing major difficulties, and that these tools can so far produce only header files or code skeletons. As an alternative, they introduced the notion of the totally complete models in an attempt to completely eliminate programming. The models in StateWORKS are sets of virtual finite state machines (VFSMs) that run on top of the VFSM Executor, which is essentially an interpreter.

This book has a similar but different approach. We try to shrink the gap between communication protocol modeling and programming, both by making detailed models and by providing the FSM library, which forces programmers to transform models into code in a uniform way. This methodology makes forward engineering well defined. As already mentioned in the previous chapter, the FSM library provides two main classes, namely *FiniteStateMachine* and *FSMSystem*. The former is used to model and implement individual FSMs and the latter is used as their execution platform, which comprises common services and an event (message) interpreter.

When it comes to programming interpreters and FSM-related libraries, a broad spectrum of possible implementations exists, starting with the traditional structural or procedural solution, continuing with a series of mixed solutions, and ending with the object-oriented solutions of both static and dynamic type. This situation is justified with the fact that the implementation style depends highly on the type of the target architecture. For example, if we consider a microcontroller as the target architecture, we are naturally forced to select a structural solution in the C/C++ programming language. If we consider more powerful architectures, in terms of resources, we may also take into consideration the object-oriented approaches supported by the C++ and Java programming languages.

In the next section, we introduce the component diagrams, which are the means of making implementation models. We then illustrate a spectrum of possible finite state machine implementations, including the catalogued state design pattern (Gamma et al., 1995), which is explained in the separate section. After that, we cover the concepts and most important design and implementation details of the FSM library (its reference manual is given in Chapter 6). We conclude this chapter with two implementation examples.

4.1 Component Diagrams

In the previous chapter, we were dealing with the abstractions in the conceptual world. The design phase typically starts with exploration in the realm of interaction diagrams, where we try to get a better feeling of the system. We finish the design phase by defining the static structure and the complete behavior of the system in the corresponding class and activity or statechart diagrams, respectively. At the end of the design phase, we also specify the deployment of individual software components by rendering the corresponding deployment diagrams.

In the implementation phase we are materializing the design abstractions, such as classes, interfaces, and collaborations, into the components that live in the physical world. As already mentioned, a component is a physical and replaceable part of the system that realizes the given set of interfaces. What we actually do at the beginning of the implementation phase is pack the design abstractions into packages with well-defined interfaces, referred to as components. The examples of such packages are traditional binary object libraries, dynamically linkable libraries (DLLs), and executables, but also tables, files, and documents.

The components and classes are very much alike. Both can:

- Realize a set of interfaces
- Participate in relations (dependencies, generalizations, and associations)
- Be nested
- Have instances
- Participate in interactions

The differences between the components and the classes are the following:

- The former represent physical entities, whereas the latter are conceptual abstractions, so they exist on different levels of abstraction.
- The former only have operations that are accessible through their interfaces, whereas the latter may have both operations and attributes.

The most important feature of the component is that it is replaceable. This means that we can substitute a component with another one without any influence on the system as a whole. This replacement is completely transparent to the users of the replaced component. A new component provides the same or perhaps even better services through the exact same interfaces.

We distinguish the following three types of components:

- The deployment components (already introduced in the context of deployment diagrams). These are the parts of the executable system, such as executables and DLLs.
- The work product components. These are the artifacts of the development process, such as project settings, source code, and data files, that are used to build the deployment components.
- The executable components. These are the parts of the run-time system, e.g., DCOM and CORBA components.

We make the implementation models by rendering the component diagrams. The set of graphical symbols that are available for rendering component diagrams is shown in Figure 4.1. As usual, we select a symbol from the set of available symbols, drag and drop it onto the working sheet, and fill in the data related to its properties. The set of symbols available for rendering component diagrams is obviously a subset of the set of symbols available for rendering deployment diagrams. The properties of these symbols are explained in the previous chapter (see the section on deployment diagrams).

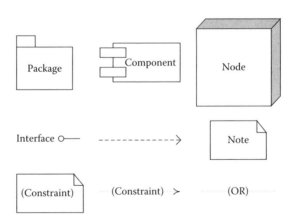

FIGURE 4.1
The set of symbols available for rendering component diagrams.

Implementation

In communication protocol engineering, we are mainly using component diagrams for:

- Modeling APIs
- Modeling executables and libraries
- Modeling source code

Well-defined application programming interfaces (APIs) are some of the most important features of the well-structured software system. An API is an interface that is realized by one or more components. Being an interface, it actually defines a set of services. It represents a clear demarcation line between the service users and the service providers. The former just get the service without caring who is providing it. The same also holds true in the opposite direction, as it is completely transparent for the service providers for whom they actually provide the service.

We may think of APIs as programmatic seams of the system. We use them to connect more components together to create more complex systems. Each component is replaceable. We can replace it with another component whenever there is a need. The developers of the component that uses some APIs do not care who or how it will be provided. They only care about how to fulfill the requirements for the component they are working on currently. Alternately, the system integrator must care that all of the needed components are provided and that they are compliant with their APIs.

Figure 4.2 illustrates the modeling of APIs by means of a very simple example. Imagine that we have been provided with the TCP/IP protocol stack packed as a dynamically linkable library, named *tcpipstack.dll*. It defines the API that comprises three interfaces, namely, *TCPSockets*, *UDPSockets*, and *IPInterface*. The first provides communication services over TCP ports, the second over UDP ports, and the third directly over IP.

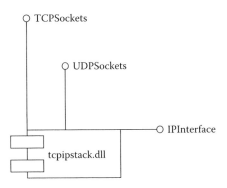

FIGURE 4.2
An example of a simple API.

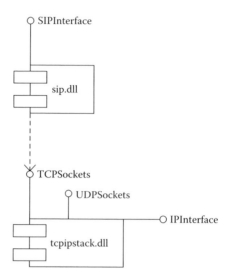

FIGURE 4.3
An example of a simple API user.

Provided with such a component, we are now able to create a new component that uses it. For example, we can create the DLL *sip.dll* (Figure 4.3). This new component provides the SIP services through the interface *SIPInterface*. The fact that *sip.dll* uses services provided through the interface *TCPSockets* is modeled by connecting these two with the dependency relation.

Besides modeling APIs, we can use component diagrams to model executables and libraries. Generally, if the system under development comprises more executables and associated object libraries, it may be wise to make a model that illustrates their relationships. This is especially important if we want to keep versioning and configuration management during the system lifetime under control.

Modeling of executables and libraries can help in making the decision regarding physical partitioning of the system. The issues that affect this decision making are the following:

- Technical issues
- Configuration management issues
- Reusability issues

Figure 4.4 shows the model of a simple executable, named *softphone.exe*. This executable uses the DLL *sip.dll* through the API that comprises the single interface *SIPInterface*. Farther down the hierarchy, *sip.dll* gets communication service that is provided by the DLL *tcpipstack.dll* through the interface *TCPSockets*.

Implementation

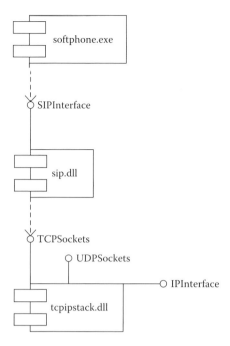

FIGURE 4.4
The model of a simple executable.

Each library and executable is built in the environment of a separate software project. Generally, a software project comprises the project configuration (settings) files, the source code files, and the object libraries. The source code files typically include the module declaration (header and include) files and the module definition files. The developers try to logically organize these files into a file system structure by placing the related files into the same directory (folder).

In the case of complex projects, the corresponding directory tree can get rather ramified, and sometimes it may not be clear where to put new software modules. This can be especially confusing for the new members of the development team. Things get even worse when we must manage splitting and merging of groups of files as development paths fork and join.

In such cases, it is advisable to make a model of the software project, also referred to as the source code model. An example of such a model is shown in Figure 4.5. The executable *Main.exe* is built in accordance with the project definition file *Main.dsw*. Because the project comprises all the module headers and module definition files, the file *Main.dsw* has a dependency relation with all of them. (For clarity, only some of these dependencies are shown in Figure 4.5.)

Farther down the hierarchy, the source code files *AutomataA.cpp* and *AutomataB.cpp* use the header files *AutomataA.h* and *AutomataB.h*, respectively. Both of these header files use the header file *Constants.h*. Finally, all

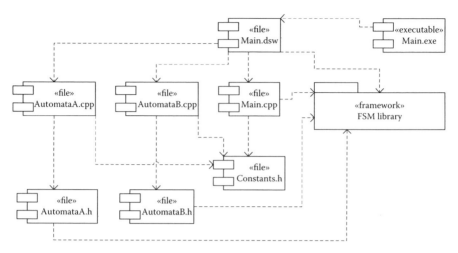

FIGURE 4.5
The model of a simple project.

of the header and source code files, except *Constants.h*, use the framework *FSMLibrary*.

4.2 The Spectrum of FSM Implementations

As mentioned in the previous chapter, we model communication protocols as finite state machines (FSMs). A broad spectrum of various solutions exists for the implementation of FSMs. This section contains a short overview of only three, perhaps the most representative approaches to the implementation of FSMs. The complete treatment of all methodologies and corresponding tools is outside the scope of this book, and as an alternative we simply want to develop ideas by exploring different implementations of a simple FSM (counter by modulo 2). The goal is to familiarize the reader with this subject by showing what the problems are and how they can be tackled.

The three approaches to FSM implementation are illustrated by simple implementations of modulo 2 counters in the Java programming language. As already mentioned, communication protocol developers today mainly use C/C++ and Java, and the selection of the programming language for a certain project mainly depends on the target platform. By mixing examples in Java and C/C++, we want to show that all these languages are applicable in the area of communication protocol engineering, and that the selection of a programming language is not the highest priority issue. Actually, we start with Java in this and the next section, and later we switch to C++.

The state design pattern is a particular FSM implementation type that is special because it was catalogued by Gamma et al. in 1995. Because of that,

Implementation 173

it receives a special treatment in the separate section. However, none of these four approaches is used later in the book. Instead, later we introduce the FSM Library-based implementation paradigm, which is more like the state-of-the-art paradigm. More clearly, first we show what is possible, and perhaps what is next, and then we turn to the current practice in communication protocol engineering.

Let us turn our attention to the subject of the implementation, a communication protocol. As already mentioned in Chapter 1, the communication protocol is defined with the syntax of its messages, the set of procedures (actions) that process the messages, and the set of reactions to exceptional events (timer and error management). In the programming world, they are modeled as finite state machines, also referred to as automata. Mathematically, the abstract automata are defined as:

$$A = (X, Y, S, t, o, S_0)$$

where
$X = \{X_1, X_2, \ldots X_n\}$ is a set of input signals (input alphabet)
$Y = \{Y_1, Y_2, \ldots Y_m\}$ is a set of output signals (output alphabet)
$S = \{S_1, S_2, \ldots S_k\}$ is a set of states (state alphabet)
S_0 is the initial state
t is the transition function, which maps the Cartesian product of SxX to S
o is an output function, which maps the Cartesian product of SxX to Y

Abstract automata are typically illustrated in the form of a state transition graph. The example of the state transition graph in Figure 4.6 illustrates the counter by modulo 2, which is actually the example of a finite state machine we want to implement in Java. It is formally defined as follows:

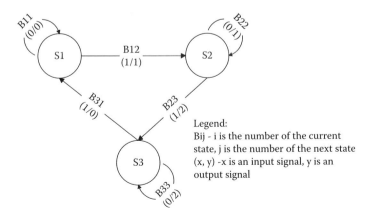

FIGURE 4.6
The counter by modulo 2 state transition graph.

TABLE 4.1
The Counter by Modulo 2 Transition Table

Next State//Output Signal	Input Signal 0	Input Signal 1
State S1	1/0	2/1
State S2	2/1	3/2
State S3	3/2	1/0

$$C = (X, Y, S, t, o, S_0)$$

where
$X = \{0, 1\}$
$Y = \{0, 1, 2\}$
$S = \{S1, S2, S3\}$
$S_0 = S1$

The functions t and o are defined in Table 4.1.

The input and output alphabets comprise the signals {0, 1} and the signals {0, 1, 2}, respectively. The automata can take one of the three possible states, namely, S1, S2, and S3. The initial state of the automata (S_0) is the state S1. Both transition and output functions are defined in Table 4.1. The rows of this table correspond to the automata states (S1, S2, and S3), whereas the columns correspond to the input signals (0 and 1). The elements of Table 4.1 have the format s/y, where s corresponds to the next state number and y corresponds to the output signal.

The same information about the next state and the output signal is shown differently in the state transition graph (Figure 4.6). The arcs of the state transition graph are labeled as $B_{ij}(x/y)$, where i is the number of the current state, j is the number of the next state, x is the input signal that triggers the transition, and y is the output signal generated by the transition. The corresponding statechart diagram is shown in Figure 4.7.

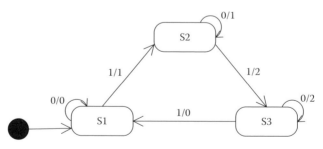

FIGURE 4.7
The counter by modulo 2 statechart diagram.

Implementation

The simplest but perhaps still the most frequently used FSM implementation is based on the structural or procedural approach. This implementation is made in the form of nested selection statements in higher-level programming languages. In the programming languages C/C++ and Java, we typically use *switch-case* statements for this purpose because the control flow structures made with *if* and *else-if* statements are less readable.

Typically, the outermost *switch-case* statement selects a case that corresponds to the automata current state. In the code paragraph that defines the processing of the current state, normally we use the second, nested *switch-case* statement, which selects the case that corresponds to the input signal. The program paragraph that corresponds to that input signal effectively performs the transition by creating the corresponding output signals and evolving to the next state. This evolution is made simply by updating the content of a variable that holds the identification of the current state (most frequently, this is just the index of the state).

Actually, the structure of the resulting program code is very similar to the program representation of SDL (SDL-PR), which was introduced in the previous chapter, and this fact is also mentioned there. Generally, communication protocol implementation based on nested *switch-case* statements looks like the following:

```
switch(state) {
  case STATE_1:
    switch(message_code) {
      case MESSAGE_CODE_1:
        // processing of the message code 1 in the state 1
        break;
      case MESSAGE_CODE_2:
        // processing of the message code 2 in the state 1
        break;
      case MESSAGE_CODE_3:
        // processing of the message code 3 in the state 1
        break;
      ...
      default:
        // processing of the unexpected message in the state 1
        break;
    }
  case STATE_2:
    switch(message_code) {
      case MESSAGE_CODE_1:
        // processing of the message code 1 in the state 2
        break;
      case MESSAGE_CODE_2:
        // processing of the message code 2 in the state 2
        break;
      ase MESSAGE_CODE_3:
        // processing of the message code 3 in the state 2
        break;
      ...
      default:
```

```
      // processing of the unexpected message in the state 2
      break;
  }
  ...
  case STATE_N:
  ...
}
```

We illustrate this general scheme by applying it to the implementation of the counter by modulo 2 in Java. The three states of the counter are labeled as *S1*, *S2*, and *S3* in the program code. The input signals 0 and 1 are labeled as *M1* and *M2*, respectively. The demonstration program reads the actual input signals from the standard input file (by default, this is the keyboard). The generated output signal is represented by a simple printout on the standard output file (by default, this is the monitor). The demo program code is the following:

```
package automata;
import java.util.*;
import java.io.*;
public class Environment1 {
  public static void main(String[] args) throws IOException {
    char ch = '0';
    Automata1 a1 = new Automata1();
    System.out.println("This is the example of counter by modulo 2.");
    System.out.println("Automata evolution has started...");
    while(true) {
      System.out.print("Enter input signal (0/1 and <ENTER>):");
      ch = (char)System.in.read();
      System.in.skip(2);
      if(((ch!='0') && (ch!='1'))) break;
      a1.processMsg(ch);
    }
  }
}
```

The demo program initially creates the object *a1*, an instance of the class *Automata1*, which is the structural and procedural implementation of the counter by modulo 2. After printing two welcome messages, it falls into an infinite *while* loop in which it prompts the user for the input signal and reads it. If the input signal is neither 0 nor 1, the demo program breaks the loop and terminates. Otherwise, it performs one step of the automata evolution by calling the procedure *processMsg()* of the object *a1*.

The Java code for the class *Automata1* is the following:

```
package automata;
public class Automata1 {
  private static final int S1 = 0;
  private static final int S2 = 1;
  private static final int S3 = 2;
  private static final char M1 = '0';
```

Implementation

```
    private static final char M2 = '1';
    private int state=S1;
    public void processMsg(char msg) {
      switch(state) {
        case S1:
          switch(msg) {
            case M1:
              System.out.println("Output signal: 0");
              break;
            case M2:
              System.out.println("Output signal: 1");
              state = S2;
              break;
            default:
              break;
          }
          break;
        case S2:
          switch(msg) {
            case M1:
              System.out.println("Output signal: 1");
              break;
            case M2:
              System.out.println("Output signal: 2");
              state = S3;
              break;
            default:
              break;
          }
          break;
        case S3:
          switch(msg) {
            case M1:
              System.out.println("Output signal: 2");
              break;
            case M2:
              System.out.println("Output signal: 0");
              state = S1;
              break;
            default:
              break;
          }
          break;
        default:
          break;
      }
    }
}
```

The implementation above starts with the definition of the symbolic constants that correspond to the possible automata states, namely *S1*, *S2*, and *S3*, and valid input signals *M1* and *M2* (input signals 0 and 1). Next, we define the variable state that holds the current automata state and we set it to the value *S1* (the automata initial state).

The method *processMsg* starts with the *switch-case* statement that selects the further execution path depending on the content of the variable state (i.e., the current automata state). Three possible cases are found that are defined by the corresponding case clauses. Each of these clauses contains a further *switch-case* statement that distinguishes between two valid input signals, namely *M1* and *M2*. The nested *case* clause that corresponds to the particular input signal prints the message, which corresponds to the output signal, and updates the variable *state*, if the current state of the automata changes.

This example demonstrates the main advantage of the structural or procedural approach, and that is simplicity, which yields greater performance in terms of execution speed. Another advantage is that we can easily construct a compiler or a code generator that generates such implementations (a good example that justifies this claim is SDL-PR). The main disadvantage of this approach is its bad scalability, which becomes evident in the case of large-scale implementations, i.e., implementations of automata that have a large number of states and state transitions.

The code size for such program implementations increases linearly with the number of states and the number of state transitions. Another disadvantage of this approach is that it is monolithic, which implies static regarding the need to change the automata, either by adding new, or deleting the existing states, or by adding or deleting state transitions.

In this type of implementation, the structure of the automata (its vertex and arcs) is built into the machine code of the implementation (hard-coded). We say that the input signal processing flow is governed by the structure of the machine code. If we want to add or delete a state or a state transition, we must change the program code, recompile it, and install the new version on the target platform. Most frequently, the installation procedure requires the system to be restarted at its end. Restarting the system means that effectively it will not be operational for a certain short interval of time. The problem is that some types of systems, such as nonstop systems, may not tolerate restarts no matter how short the time interval is.

Some systems try to make restarts allowable by providing processor tandem configurations. Typically in such a system, one of the processors continues the normal operation while the other is restarting after an update. In that case, we have a synchronization problem, which of course can be solved but it could be rather complex. Generally, system restarts are problematic and should be handled with special care.

On the other end of the spectrum of FSM implementations, we have the diametrical approach to FSM implementation in which the structure of the automata is not defined by the program control flow but rather with the corresponding data structure. The simple interpreter uses this data structure to process the incoming events (messages), therefore it is referred to as an event interpreter. The data structure implementations in assembler and C programming language are built from lists and lookup tables.

Implementation

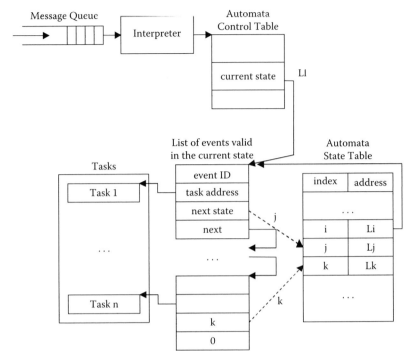

FIGURE 4.8
The event interpreter and the data structure that defines the FSM structure.

The automata evolution is driven by the incoming events. Each input event triggers one step of the evolution. The event interpreter carries out the evolution step by traversing the data structure to determine the current state and the state transition that corresponds to the input event type. In contrast to this common part of the message processing flow — which is directed by the data structure — program parts that correspond to particular reaction tasks are dedicated routines that perform specific functions, which cannot be generalized.

Figure 4.8 illustrates the FSM implementation based on the event interpreter and the data structure that defines the FSM structure (essentially, the state transition graph). New incoming events (messages) are added at the end of the message queue (see the top left corner of Figure 4.8). The interpreter takes the messages from the head of the message queue and processes them by using the data structure, which comprises:

- Automata control table
- Automata state table
- List of valid events (one such list exists for each automata state)

The automata control table is assigned to automata to store its current state and optionally some of its additional attributes. The automata state table is a lookup table that maps the state index into the address of the corresponding list of valid events in that state. The elements of this list contain the complete information necessary and sufficient to perform the state transition from the current state to the next state, which is determined by the event type. This information is stored in the following fields:

- *event ID*: holds the event type to which this element corresponds
- *task address*: contains the pointer to the corresponding routine (procedure)
- *next state*: stores the index of the next state
- *next*: contains the pointer to the next element in the list

The event interpreter processes the message through the following steps:

- Get the message from the head of the message queue.
- Locate the automata control table by examining the content of the message header (message destination field, in particular).
- Read the current state and locate the corresponding list of valid events by looking up the automata state table.
- Determine the event type by examining the content of the message header (message code field, in particular) and locate the corresponding element in the list of valid events (ignore the event if such an element does not exist).
- Perform the task by calling the corresponding task routine as a subroutine (procedure).
- Read the index of the next state from the field *next state*.
- Update the field *current state* by storing the next state index to it.

The advantage of this approach is that we can construct a compiler that transforms the design FSM model into the corresponding data structure and the set of task routines. The automatic translation performed by the compiler increases the probability that the implementation is compliant with the design model and, therefore, that it is correct. Moreover, the routine performed by the event (described above) is fairly simple and short. The price that is paid for the correctness and simplicity is poor performance. The decrease in the processing throughput is proportional to the number of memory accesses to the corresponding elements of the data structure.

Two characteristics of this approach are not obvious from Figure 4.8 and require further explanation. The first characteristic is universality. Since the FSM structure is built into the corresponding data structure, the event interpreter routine is completely independent from it. The event interpreter

Implementation

always repeats the same routine. This is the same for all FSMs. Therefore, this routine is universal in contrast to the implementation with nested *switch-case* statements, which implement just one particular FSM. This characteristic is especially important from the point of software maintenance. If we want to change the FSM structure by adding or deleting states or state transitions, we must update the data structure. There is no need to change the simple interpreter routine at all.

The second characteristic of the event interpreter-based approach is that it enables sharing of common tasks between more state transitions. In principle, this is also possible in the nested *switch-case*-based approach by introducing common functions, which are called from the corresponding case program clauses, but this is seldom used by their practitioners. In the event interpreter-based approach, this possibility becomes more apparent and, therefore, really used because tasks are already specified as procedures (subroutines) rather than *case* program clauses.

Because of task sharing, the number of tasks may generally be smaller than the number of state transitions. We can also organize tasks hierarchically, such that higher-level tasks call their subordinate tasks. This makes it possible to implement more complex tasks by using simple primitives. Such organization has the following advantages:

- Better performance in terms of code size
- Enables dynamic mutation of tasks

By exploiting these characteristics in environments with dynamic loaders, such as Java, we can implement dynamically reconfigurable automata. The automata in such environments change during normal system operation and those changes do not demand any system restarts. In such environments, it is desirable to use the object-oriented approach and to define the FSM structure with the set of objects rather than with a data structure, such as the one previously described. The event interpreters in such implementations interact with the objects that materialize the FSM structure instead of using the traditional data structures.

The following code illustrates FSM structure modeling with the group of classes written in Java:

```
package automata2;
import java.util.*;
import java.io.*;

class Task {
  public int id;
  public Task(int ident) {id=ident;}
  public void processMsg() {System.out.println(id);}
}

class Branch {
  private String msgcode;
```

```java
    private Task task;
    private String nextstateid;

    public Branch(String msg, Task tsk, String nextsts) {
      msgcode=msg;
      task=tsk;
      nextstateid=nextsts;
    }
    public String getMsgCode() {return msgcode;}
    public Task getTask() {return task;}
    public String getNextStateId() {return nextstateid;}
}

class State {
  private String stateid;
  public Set setofbranches;

  public State(String id,Set branches){
    stateid=id;
    setofbranches=branches;
  }
  public String getStateId() {return stateid;}
  public Set getSetOfBranches() {return setofbranches;}
}

class AStructure {
  private String automataid;
  private Set setofstates;

  public AStructure(String id,Set states) {
    automataid=id;
    setofstates=states;
  }
  public String getAutomataId() {return automataid;}
  public Set getSetOfStates() {return setofstates;}
}

class Automata {
  protected AStructure structure;
  protected String stateId;
  protected State initial;

  public Automata(AStructure str,String id,State s) {
    structure = str;
    stateId=id;
    initial=s;
  }
  public void processMsg(String msg) {
    State currentS = initial;
    Iterator iterA = structure.getSetOfStates().iterator(); while(iterA.hasNext()) {
      State eachS = (State)iterA.next();
      if(eachS.getStateId().equals(stateId)) {
        currentS=eachS;
        break;
```

Implementation

```
      }
    }
    Iterator iterS =
currentS.getSetOfBranches().iterator(); while(iterS.hasNext()) {
      Branch eachB = (Branch)iterS.next();
      if(eachB.getMsgCode().equals(msg)) {
        Task t=eachB.getTask();
        t.processMsg();
        stateId=eachB.getNextStateId();
        break;
      }
    }
  }
}
```

The class *Task* models the task that is performed during the transition from the current state to the next state. The task identification is stored in the class field *id*. The user of the class *Task* specifies the particular task identification as the parameter of the class constructor. The default message processing function, named *processMsg()*, just prints the task identification to the standard output file.

The class *Branch* models the arc of the state transition graph. The attributes of the state transition are the message code that triggers the state transition, the task that is performed during the state transition, and the identification of the next stable state. The corresponding fields are named *msgcode*, *task*, and *nextstateid*, respectively. These fields are set by the class constructor. The current content of these fields is returned by the functions *getMsgCode()*, *getTask()*, and *getNextStateId()*, respectively.

The class *State* models a single FSM state. The state attributes are the state identification and the set of the outgoing state transitions (the target state is irrelevant; it can be this state or some other state). The corresponding class fields are named *id* and *branches*, respectively. Their content is set by the class constructor and returned by the functions *getStateId()* and *getSetOfBranches()*, respectively.

The class *AStructure* models the FSM structure. Its attributes are the automata identification and the corresponding set of states. The corresponding class fields are *automataid* and *setofstates*. The class constructor gets particular values for these fields through its parameters. The functions *getAutomataId()* and *getSetOfStates()* return the current values of these fields.

Finally, the class *Automata* models the complete FSM. Its attributes are the FSM structure (essentially the set of sets of state transitions), the current state identification, and the initial state identification. The corresponding class fields are named *structure*, *stateId*, and *initial*, respectively. These fields are set by the class constructor.

The function *processMsg(String msg)* is the event interpreter. The input argument *msg* is the message, which triggered the state transition. The interpretation starts with the iteration through the set of states to locate the object that corresponds to the FSM current state (its identification is stored in the field *stateId*). This is a typical object-oriented approach, which avoids unpopular *switch-case* and similar selection statements. Principally, this first iteration is really not needed and can be easily eliminated by saving the current state object instead of the current state identification. However, the first iteration is intentionally kept to make the example more informative by showing how we can use two subsequent iterations to search through the set of sets of state transitions.

The second iteration searches through the set of state transitions that correspond to the current state to locate the state transition that corresponds to the input message *msg*. After locating the state transition, it gets the object that corresponds to the state transition task and calls its *processMsg()* functions, which in its turn prints the task identification to the standard output file.

From the program code given above, the classes *Task, Branch, AStructure,* and *Automata* are obviously generic and can be used for the construction of any FSM. Besides that, this solution enables the design and implementation of dynamically reconfigurable FSMs because sets in Java can be dynamically updated and the corresponding task object dynamically loaded and unloaded.

We illustrate the applicability of this set of classes with the following implementation of the counter by modulo 2 in Java (the corresponding overall class architecture is shown in Figure 4.9):

```
class Task0 extends Task {
  public Task0(int ident) {super(ident);}
  public void processMsg() {System.out.println("0");}
}

class Task1 extends Task {
  public Task1(int ident) {super(ident);}
  public void processMsg() {System.out.println("1");}
}

class Task2 extends Task {
  public Task2(int ident) {super(ident);}
  public void processMsg() {System.out.println("2");}
}

class Automata2 {
  public static void main(String[]args) throws IOException {
    Automata a2 = makeAutomata();
    char ch;
    String msg;
    System.out.println("This is the example of counter by modulo 2.");
    System.out.println("The automata evolution has started...");
    while(true) {
      System.out.print("Enter input signal (0/1 and <ENTER>): ");
      ch = (char)System.in.read();
      System.in.skip(2);
      if(((ch!='0') && (ch!='1'))) break;
```

Implementation

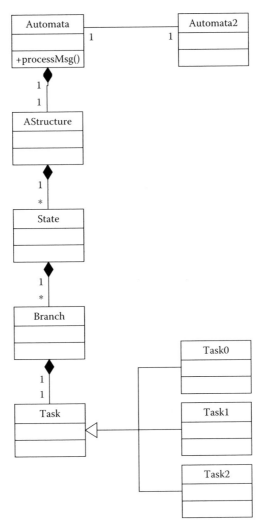

FIGURE 4.9
The static structure used in the second approach to the FSM implementation.

```
    if(ch=='0') msg="0"; else msg="1";
    a2.processMsg(msg);
  }
}

private static Automata makeAutomata() {
  Branch b11 = new Branch("0",new Task0(0),"0");
  Branch b12 = new Branch("1",new Task1(1),"1");
  Set s1 = new HashSet();
  s1.add(b11); s1.add(b12);
  state S1 = new State("0",s1);

  Branch b22 = new Branch("0",new Task1(1),"1");
```

```
    Branch b23 = new Branch("1",new Task2(2),"2");
    Set s2 = new HashSet();
    s2.add(b22); s2.add(b23);
    State S2 = new State("1",s2);

    Branch b33 = new Branch("0",new Task2(2),"2");
    Branch b31 = new Branch("1",new Task0(0),"0");
    Set s3 = new HashSet();
    s3.add(b33); s3.add(b31);
    State S3 = new State("2",s3);

    Set a = new HashSet();
    a.add(S1); a.add(S2); a.add(S3);
    AStructure as = new AStructure("0",a);

    Automata au = new Automata(as,"0",S1);
    return au;
  }
}
```

At the beginning of this example, we define the application specific tasks, namely, *Task0*, *Task1*, and *Task2*, which are responsible for printing the counter by modulo 2 outputs (0, 1, and 2, respectively). Note that the number of tasks (three) is smaller than the number of state transitions (six) in this particular example. The application specific *processMsg()* functions are defined by overriding the default functions.

The definitions of the classes *Task0*, *Task1*, and *Task2* are followed by the definition of the class *Automata2*, which comprises two public functions, the functions *main()* and *makeAutomata()*. The function *main()* starts by calling the function *makeAutomata()*, which in its turn returns the counter by modulo 2 object, named *a2*. After that, it falls into an infinite *while* loop in which it reads the standard input file. If the input character is neither "0" nor "1," it breaks the loop and the program terminates. Otherwise, it converts an input character into the corresponding string ("0" and "1," respectively) and passes it as an input event to the event interpreter.

The function *makeAutomata()* constructs individual state transitions (instances of the class *Branch*), individual states (instances of the class *State*), counter by modulo 2 structure (an instance of the class *AStructure*), and the counter by modulo 2 itself (an instance of the class *Automata*). It first constructs the state transition *b11*, which for the input "0" moves the FSM from the state *S1* to the same state, and during that transition it performs the task *Task0*. Similarly, it constructs the state transition *b12*, which for the input "1" moves the FSM from the state *S1* to the state *S2*, and during that transition it performs the task *Task1*. Next, it constructs the set of state transitions *s1* and the state *S1*.

Likewise, this function constructs the state transitions *b22* and *b23* and the state *S2*, as well as the state transitions *b33* and *b31* and the state *S3*. Finally, it constructs the structure of the counter by modulo 2, named *as*, and the counter by modulo 2, named *au*.

Implementation

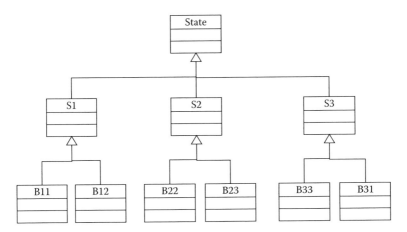

FIGURE 4.10
The counter by modulo 2 state class hierarchy.

The third approach to FSM implementation from the broad spectrum of implementations is illustrated next. In this approach, we define the FSM structure with the corresponding class hierarchy and set of lookup tables that map FSM inputs into the corresponding state transitions. This approach also uses message interpretation and is therefore universal, like the previous one, but it yields much better performance that is comparable with the performance of the first approach (nested *switch-case* statements).

The first idea behind this concept is to model each FSM stable state with the class that is derived from the basic class *State*. The second idea is to consider a state transition (represented with the corresponding arc of the state transition graph) as a transient (i.e., unstable) state. Each state transition is modeled with a class that is derived from the class that represents its originating stable state.

These two ideas lead to the class hierarchy with two hierarchical levels. The root of the class hierarchy is the basic class *State*. The first level of hierarchy defines the FSM stable states whereas the second level of hierarchy defines its unstable states, i.e., state transitions.

We illustrate this approach with the example of counter by modulo 2. The corresponding class hierarchy is shown in Figure 4.10. The first hierarchy level defines the FSM stable states *S1*, *S2*, and *S3*. All of these are derived from the basic class *State*. The second level defines FSM state transitions *B11*, *B12*, *B22*, *B23*, *B33*, and *B31*. Notice that *B11* and *B12* are derived from their originating state *S1*. Similarly, *B22* and *B23* are derived from *S2*, and *B33* and *B31* are derived from *S3*.

The third idea behind this approach is that FSM evolution takes place by traversing the class hierarchy tree and by using polymorphism, one of the most powerful abstractions of object-oriented programming. Concretely, the event interpreter performs the following steps:

- Use FSM input message (signal) and the lookup table (map), which is associated with the FSM current state, to determine the corresponding unstable state (state transition).
- Perform the application-specific task by calling the message processing function defined within the class that models the corresponding unstable state.
- Move the FSM into its next stable state.

The class hierarchy for the counter by modulo 2 is defined with the following Java module:

```
package automata;
import java.util.*;

class State {
  public State msgToBranch(String msg) {return new State();}
  public State processMsg() {return new State();}
}

class S1 extends State {
  public State msgToBranch(String msg) {
    return Structure3.getBranch("0",msg);
  }
}
class S2 extends State {
  public State msgToBranch(String msg) {
    return Structure3.getBranch("1",msg);
  }
}
class S3 extends State {
  public State msgToBranch(String msg) {
    return Structure3.getBranch("2",msg);
  }
}

class B11 extends S1 {
  public State processMsg() {
    System.out.println("Output: 0");
    return new S1();
  }
}
class B12 extends S1 {
  public State processMsg() {
    System.out.println("Output: 1");
    return new S2();
  }
}
class B22 extends S2 {
  public State processMsg() {
        System.out.println("Output: 1");
        return new S2();
```

Implementation

```
    }
  }
class B23 extends S2 {
  public State processMsg() {
    System.out.println("Output: 2");
    return new S3();
  }
}

class B33 extends S3 {
  public State processMsg() {
    System.out.println("Output: 2");
    return new S3();
  }
}
class B31 extends S3 {
  public State processMsg() {
    System.out.println("Output: 0");
    return new S1();
  }
}

public class Automata3 {
  private State state;

  public Automata3() {
    state = new S1();
  }
  public void processMsg (char chmsg) {
    String msg;
    if(chmsg=='0') msg="0"; else msg="1";
    state = state.msgToBranch(msg);
    state = state.processMsg();
  }
}
```

The basic class *State* has two default functions, *msgToBranch()* and *processMsg()*. Both functions return an instance of the class *State*. The fact that the instance of the class derived from the class *State* is also considered to be the instance of the class *State* enables the event interpreter to employ polymorphism. We will return to this point shortly.

The function *msgToBranch()* is responsible for mapping the FSM input message into the corresponding state transition object. The input message in this simple example is a one-character string ("0" or "1"). The function can return any instance of the basic class *State*, but normally in this example it should return the instance of the class *B11, B12, B22, B23, B33*, or *B31*.

The function *processMsg()* caries out the application-specific task for the given input message. It returns the FSM next stable state. The idea is that the FSM dynamically changes its behavior. The FSM is in a certain state, either stable or unstable, at any point in time, but it is always represented by a single object. That object is actually returned by one of these two functions, which are called in the course of FSM evolution.

Next, we define the classes that model the FSM stable states, namely, *S1*, *S2*, and *S3*. Each of these classes extends the basic class *State* and overrides the default function *msgToBranch()* with the application-specific one. These particular functions actually delegate their responsibility to the function *getBranch()* of the class *Structure3* by passing their identification ("0," "1," and "2" for *S1*, *S2*, and *S3*, respectively) and the input message to it. More precisely, these simple functions just return the unstable state object that is provided by the function *getBranch()* to their caller, and that is the event interpreter.

The stable state classes are followed by the classes that model the FSM unstable states, namely, *B11*, *B12*, *B22*, *B23*, *B33*, and *B31*. Each of these classes extends the corresponding stable state class and overrides the default function *processMsg()*, which it inherits from the basic class *State*, with the application-specific one. These particular functions perform the application-specific tasks and return the corresponding next stable state object (*S1* for *B11* and *B31*, *S2* for *B12* and *B22*, and *S3* for *B23* and *B33*). The application-specific tasks in this simple example are implemented as the corresponding print statements to the standard output file.

The FSM is modeled with the class *Automata3*. This class has a single attribute named *state*, which is set by the class constructor to the FSM initial stable state, namely *S1*. Later during the FSM evolution it changes and can become any FSM state, either stable or unstable.

The class *Automata3* has a single function, named *processMsg()*, that is the FSM event interpreter. This function performs one state transition in two steps. In the first step, it calls the function *msgToBranch()* of the FSM current stable state object. This effectively starts the state transition by moving the FSM from its current stable state to the unstable state that corresponds to the input message. In the second step, the event interpreter calls the function *processMsg()* of the FSM unstable state, which performs the application-specific task and returns the FSM next stable state object. This effectively completes the state transition. Interestingly, the state class hierarchy in this approach is completely application-specific whereas the event interpreter is very simple and generic and therefore can be reused in implementations of other FSMs.

The following utility classes support mapping of input messages to the corresponding state transitions (unstable state objects):

```
package automata;
import java.util.*;

class MapContainer {
  private String identification;
  private Map map;

  public MapContainer(String id,Map m){
    identification = id;
    map = m;
  }
  public String getId() {return identification;}
```

Implementation

```
    public Map getMap() {return map;}
}

public class Structure3 {
  private static Set maps;

  public void setMaps(Set m) {
    maps = m;
  }
  public static State getBranch(String id,String msg) {
    Map m = new HashMap();
    Iterator iter = maps.iterator();
    while(iter.hasNext()) {
      MapContainer each = (MapContainer)iter.next();
      if(each.getId().equals(id)) {
        m = each.getMap();
        break;
      }
    }
    return (State)m.get(msg);
  }
}
```

The class *MapContainer* stores the map identification and the map itself in the attributes *identification* and *map*, respectively. These attributes are set by the class constructor. Their current content is available through the corresponding *get* functions.

The class *Structure3* contains a set of maps for all FSM stable states. This set is established by the function *setMaps()* and searched by the function *getBranch()*. The input parameters of the function *getBranch()* are the map (i.e., stable state) identification and the input message. The function *getBranch()* iterates through the set of map containers, locates the one with the given identification, uses the located map to get the state transition that corresponds to the input message, and returns it to its caller.

An important feature of this approach is that it is based on Java sets and maps, which makes it an ideal environment for making dynamically reconfigurable FSMs as Java sets and maps can be dynamically updated. For example, if we want to add a new state transition *B21*, it would be sufficient to write, compile, and dynamically load a new class *B21* that represents it and to add the corresponding entry in the map that is associated to the FSM stable state *S2*.

Because the current Java version does not support a map of maps, the solution for mapping input events to the corresponding state transitions presented here is based on the usage of a set of maps. Worth mentioning is the fact that an environment with a map of maps would enable top performance implementations based on two connected mappings. The key for the first mapping would be the FSM current stable state whereas the key for the second mapping would be the input message. The performance of such implementations would be even better than the performance of the implementations based on nested *switch-case* statements.

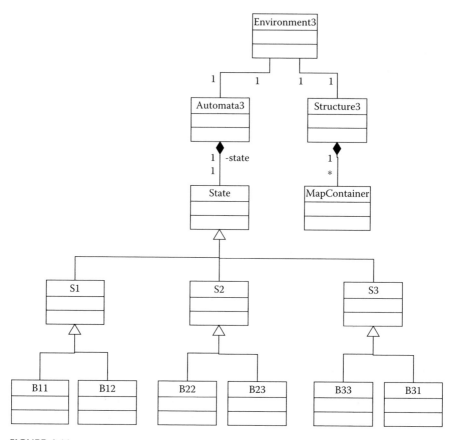

FIGURE 4.11
The static structure used in the third approach to the FSM implementation.

The class *Environment3* uses the previously defined classes and demonstrates their usability. The corresponding Java code is the following (the overall class architecture is shown in Figure 4.11):

```
package automata;
import java.util.*;
import java.io.*;

public class Environment3 {
  public static void main(String[] args) throws IOException {
    char ch = '0';
    Automata3 a3 = new Automata3();

    Map m1 = new HashMap();
    m1.put("0",new B11()); m1.put("1",new B12());
    MapContainer M1 = new MapContainer("0",m1);

    Map m2 = new HashMap();
    m2.put("0",new B22()); m2.put("1",new B23());
```

Implementation

```
    MapContainer M2 = new MapContainer("1",m2);

    Map m3 = new HashMap();
    m3.put("0",new B33()); m3.put("1",new B31());
    MapContainer M3 = new MapContainer("2",m3);

    Set maps = new HashSet();
    maps.add(M1); maps.add(M2); maps.add(M3);

    Structure3 st3 = new Structure3();
    st3.setMaps(maps);

    System.out.println("This is the example of counter by modulo 2.");
    System.out.println("The automata evolution has started...");
    while(true) {
      System.out.print("Enter input signal (0/1 and <ENTER>): ");
      ch = (char)System.in.read();
      System.in.skip(2);
      if(((ch!='0') && (ch!='1'))) break;
      a3.processMsg(ch);
    }
  }
}
```

The function *main* starts by creating the object *a3*, an instance of the counter by modulo 2. It then creates all the necessary maps and map containers, the set of maps named *maps*, the object *st3*, an instance of the class *Structure3*. After this, it sets the set of maps by calling the function *setMaps()* and falls into an infinite *while* loop in which it reads FSM input messages and calls the event (message) interpreter until the user enters a signal that is neither "0" nor "1."

The keys for searching Java maps in this simple example are just simple strings ("0" and "1"). This Java map is a rather powerful abstraction because its key may be any class whose instances are comparable. This makes it possible to model real communication protocol messages with such classes and to build Java maps for them. Once we model the messages by the corresponding objects, FSM objects can interact with them in an object-oriented fashion.

If we want to provide a full object-oriented treatment of communication protocol messages, we must provide the corresponding serialization functions. Two types of these functions are actually used. The first type is used for converting an object into a series of octets that can be transported over the communication line. The second type performs the reverse operation by converting the received series of octets into the corresponding object. If we do not provide these serialization functions, we are forced to operate directly on numbers and use *switch-case* and similar statements unpopular in the object-oriented world.

4.3 State Design Pattern

The State design pattern is one of the approaches to FSM implementation. As already mentioned, the State pattern is shown in a separate section because it was catalogued by Gamma et al. and therefore it is not just another example but rather a well-defined and proven concept. The reader may find the complete description of the State pattern in the original book on design patterns (Gamma et al., 1995). Here we present just a brief overview and an example that demonstrates the State pattern applicability.

The original motivation to introduce this design pattern was to support objects that change their behavior as their state changes, exactly what the FSMs do. For example, when counter by modulo 2 (Figure 4.6) is in its initial state $S1$, it produces the output 0 for the input 0, but when its state changes to $S2$ or $S3$, it produces different outputs for the same input (1 in the state $S2$, and 2 in the state $S3$). Similarly, the input 1 yields the output 1 in the state $S1$, the output 2 in the state $S2$, and the output 0 in the state $S3$.

The key idea of this design pattern is to separate the FSM appearance from its behavior. We define the FSM appearance with the FSM wrapper class, which is referred to as a context. The **context** defines the user interface (a set of operations accessible by the FSM users) and contains the current FSM state object, which is one of the concrete FSM state objects.

The FSM behavior is defined with the wrapped state hierarchy. The root of this hierarchy is the generic state class, which actually defines an interface for the concrete states of the context. Each concrete state class is derived from the generic state class and it provides the state-specific behavior of the context (FSM).

The State pattern revolves around polymorphism. Essentially, context (FSM) delegates the state-specific requests to the current state object. More precisely, each operation defined within the user interface simply calls the corresponding operation on the current state object (these operations usually have the same name). The context can pass itself as a parameter to the called operation and thus make itself accessible to the concrete state, if needed.

Typically, clients initially configure the context with state objects. Later, during normal system operation, clients do not deal with state objects directly. Notice that either the context class or the concrete state subclass can change the context current state. Therefore the FSM transition logic can be centralized, distributed, or hybrid.

According to the authors, the State pattern consequences are the following:

- It localizes state-specific behavior.
- It makes state transitions explicit.
- State objects can be shared.

Implementation

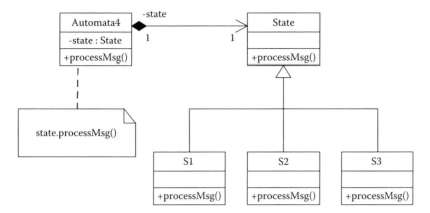

FIGURE 4.12
The static structure used by the State design pattern.

At the end of this short overview of the State pattern, we illustrate its applicability with the simple example — a State pattern-based implementation of the counter by modulo 2. The corresponding class diagram is shown in Figure 4.12. The context in this example is the class *Automata4*. The attribute *state* holds the current FSM state object. The key function *processMsg()* delegates message processing to the current FSM state object by calling its function *processMsg()*.

The generic state class *State* defines a simple interface, which comprises a single function, *processMsg()*. Generally, such a function would define the default FSM behavior, which can then be overridden in the concrete substate classes. In this simple example, as we will shortly see, no such a behavior is allowed and therefore the corresponding operation is simply empty.

The concrete substate classes *S1*, *S2*, and *S3* are derived from the generic state class *State*. Each of these classes provides a state-specific behavior by overriding the function *processMsg()* with its own particular definition. The corresponding code in Java is the following:

```
package automata4;
import java.util.*;

public class Automata4 {
  private State state;
  public Automata4() {state = new S1();}
  public void setState(State s) {state = s;}
  public void processMsg(char msg) {
    state.processMsg(this,msg);
  }
}

class State {
  public void processMsg(Automata4 a,char ch) {
  }
}
```

```
class S1 extends State {
  public void processMsg(Automata4 a,char ch) {
    if(ch=='0') {
      System.out.println("Output 0");
      a.setState(new S1());
    } else {
      System.out.println("Output 1");
      a.setState(new S2());
    }
  }
}

class S2 extends State {
  public void processMsg(Automata4 a,char ch) {
    if(ch=='0') {
      System.out.println("Output 1");
      a.setState(new S2());
    } else {
      System.out.println("Output 2");
      a.setState(new S3());
    }
  }
}

class S3 extends State {
  public void processMsg(Automata4 a,char ch) {
    if(ch=='0') {
      System.out.println("Output 2");
      a.setState(new S3());
    } else {
      System.out.println("Output 0");
      a.setState(new S1());
    }
  }
}
```

The definition of the class *Automat4* begins with the definition of the field *state*, which is used to store the FSM current state object. The class constructor sets this field to the FSM initial state object, which is an instance of the class *S1*. The function *setState()* is used by the FSM concrete state objects to change the FSM state (an example of distributed transit logic). The function *processMsg()* simply calls the corresponding function on the FSM current state object.

The class *State* defines a simple state interface with just one function — *processMsg( )* — which is empty because this example has no default behavior. The class *S1* is an example of a concrete substate class. It defines the *S1*-specific FSM behavior by overriding the function *processMsg()* that it inherits from the base class *State*. This function checks whether the input signal is 0 or 1, prints the corresponding output signal, and changes FSM state by calling the function *setState()*. We made the context accessible by passing it as a parameter to the function *processMsg()*.

The following Java code creates the working environment for this example (given without the comments because similar code is already explained in a previous section):

```
package automata4;
import java.util.*;
import java.io.*;

public class Environment4 {
  public static void main(String[] args) throws IOException {
    char ch = '0';
    Automata4 a4 = new Automata4();
    System.out.println("This is the example of counter by modulo 2.");
    System.out.println("The automata evolution has started...");
    while(true) {
      System.out.print("Enter input signal (0/1 and <ENTER>): ");
      ch = (char)System.in.read();
      System.in.skip(2);
      if(((ch!='0') && (ch!='1'))) break;
      a4.processMsg(ch);
    }
  }
}
```

4.4 Implementation Based on the FSM Library

In the previous two sections, we have explored various approaches to the FSM implementations by the means of simple examples. The reader should be much more familiar with the FSM implementation by now, but for the serious communication protocol engineering we need much more. We need a well-established working environment that will enable productive and repeatable development processes that yield maintainable products (communication protocols) of high quality.

The main measure (metrics) of quality in the context of communication protocols is their reliability, which is considered to be proportional to the number of remaining software bugs. Another important quality measure is the product performance measure with its throughput (the number of messages processed in the given interval of time) and hardware resources needed to achieve that throughput (RAM and ROM size and processor speed measured in MIPS or MHz). Generally, one of the key factors to successful software quality assurance is the quality of the software tools used in the development process. Communication protocol engineering is by no means an exception in this respect.

In this section, we present an example of the state-of-the-art working environment for the productive development of communication protocols. The environment is effectively created by an integrated development environment, which includes a C++ compiler and the domain-specific C++ library, named FSM Library. As already mentioned, the FSM Library includes two fundamental classes, *FSMSystem* and *FiniteStateMachine*. The former creates the execution platform for a group of FSMs whereas the latter is the base class for implementing individual FSMs.

The FSM Library API comprises two interfaces, which are defined by the class *FSMSystem* and *FiniteStateMachine*. The complete FSM Library programmer reference manual is given in Chapter 6. The reference manual also includes two representative implementation examples. In this section, we focus on the FSM Library concepts and internals.

The key concept behind the FSM Library is to enable productive implementations of FSMs in a uniform way. The main task of the FSM Library user is to implement the FSM state transition functions. The user does this by translating the design artifacts (statechart diagram, activity diagram or SDL diagram) into the corresponding C++ class function members. This translation can be done manually or with a software tool (typically used if the product performance is not critical).

The process of translation is both productive and uniform because the FSM Library provides all the functions needed to effectively construct an FSM state transition. These functions can be classified into the following function groups:

- Message handling functions (both message header and message payload handling functions). These functions support both message coding and decoding (i.e., message synthesis and analysis).
- Message sending functions.
- Timer handling functions (essentially, start, stop, and restart timer).

The reader may be puzzled by the fact that the list given above does not include any message receiving functions. The FSM Library is specific in this respect. The developer does not need to explicitly call a function that receives a message (signal). Rather, the FSM execution platform (provided by the class *FSMSystem*) routes all sent messages toward their destination automata, locates the state transition function that corresponds to the message type (determined by the content of the corresponding message header field), and calls it as its subroutine. We will see shortly that the function that performs the message routing and processing (named *Start*) is actually the event interpreter.

Therefore, the FSM Library completely supports the message handling style present in the design artifacts (statecharts, activity diagrams, and SDL diagrams), which just name the input event (message) without taking care of how that event is effectively recognized (received). The FSM Library provides the class *FSMSystem* to support the straightforward implementations of design artifacts. Once provided with the class *FSMSystem*, the

Implementation 199

developers do not care how the message is received; they simply write the C++ function that performs the state transition when the message is received.

Other FSM Library specifics are the following:

- The FSM implementation is independent from the underlying real-time kernel.
- The FSM Library provides the mechanism to send messages to the dynamically allocated automata instances, which are referred to as unknown automata instances.
- The FSM Library provides public mailboxes, which can be used as message queues with different priorities.
- The FSM Library separates the message handling functions from the real-time kernel. This feature is referred to as the encapsulation of the message handling functions.
- The FSM Library treats timers as special messages, which are distinguished from the communication protocol messages by the code that determines the message type.
- The logging system provided by the FSM Library is based on the test version of the real-time kernel, which is derived from the target (final) real-time kernel.
- The FSM implementation is independent from the concrete formats of the communication protocol messages.
- The FSM Library provides automatic message buffer reallocation in cases where current buffer capacity becomes insufficient for storing additional message parameters.

The following paragraphs provide short comments on each of these FSM Library specifics. We proceed through the list of specifics from its beginning toward its end.

An important design decision was to make the FSM Library independent from the underlying run-time kernel. This decision is important because it enables easy porting of the FSM implementations to various target platforms (bare machine, UNIX, Windows NT). The internal class *KernelAPI* facilitates this independence. It represents a clean interface between the FSM implementation and the run-time system. The kernel developer must derive a new class from the class *KernelAPI* and write its real member functions by taking into account the details of the particular target platform. An example of such implementation is shown later in this section.

The second FSM Library-specific feature is related to the beginning of the communication between two FSMs, namely, FSM A and FSM B, where the former has the active role and the latter is passive. The problem is simple if A always communicates with the same B, but it becomes more complex if B is not known in advance (B is an unknown FSM). Consider a pool of FSMs, where each is capable of performing the same task. FSM A is principally interested in engaging any instance from the pool that is free.

The FSM Library facilitates the communication with the unknown automata by placing all relevant data into the header of the message that is sent to it. The message destination is set to the special code, named UNKNOWN_AUTOMATA. The function member *Start* of the class *FSMSystem* recognizes this code and dynamically allocates an automata instance, which will be the message destination and therefore involved in the further communication with the message originator. In the case when there are no free automata instances available in the pool, the function *Start* calls the special function *NoFreeInstances*, which is responsible for the recovery procedure. Typically, this function informs the message originator about the automata instance outage by sending it an appropriate signal, such as *NAK*, *DISCONNECT*, and so on.

The third FSM Library-specific feature is the provision of general purpose mailboxes, which can be used both as public mailboxes and private mailboxes. The former are actually FIFO message queues that contain messages for various destinations whereas the latter contain messages for a single destination, which is an FSM that owns the private mailbox. Generally, we can use only a single public mailbox to enable the communication between all FSMs present in the system. Such a solution can suffice in the case of simple systems with a small number of FSMs and soft real-time requirements. However, a single public mailbox may not be sufficient in the case of more complex systems because the FSM Library mailbox is just a FIFO message queue without any support for message prioritization.

The absence of message prioritization can lead to a case where an FSM processes an outdated message instead of processing the corresponding timeout message just because the outdated message is ahead of the timeout message in the public mailbox. Such cases can lead to dysfunctional behaviors that are not caused by design oversights but rather with the inappropriate implementation.

The regular method of supporting message prioritization in the FSM Library-based implementations is to use more public mailboxes that are assigned different priorities. For example, we can use three public mailboxes for three different priorities. These three public mailboxes are effectively treated as three FIFO message queues with different priority (e.g., high, medium, and low). We can select a strategy of using private mailboxes instead. We can also mix public and private mailboxes if we wish. Actually, the function *Start* (the member of the class *FSMSystem*) treats them equally. In its loop, it searches all the mailboxes for messages. The effective mailbox priority is determined by the order of that search (i.e., it starts from the mailbox index 0).

The fourth FSM Library-specific feature is the encapsulation of the message handling functions. Generally, real-time kernels can store the message source and destination information in the message header or in the separate data structure. By separating the message handling functions into a group that

Implementation

handles the message header and a group that handles the message payload, the FSM Library provides complete FSM implementation independence from the message source and destination information location.

An additional enhancement related to the message destination provided by the FSM Library is the support for sending messages to the *left* or to the *right* FSM. The abstraction of the *left* and *right* FSM originally comes from SDL. If the SDL symbol for sending a message points to the left, we say that the message is sent to the *left* FSM. Similarly, if the symbol points to the right, we say that the message is sent to the *right* FSM.

The internal class *KernelAPI* provides the functions *SendMessageLeft* and *SendMessageRight*, which are inherited by the class *FiniteStateMachine*, to support this abstraction. These two functions enable the direct coding of the corresponding parts of SDL diagrams, and the resulting C++ code has a great similarity with the original SDL diagrams. For example, consider the following snippet of C++ code that corresponds to a state transition:

```
StopTimer(FE4_TIMER1);
DisconnectRingTone();
PrepareNewMessage(0x00,r2_SetupRespConf);
SendMessageLeft();
StartChargingIncoming();
Connect();
SetState(FE4_ACTIVE);
```

The call of the function *SendMessageLeft()* above is a direct encoding of the corresponding left-pointing SDL graphical symbol. This snippet of code is a typical state transition implementation based on the FSM Library, which is rather short and easy to read and map to the original design model. These are two key implementation features that ensure productivity and quality.

The fifth FSM Library-specific feature is that it treats timers as special messages, distinguished from the communication protocol messages by the code that determines the message type. Some of the message header parameters are meaningless for timers. The corresponding message header fields are used by the FSM Library API functions related to timers to store the data specific for individual timers, such as timer duration.

All timers used by a certain FSM type must be initialized in the FSM class function member *Initialize()* by calling the function *InitTimerBlock()* (see Section 6.8.74). The parameters of this function are the timer identification, the timer duration, and the identification of the message to be sent when the timer expires. In response to a series of *InitTimerBlock()* calls, the system creates the corresponding array of timers. The identification of a timer effectively becomes the index of this array.

Once initialized, the timer can be started by the function *StartTimer()*, stopped by the function *StopTimer()*, restarted by the function *RestartTimer()*, or checked by the function *IsTimerRunning()*. All these functions have a single parameter, the identification of the timer. Therefore, the resulting C++ code resembles the original design model to a great extent. Moreover, when the timer expires, the corresponding message is automatically sent to the FSM that started it, which processes this message in the same fashion as all other messages. This feature also contributes to the similarity of the resulting C++ code and the original design model.

The sixth FSM Library-specific feature is that the logging subsystem provided by the FSM Library is based on the test version of the real-time kernel, which is derived from the target (final) real-time kernel. The logging subsystem is important in communication protocol engineering because certain design oversights or implementation errors become evident only in complex circumstances, which can happen only after long run-time periods. Typically, such circumstances are difficult to repeat and therefore developers normally use log files to backtrack the sources of errors once they occur.

The FSM Library provides a complete logging subsystem that is used both during system testing and normal system exploitation. The internal class *LogAutomata* defines the necessary set of functions. FSM tracing is based on the interception of all relevant internal functions, such as FSM state updating, message processing, timer management functions, and so on. Automatic logging of various events makes the resulting log file outlook uniform, and thus easy to read by any member of the development team. All logging events are prioritized, which helps developers to easily define exactly which events they want to trace.

Traditionally, log files are located on mass storage devices such as hard disks or flash memory. The FSM Library introduces an enhancement in this respect. The internal class *LogInterface* defines the interface between the system implementation and the concrete logging media, such as the conventional log file, the TCP/IP connection to the logging server, and so on. Logging to the concrete media is provided by a subclass that is derived from the base class *LogInterface*. Examples of such classes are the classes *LogFile* and *LogTCP*.

The seventh FSM Library-specific feature is that the FSM implementation is independent of the concrete formats of the communication protocol messages. The feature is facilitated by the internal class *MessageHandler*, which provides a set of generic functions for manipulating message parameters. Basically, two families of these functions exist, namely, *get* and *add*. The former return the value of the given parameter whereas the latter add the given message parameter to the message. The parameter is specified with its identification (code) and its value.

The class *MessageHandler* uses the class *MessageInterface*, which is an abstract class that defines the interface for the abstract message format. Normally, the developer derives a class from the class *MessageInterface* for each concrete message format and writes its function in accordance with the format-specific details. An example of such a class is the class *StandardMessage*, which models a message that comprises a sequence of octets (characters). Such an approach centralizes message handling functionality. This centralization eliminates code redundancy and increases code coverage.

Implementation

Additionally, development team productivity is increased because message handling functions and FSMs can be developed in parallel.

The eighth and last FSM Library-specific feature is that it provides automatic message buffer reallocation in cases where the current buffer capacity becomes insufficient for storing additional message parameters. Although this functionality is rather easily implemented, it is important because it makes the process of message creation completely transparent. The programmer just adds parameters to the new message as needed, without having to take care about the size of the free space in the corresponding buffer. This detail is completely hidden by the message handling functions.

4.4.1 Using the FSM Library

Using the FSM Library is rather easy. It helps a lot in both the design and implementation phases of the development process. The author's experience shows that both students and engineers working in the industry can start using it only after a couple of days of training. Actually, it does not take more than writing one example based on the FSM Library to start using it. Besides that, it is a well-established working environment that has been used in a series of the real-world projects for the industry.

When it comes to design, the FSM Library greatly simplifies matters by providing two fundamental classes, *FSMSystem* and *FiniteStateMachine*. The existence of these two classes makes the system static structure well known from the start (Figure 3.5). Each protocol is modeled by the subclass derived from the base class *FiniteStateMachine*. The resulting FSM is executed by the event interpreter, which is hidden inside the class *FSMSystem*. These two classes practically encapsulate all domain-specific design patterns needed for designing a communication protocol.

The overall result is that the class diagram is almost not needed at all, at least not for realistic communication systems that comprise less then a dozen communication protocols. Even for very complex communication systems based on the FSM Library, the class diagram can be used more as an accompanying document. The most informative part of such a class diagram would be the one that specifies the mailboxes present in the system, as well as the timers used by individual FSM types.

Real valuable design artifacts for the paradigm based on the FSM Library are the complete models of the system behavior in the form of the activity, statechart, or SDL diagrams. This is the case because the FSM Library *de facto* specifies the skeleton of the system static structure but it does not (and cannot) specify the complete system behavior. It provides only primitive behavior from which we can build more complex behavior, in particular, the state transitions.

Once we have finalized the detailed design diagrams (activity, statechart, or SDL diagram), we are ready to proceed to the implementation phase of the development process. The main task of implementing FSMs by using the FSM Library, besides writing the initialization function and a couple of simple auxiliary functions, is the encoding of state transitions by using the set of primitives provided within the FSM Library application programming interface (see Section 6.8). A good thing about these primitives is that they provide mapping of SDL steps in almost a one-to-one manner. The names of the primitives are almost self-documenting, at least after the short experience you get by using them. The code resembles the original design artifacts (especially SDL diagrams). All these attributes helps any member of the development team to read, understand, and continue the work that was done by some other member of the development team, especially if they have the design artifact at their disposal.

Also worth mentioning is that besides forward engineering, the FSM Library helps backward engineering, too. This is especially true if the backward engineering is done by hand. Using software tools for that purpose is also possible if the development team strictly obeys certain coding guidelines. The key for the successful forward and backward engineering is the well-defined API (see Section 6.8).

We demonstrate the usage of the FSM Library API by the examples at the end of this chapter, as well as with the examples at the end of Chapter 6.

4.4.2 FSM Library Internals

This section describes the FSM Library internals. The main FSM Library components are the following:

- The class *FSMSystem*
- The class *FiniteStateMachine*
- The real-time kernel

The class *FSMSystem* provides the following functionalities:

- Initialization of the FSM objects. The result is a set of the corresponding transition tables, which determine which state transitions are triggered by the individual events (messages).
- Routing of messages. This component locates the message destination FSM, looks up its state transition table to find the state transition that corresponds to the message type, and calls the corresponding function as its subroutine.
- Public mailbox prioritization. The public mailbox priority decreases as its identification increases. The identification is actually the index of the corresponding mailbox array. The public mailbox with the identification 0 has the highest priority.

Implementation 205

- Allocation of FSMs from the pool of FSMs. If the message destination is an unknown object of a certain type, a free FSM from the corresponding pool is allocated to process that message.

The class *FiniteStateMachine* provides the following functionalities:

- Maintaining the current state variable (the field member of this class)
- Maintaining the state transition table
- FSM evolution support by providing the address of the state transition function that corresponds to the incoming message type
- Message handling (message checking, parsing, and creation)
- Message exchange (the message send operation is explicit whereas the message receive operation is implicit)
- Memory management (supports requesting and releasing buffers for messages)
- Timer management (supports starting, stopping, restarting, and testing timers)

The functionalities provided by the real-time kernel are inherited by the class *FiniteStateMachine* (message exchange, buffer, and timer management). The following subsections describe the internals of these three components.

4.4.2.1 FSMSystem Internals

As already mentioned, the class *FSMSystem* provides the execution platform for all FSMs present in the system. The list of concrete functionalities provided by this class is already given in the previous section. The heart of the class *FSMSystem* is the function *Start*, which actually provides all the listed functionalities. Essentially, it is the event (message) interpreter. Its program code in C++ is the following:

```
void FSMSystem::Start(){
  SystemWorking = true;
  while(SystemWorking) {
    Sleep(1);
    for(uint8 i=0; i<NumberOfMbx; i++) {
      uint8 *msg = GetMsg(i);
      if(msg == NULL){
        continue;
      }
      uint8 automataType = GetMsgToAutomata(msg);
      if(((automataType > NumberOfAutomata) ||
         (NumberOfObjects[automataType] == 0))){
        // Error handling
        DiscardMsg(msg);
        continue;
      }
```

```
      uint32 objNum = GetMsgObjectNumberTo(msg);
      if(objNum == UNKNOWN_AUTOMATA){
        ptrFiniteStateMachine object =
          FreeAutomata[automataType].Get();
        if(object != 0) object->Process(msg);
        else
          (Automata[automataType][0])->NoFreeObjectProcedure(msg);
        continue;
      }
      else if(objNum > NumberOfObjects[automataType]) {
        // Error handling
        DiscardMsg(msg);
        continue;
      }
      else {
        (Automata[automataType][objNum])->Process(msg);
      }
    }
  }
}
```

The function *Start* initially sets its field member *SystemWorking* to the value *true* and enters the loop, which is executed while *SystemWorking* has the value *true*. Once this variable is set to the value *false* (this is exactly what the API function *StopSystem()* does), the function *Start* exits the loop and terminates. Because this function is the FSM event interpreter, once it stops the whole system stops.

Inside the *while* loop, this function enters the nested *for* loop in which it checks all mailboxes for messages. This *for* loop starts from the mailbox with the identification (index) 0, thus making it the highest priority mailbox. As it proceeds toward the identification *NumberOfMbx*, the priority of the corresponding mailboxes decreases.

Once it finds a message in the mailbox, it exits the nested *for* loop and continues with determining the destination automata (FSM) type identification by calling the function *GetMsgToAutomata()*. If the identification is invalid (greater than the configuration parameter *NumberOfAutomata*) or if no instances of that type are found, the function discards the message by calling the function *DiscardMsg()* and continues the main loop.

If the automata type identification is valid and at least one instance of that type is found, the function *Start* determines the destination object identification by calling the function *GetMsgObjectNumberTo()*. If this identification is equal to *UNKNOWN_AUTOMATA*, the function *Start* tries to allocate an object from the pool of objects of the given type by calling the function *Get()* on the object of that type.

If at least one free object is found in the pool (actually an array of objects of the given type), the function *Get()* will return the identification (array index) of the first one and in its turn, the function *Start* will call its function *ProcessMsg()*. Behind the scenes, the function *ProcessMsg()* locates the state transition that corresponds to the message type, calls it as its subroutine,

Implementation 207

and continues the main loop. If no free objects are in the pool, the function *Start* discards the message and continues the main loop.

Finally, if the message destination is a known object (its identification is not equal to UNKNOWN_AUTOMATA), the function Start checks if its identification is valid (not greater than the configuration parameter NumberOfObjects[automataType]). If the object identification is valid, the function Start calls object function processMsg() and continues the main loop.

4.4.2.2 FiniteStateMachine Internals

The class *FiniteSateMachine* is at the top of the FSM Library class hierarchy (Figure 4.13). It hides the details of the FSM Library internal static structure from its user. The class *FiniteStateMachine* inherits logging-related functionality from the class *LogAutomata* (shown as the left branch of the class hierarchy in Figure 4.13). Alternately, the class *FiniteStateMachine* inherits the buffer, timer, and message management functionality from the class *KernelAPI* (shown as the right branch of the class hierarchy in Figure 4.13). Both *FiniteStateMachine* and *KernelAPI* inherit the message management functionality from the class *MessageHandler*.

The class *LogAutomata* conceptually uses the logging services provided through the interface created by the class *LogInterface*. The logging services are provided in run-time reality by the object that is an instance of a subclass, which is derived from the base class *LogInterface*. Figure 4.13 shows two examples of such classes, namely, *LogFile* and *LogTCP*. The former provides the recording of log events into the file located on some mass storage device. The latter uses the TCP/IP network to send log events packed into messages to the logging server, which in its turn writes the log events to a file, perhaps located on its hard disk.

Similarly, the class *MessageHandler* uses services of the abstract interface provided by the class *MessageInterface*. The real providers of the message handing services are subclasses derived from the base class *MessageInterface*. Figure 4.13 shows three examples of such classes, namely, *StandardMessage*, *H323Message*, and *SS7Message*. In the examples in this book, we use the class *StandardMessage*, which creates the abstraction of the message comprising a series of octets (characters) that can be partitioned into an arbitrary number of message fields (carrying message parameters) of arbitrary size (given as a number of octets).

In the text that follows, we cover the most important details of the class *FiniteStateMachine*, *KernelAPI*, and *MessageHandler*. The effect of this top-down approach is that we introduce first the functionality solely provided by the class *FiniteStateMachine*, then the functionality that the class *FiniteStateMachine* inherits from the class *KernelAPI*, and finally the functionality that the class *FiniteSateMachine* inherits from the class *MessageHandler*.

The class *FiniteStateMachine* comprises all attributes and operations necessary for the definition and evolution of a single FSM. The FSM state is modeled with the structure *SState*:

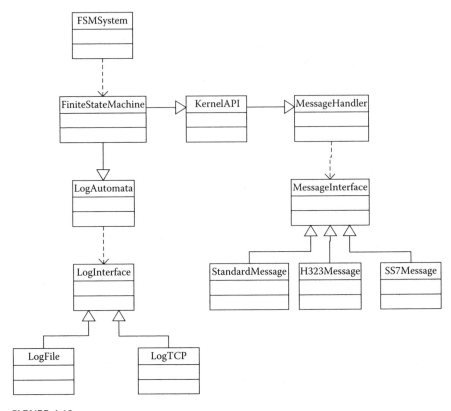

FIGURE 4.13
The internal FSM Library static structure.

```
struct SState {
  SState(uint16 maxNumOfProceduresPerState);
  ~SState();
  bool StateValid; // if true, data are valid
  unsigned short NumOfBranches; // number of branches in a state
  // procedure for processing unexpected message
  PROC_FUN_PTR UnexpectedEventProcPtr;
  SBranch* PBranch; // pointer on data for each branch
};
```

The field *NumOfBranches* contains the number of outgoing state transitions (branches) for the corresponding state. The field *UnexpectedEventProcPtr* is a pointer to the C++ function that handles the reception of unexpected messages. Finally, the field *PBranch* contains a pointer to the array of the *SBranch* instances, which model individual outgoing state transitions. The structure *SBranch* definition is the following:

```
struct SBranch {
  uint16 EventCode; // message code
  PROC_FUN_PTR ProcPtr; // message processing function
};
```

Implementation

The field *EventCode* contains the code of the event (message) that triggers this state transition. The field *ProcPtr* contains the pointer to the C++ function that performs the actions during this particular state transition.

Generally, an FSM can use a number of timers. Each timer is represented with the instance of the structure *TimerBlock*:

```
struct TimerBlock {
  TimerBlock(uint16 v, uint16 s) :
    Count(v), SignalId(s), Valid(false), TimerBuffer(0){}
  TimerBlock() :
    Count(INVALID_32), SignalId(INVALID_16), Valid(false),
    TimerBuffer(0) {};
  uint32 Count;       // in time slices
  uint16 SignalId;    // message code
  bool Valid;         // if true, data is valid
  ptrBuff TimerBuffer; // Ptr to timer buffer
};
```

The field *Count* defines the timer duration, the field *SignalId* defines the code of the message (signal) that is generated when the timer expires, the field *Valid* is set if the timer is running, and the field *TimerBuffer* contains the pointer to the buffer used by the timer expiration message.

The main private field members of the class *FiniteStateMachine* are the following:

```
class FiniteStateMachine : public KernelAPI, LogAutomate {...
  private:
    uint16 NumOfStates;  // Number of FSM states
    uint16 NumOfTimers;  // Number of timers
    uint16 MaxNumOfProcPerState;  // Max. no. of branches
    SState *States[MAX_STATE_NO];  // State data
    uint32 ConnectionId;  // Current connection
    uint32 CallId;  // Current call
    uint8 State;  // Current state
```

The fields *NumOfStates*, *NumOfTimers*, and *MaxNumOfProcPerState* are the dimensions of the corresponding arrays. They define the number of FSM states, the number of timers it uses, and the maximum number of branches, respectively. The field *States* is an array of pointers to the instances of the structure *SState* that contains pointers to arrays of instances of the structure *SBranch*. This data structure corresponds to the FSM state transition table.

The field *ConnectionId* carries the domain-specific name but actually contains the FSM object identification that is unique within the scope of objects of the same type. During the system initialization, the class *FSMSystem* creates the array of FSM objects of the same type. The index of the object in that array is written into this field at that time. This identification can be used as appropriate for the application at hand. The FSM Library user can take advantage of the fact that all message sending functions automatically copy the content of this field into the object identification field of the message header.

The field *CallId* carries another domain-specific name but it can be used for various purposes in various applications. In contrast to the field *ConnectionId* whose uniqueness is limited to the scope of a single FSM type, the value of the field *CallId* is unique in the scope of the whole system. Traditionally, it has been used to identify a single call, but generally it can be used to identify any communication process of interest. Like the field *ConnectionId*, this field is also copied by the message sending functions to the message header automatically.

Finally, the field *State* is the FSM current state identification, which is the value of the index of array defined in the field *States*. This field defines the context of the FSM.

As already mentioned, the FSM Library supports the abstraction of the *left* and *right* FSM. The message sending functions, namely *SendLeftAutomata()* and *SendRightAutomata()* — originally defined in the class *KernelAPI* — require the data about the *left* and *right* FSM. Relevant *FiniteStateMachine* attributes are the following:

```
// Left automata data
uint8   LeftMbx;       // left mbx id
uint8   LeftAutomata;  // left automata
uint8   LeftGroup;     // left group
uint32  LeftObjectId;  // left object
// Right automata data
uint8   RightMbx;       // right mbx id
uint8   RightAutomata;  // right automata
uint8   RightGroup;     // right group
uint32  RightObjectId;  // right object
```

We finish the overview of the *FiniteStateMachine* internals with its initialization and control functions:

```
FiniteStateMachine(
  uint16  numOfTimers = DEFAULT_TIMER_NO,
  uint16  numOfState = DEFAULT_STATE_NO,
  uint16  maxNumOfProceduresPerState = DEFAULT_PROCEDURE_NO_PER_STATE);
virtual void Initialize(void) = 0;
void InitEventProc(uint8 state, uint16 event, PROC_FUN_PTR fun);
void InitUnexpectedEventProc(uint8 state, PROC_FUN_PTR fun);
PROC_FUN_PTR GetProcedure(uint16 event);
virtual void NoFreeInstances() = 0;
virtual void Process(uint8 *msg);
void FreeFSM();
```

The class constructor first sets the number of timers, the number of states, and the maximal number of branches per state. It then calls the function *Initialize()*, provided by the user. This function typically uses a series of calls to functions *InitEventProc()* and *InitUnexpectedEventProc()*. The former defines the state transition function for the given state and message type

Implementation 211

whereas the latter defines the unexpected message handler for the given state.

The function *GetProcedure()* is a control function that returns the address of the state transition function for the given message type in the current state. The function *NoFreeInstances()* is a recovery function that is called in cases where no more free objects of this type are found. The function *Process()* is the prototype of the state transition function. The function *FreeFSM()* releases the FSM object by returning it to the pool of objects of this type.

The class *KernelAPI* provides the following groups of functions:

- Initialization functions
- Memory management functions
- Message management functions
- Timer management functions

The initialization functions provided by the class *KernelAPI* are its constructors (see Section 6.8) and the function *setKernelObjects*, whose prototype is the following:

```
void setKernelObjects(TPostOffice *o, TBuffers *b, CTimer *t);
```

The parameters of this function are the pointers to the objects that comprise the system mailboxes, buffers, and timers. These objects will be described in the next section.

The memory management functions provided by the class *KernelAPI* are the following:

```
uint8  *GetBuffer(uint32 length);
void   RetBuffer(uint8 *buff);
bool   IsBufferSmall(uint8 *buff, uint32 length);
uint32 GetBufferLength(uint8 *buff);
```

The function *GetBuffer()* returns the pointer to the buffer of the sufficient size (not less than specified by its parameter). The function *RetBuffer()* releases the given buffer. The function *IsBufferSmall()* checks the size of the given buffer. The function *GetBufferLength()* returns the size of the given buffer.

The message management functions provided by the class *KernelAPI* are the following:

```
void Discard(uint8* buff);
void SetMessageFromData();
void SendMessage(uint8 mbxId);
void SendMessage(uint8 mbxId, uint8 *msg);
void SendMessageLeft();
void SendMessageRight();
void ReturnMsg(uint8 mbxId);
```

The function *Discard()* releases the given message. The function *SetMessageFromData()* copies the data about this FSM (type, group, and instance identifications) to the corresponding fields of the new message header. According to the FSM Library terminology, the current message is the one that has been received and processed whereas the new message is the message that is currently under construction (and will be subsequently sent).

The function *SendMessage(uint8 mbxId)* sends the new message to the given mailbox. The function *SendMessage(uint8 mbxId, unit8 *msg)* sends the given message to the given mailbox. The functions *SendMessageLeft()* and *SendMessageRight()* send the new message to the left and right automata, respectively. The function *ReturnMsg()* sends the current message to the given mailbox.

The timer management functions provided by the class *KernelAPI* are the following:

```
uint8 *StartTimer(uint16 code, uint32 count, uint8 *info=0);
void  StopTimer(uint8 *timer);
bool  IsTimerRunning(uint8 *timer);
```

The function *StartTimer()* starts the given timer by setting its duration and the corresponding message buffer. The function *StopTimer()* stops the given timer. The function *IsTimerRunning()* checks if the given timer is running.

The interface defined by the class *MessageHandler* comprises the following two parts:

- Message header handling
- Message payload handling

The message header handling part provides getting and setting functions for the individual message header fields. The main message header fields are the following:

- *MSG_FROM_AUTOMATA* : the identification of the originating FSM type
- *MSG_TO_AUTOMATA* : the identification of the destination FSM type
- *MSG_CODE* : the identification of the message type
- *MSG_OBJECT_ID_FROM* : the identification of the originating FSM object
- *MSG_OBJECT_ID_TO* : the identification of the destination FSM object
- *MSG_CALL_ID* : the identification of the application-specific communication process
- *MSG_INFO_CODING* : the identification of the message format type
- *MSG_LENGTH* : the message payload length in octets

Implementation 213

The timer message is a special message. If the timer expires, it is sent to the same FSM that created it. Because of this, the message header fields *MSG_FROM_AUTOMATA* and *MSG_OBJECT_ID_FROM* are not needed, and thus can be used to hold the information about the timer duration and the destination mailbox identification.

The class *MessageInterface* defines the set of abstract functions that handle the message payload. The key idea behind the abstraction introduced by the class *MessageInterface* is the generic message parameter definition, which is independent from the particular message format. Each message parameter is uniquely defined by the following data:

- The message parameter identification
- The message parameter length (size)
- The message parameter value (content)

Depending on the message format type, the first and the second items listed may be implicit or explicit. Some of the messages carry the message parameter identification and length and some do not. However, all three items must be known to the message handling functions.

Another important fact related to the message format is that particular message formats can be disassembled to a series of primitive elements of the following type:

- Byte (1 byte)
- Word (2 bytes)
- DWord (4 bytes)
- Sequence of bytes (n bytes)

Therefore, the class *MessageInterface* includes the functions that provide access to these primitive types of information. These functions can be partitioned into the following two groups:

- Current message handing functions
- New message handling functions

The current message handling functions are the following:

```
uint8 *GetParam(uint8 paramCode);
bool  GetParamByte(uint8 paramCode, BYTE &param);
bool  GetParamWord(uint8 paramCode, WORD &param);
bool  GetParamDWord(uint8 paramCode, DWORD &param);
```

The first function returns a pointer to the parameter (sequence of octets) whose identification (*paramCode*) is given. The next three functions return the requested parameter of the size *Byte*, *Word*, and *DWord*, respectively. The new message handling functions are the following:

```
uint8  *AddParam(uint8 paramCode, uint8 paramLength, uint8 *param);
uint8  *AddParamByte(uint8 paramCode, BYTE param);
uint8  *AddParamWord(uint8 paramCode, WORD param);
uint8  *AddParamDWord(uint8 paramCode, DWORD param);
bool   RemoveParam(uint8 paramCode);
```

The first four functions add the given sequence of octets, *Byte*, *Word*, and *DWord* parameter to the new message, respectively. The function *RemoveParam()* removes the parameter — whose identification is given — from the message.

Each message handling function consists of two parts, a preparation part and an operation part. The preparation part of the current message handling functions includes preparing temporary data and message parsing. In case of message syntax errors, message handling functions report an error by returning the *value false*. The preparation part of the new message handling functions includes allocation of the message buffer and initialization of the message header fields *MSG_CODE*, *MSG_INFO_CODING* and *MSG_LENGTH* (initially set to 0).

4.4.2.3 Kernel Internals

As already mentioned, the class *FiniteStateMachine* is made independent of the particular real-time kernel with the introduction of the API defined by the class *KernelAPI*. Generally, the class *FiniteStateMachine* can use services provided by any real-time kernel that is a subclass of the class *KernelAPI*. In this section, we cover the internals of one such kernel (a default one), which is simply referred to as *Kernel*.

Figure 4.14 shows the static structure of *Kernel*. The root of the structure is the class *KernelAPI*, which acts as the wrapper of *Kernel*. This class contains pointers to the following three main parts of *Kernel*:

- Memory manager
- Message manager
- Time manager

The interfaces to these three resource managers are defined by the classes *TBuffers*, *TPostOffice*, and *CTimer*, respectively. The memory manager comprises the class *TBuffers* and a set of instances of the class *TBufferQueue*. The message manager consists of the class *TPostOffice* and a set of instances of the class *TMailBox*. The time manager is implemented by the class *CTimer* itself.

The class *TBuffers* creates the abstraction of a set of buffer pools. The size of the buffers in the pool is the same, but these sizes are different between the pools. For example, we can have three pools with three different sizes,

Implementation

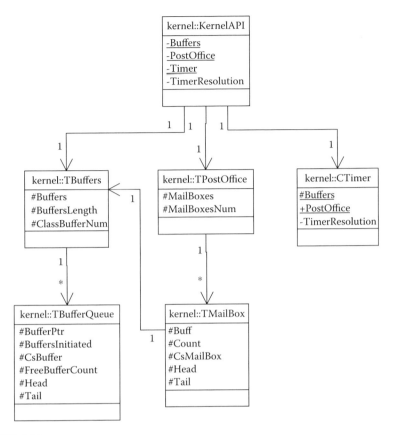

FIGURE 4.14
The internal *Kernel* static structure.

namely, small, medium, and large. The class *TBufferQueue* models one such a pool.

The constructor of the class *TBufferQueue* initially allocates an array of bytes (uint8), which is the actual memory space that accommodates the memory pool:

```
// calculate memory size for all buffers and get memory for them
memSize = bufferLength + BUFF_HEADER_LENGTH;
memSize *= buffersNo;
BufferPtr = new uint8[memSize];
```

This memory space is then partitioned into individual memory buffers that are added to the list of free buffers that actually represent the buffer pool. A buffer consists of the buffer header and the space for useful data. The buffer header comprises the pointers to the previous and to the next element in the list and the buffer code that indicates buffer size. Each buffer pool is defined with the pointer to the list of free buffers and the size of the buffers in that list. The class *TBuffers* holds the array of pointers to the instances of the class *TBufferQueue* (in the field member *Buffers*), as well as the array of the corresponding buffer sizes (in the field member *BuffersLength*).

The function *GetBuffer()* provided by the class *KernelAPI* first searches the field *BuffersLength* to find the pool of buffers of the sufficient size. It then gets the buffer from the head of the list of free buffers and returns the pointer to it. The function *RetBuffer()* uses buffer code from its header to return the buffer by adding it to the end of the corresponding list.

The class *TPostOffice* stores the array of pointers to the corresponding mailboxes. A mailbox is implemented as an instance of the class *TMailBox*. Actually, the class *TMailBox* is very similar to the class *TBufferQueue*. The main difference between them is that the former provides atomic (uninterruptible) access to the list of messages. This feature is needed because the list of messages is a resource shared by two concurrent processes, namely the event interpreter and the time interrupt routine.

The atomic mailbox access is ensured by two virtual functions, namely *MbxLock()* and *MbxUnlock()*. The former function locks the mailbox and the latter unlocks it. These functions ensure the FSM Library's portability. They can be implemented by the use of semaphores provided by the local operating systems. (The FSM Library supports OS Linux® and Windows® NT at the moment.)

The class *CTimer* is the most target-platform dependent part of *Kernel*. It consists of two parts, a platform-dependent part and a platform-independent part. The platform-dependent part comprises the time-driven routine that is periodically called by the local operating system and the routines that provide access to shared data. The platform-independent part consists of the list of running timers and routines that maintain that list. The list of running timers is implemented as a traditional delta list (the timer at the head of the list contains the absolute time interval whereas all other timers contain the time interval relative to the previous timer in the list).

To simplify timer maintenance, the function *StopTimer()* does not analyze the current status of the given timer (already expired or still running) — it simply marks the timer as expired. If the timer was still running, it will remain in the list of running timers. When it expires, it is forwarded to the given mailbox and from there it is discarded by the function member *Get()* of the class *TMailBox*.

4.4.3 Writing FSM Library-Based Implementations

Normally, we start by deriving subclasses from the base class *FiniteStateMachine*. For each such subclass, we must define the following functions (see Section 6.8 for more details):

- *GetMessageInterface()*: This function returns the pointer to the particular message interface object.

Implementation

- *SetDefaultHeader()*: This function sets the default message header parameters.
- *GetMbxId()*: This function returns the identification of the mailbox associated to this FSM type.
- *GetAutomata()*: This function returns the identification of this FSM type.
- *SetDefaultFSMData()*: This function sets default FSM data.
- *NoFreeInstances()*: This recovery function is called when the pool of objects is exhausted.
- *Initialize()*: This function initializes FSM-related data, including the state transition table.

We then write the main program, which typically follows these steps:

- Create an instance of the class *FSMSystem*.
- Initialize the real-time kernel.
- Set the system parameters.
- Register (add) all FSM objects with the instance of the class *FSMSystem*.
- Start the system by calling the function *Start()* (defined within the class *FSMSystem*).

4.5 Examples

This section includes two representative examples of the FSM Library-based implementations. The first example is the implementation of the application for reading Internet electronic mail. The second example shows the implementation of the SIP invite client transaction.

4.5.1 Example 1

This example demonstrates how an application for reading Internet electronic mail can be constructed. The application is actually an e-mail client that comprises the following three objects (see the general collaboration diagram in Figure 4.15):

FIGURE 4.15
The receive e-mail application collaboration diagram.

- *user*: a user interface
- *pop3*: the implementation of the POP3 protocol (refer to the original RFC 1939, freely available on the Internet at www.ietf.org/rfc/rfc1939.txt)
- *channel*: responsible for the direct communication with the e-mail server over the TCP protocol

As shown in Figure 4.15, the objects *user*, *pop3*, and *channel* are the instances of the classes *UserAuto*, *ClAuto*, and *ChAuto*, respectively. The object *pop3* is the central object. On its left side is the object *user* and on its right side is the object *channel*. The interaction between these objects is illustrated with three typical scenarios that are shown in Figure 4.16, Figure 4.17, and Figure 4.18. Figure 4.16 shows a successful session during which all pending e-mails are received and saved as files on a mass storage device. The flow of events from the point of view of the object *pop3* is the following:

- Triggered by the reception of the message *User_Check_Mail* from the left object, it sends the message *Cl_Connection_Request* to the right object.
- Upon the reception of the message *Cl_Connection_Accept* from the right object, it sends the message *User_Connected* to the left object. The connection with the e-mail server is successfully established at this point.
- After receiving the user name and password carried by the message *User_Name_Password* from the left object, it first sends the user name in the message *MSG(USER name)* to the right object, which is acknowledged with the message *MSG(+OK)* from the right object, and it then sends the password in the message *MSG(PASS password)* to the right object, which is also acknowledged with the message *MSG(+OK)* from the right object. The user authentication procedure is successfully finished at this point.
- It then checks the status of the pending e-mails by sending the message *MSG(STAT)* to the right object and receiving the answer in the message *MSG(+OK nn mm)*, where *nn* is the number of messages in the maildrop and *mm* is the size of the maildrop in octets.
- While pending e-mails remain, it repeats the sequence of the e-mail read procedure and the e-mail delete procedure. The e-mail read procedure starts with the message *MSG(RETR nn)* to the right object (*nn* is the order number of the e-mail message to be received). The right object in its turn sends an e-mail message in a series of *MSG(mail)* messages (the size of the last one is smaller than 255 octets). The e-mail delete procedure starts with the message *MSG(DELE nn)* sent to the right object (*nn* is the order number of the message to be deleted by the e-mail server). After reception of

Implementation

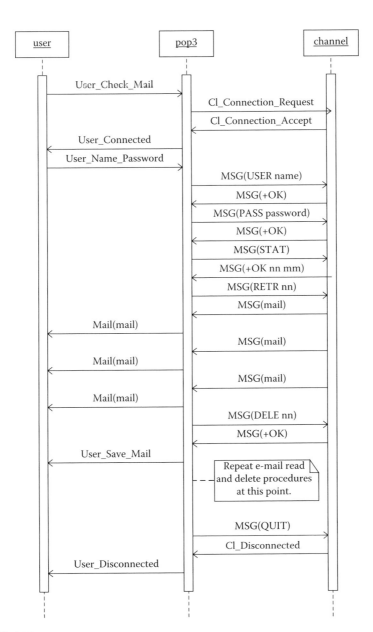

FIGURE 4.16
The successful receive e-mail session establishment scenario.

the acknowledgment *MSG(+OK)* from the right object, the left object is informed accordingly with the message *User_Save_Mail* (normally, the object user should save the current e-mail message as a file on a mass storage device at this point).

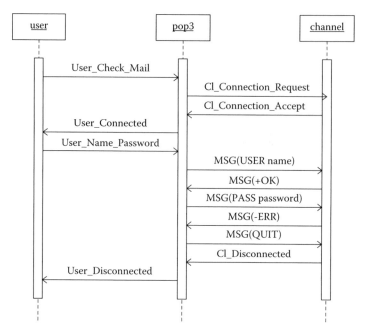

FIGURE 4.17
The invalid e-mail password processing scenario.

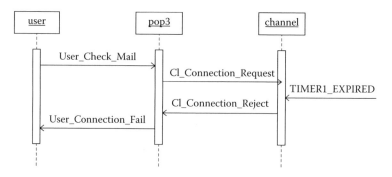

FIGURE 4.18
The unsuccessful receive e-mail session establishment scenario.

- Finally, the object *pop3* starts the session closing procedure by sending the message *MSG(QUIT)* to the right object. Then, upon reception of the message *Cl_Disconnected* from the right object, it sends the message *User_Disconnected* to the left object.

Figure 4.17 shows the invalid password processing scenario. It is the same as the previous scenario up to the point where the object *pop3* sends the message *MSG(PASS password)* to the right object. Because the password is

Implementation 221

invalid, the right object responds with the message *MSG(ERR)* and the object *pop3* immediately proceeds to the session closing procedure.

Figure 4.18 shows the unsuccessful session establishment scenario. It starts in the same way as the scenario in Figure 4.16. Assume that the TCP connection with the e-mail server cannot be established for some reason. Therefore, *TIMER1_ID* that was started by the right object expires and the associate message *TIMER1_EXPIRED* triggers the right object to send the message *Cl_Connection_Reject*. The object *pop3* in its turn sends the message *User_Connection_Fail* to the left object.

To keep this example simple enough, we focus further on the design and implementation of the key object in this application, the object *pop3*. The complete dynamic behavior of this object is specified with the SDL diagram, which is shown in Figure 4.19 and Figure 4.20. The corresponding FSM is defined with nine states (*Cl_Ready, Cl_Connecting, Cl_Authorizing, Cl_User_Check, Cl_Pass_Check, Cl_Mail_Check, Cl_Receiving, Cl_Deleting*, and *Cl_Disconnecting*), six input messages (*User_Check_Mail, Cl_Connection_Reject, Cl_Connection_Accept, User_Name_Password, MSG*, and *Cl_Disconnected*), and seven output messages (*Cl_Connection_Request, User_Connection_Fail, User_Connected, MSG, Mail, User_Save_Mail*, and *User_Disconnected*).

By convention, the names of all messages (except *Mail*) exchanged between the object *pop3* and the left object begin with the prefix *User_*. The names of the control messages exchanged between the object *pop3* and the right object begin with the prefix *Cl_*. The names of the POP3-related messages exchanged between the object *pop3* and the right object are named *MSG*. Two types of *MSG* messages are used — commands directed to the e-mail server and responses received from it.

The *MSG* commands are the following:

- *MSG(USER name)*: corresponds to the original POP3 command for specifying the name of the user mailbox
- *MSG(PASS password)*: corresponds to the original POP3 command for specifying the password for the previously specified mailbox
- *MSG(STAT)*: corresponds to the original POP3 command for inquiring about the mailbox status
- *MSG(RETR nn)*: corresponds to the original POP3 command for reading the pending e-mail message number *nn*
- *MSG(DELE nn)*: corresponds to the original POP3 command for deleting the pending e-mail message number *nn*
- *MSG(QUIT)*: corresponds to the original POP3 command for closing the current session

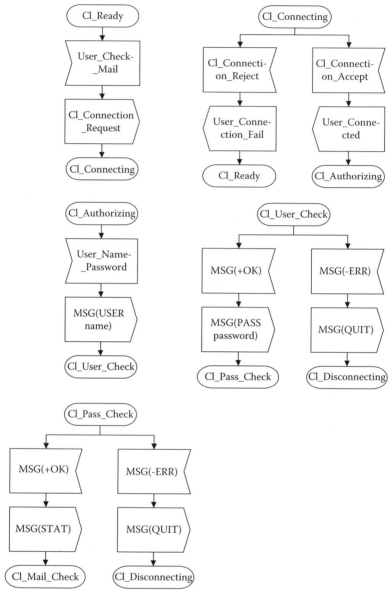

FIGURE 4.19
The POP3 client SDL diagram, part I.

The *MSG* responses are the following:

- *MSG(+OK)*: corresponds to the original POP3 acknowledgment message
- *MSG(ERR)*: corresponds to the original POP3 error message

Implementation

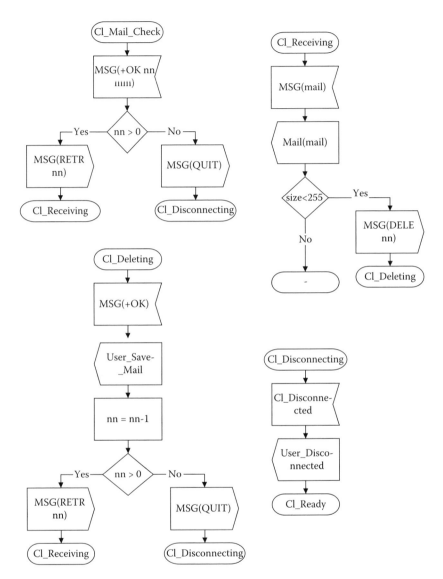

FIGURE 4.20
The POP3 client SDL diagram, part II.

- *MSG(mail)*: corresponds to the actual e-mail message that was received from the e-mail server

Figure 4.19 shows valid state transitions for the states *Cl_Ready*, *Cl_Connecting*, *Cl_Authorizing*, *Cl_User_Check*, and *Cl_Pass_Check*. The eight state transitions are shown in Figure 4.19, as follows:

- From *Cl_Ready* to *Cl_Connecting*, triggered by *User_Check_Mail*

- From *Cl_Connecting* to *Cl_Ready*, triggered by *Cl_Connection_Reject*
- From *Cl_Connecting* to *Cl_Authorizing*, triggered by *Cl_Connection_Accepted*
- From *Cl_Authorizing* to *Cl_User_Check*, triggered by *User_Name_Password*
- From *Cl_User_Check* to *Cl_Pass_check*, triggered by *MSG(+OK)*
- From *Cl_User_Check* to *Cl_Disconnecting*, triggered by *MSG(ERR)*
- From *Cl_Pass_Check* to *Cl_Mail_check*, triggered by *MSG(+OK)*
- From *Cl_Pass_Check* to *Cl_Disconnecting*, triggered by the *MSG(ERR)*

Figure 4.20 shows valid state transitions for the states *Cl_Mail_Check*, *Cl_Receiving*, *Cl_Deleting*, and *Cl_Disconnecting*. The seven state transitions are shown in Figure 4.20, as follows:

- From *Cl_Mail_Check* to *Cl_Receiving*, triggered by *MSG(+OK)* and guarded by the condition $nn > 0$
- From *Cl_Mail_Check* to *Cl_Disconnecting*, triggered by *MSG(+OK)* and guarded by the condition $!(nn > 0)$
- From *Cl_Receiving* to *Cl_Deleting*, triggered by *MSG(mail)* and guarded by the condition $mail(size) < 255$
- From *Cl_Receiving* to *Cl_Receiving*, triggered by *MSG(mail)* and guarded by the condition $!(mail(size) < 255)$
- From *Cl_Deleting* to *Cl_Receiving*, triggered by *MSG(+OK)* and guarded by the condition $nn > 0$
- From *Cl_Deleting* to *Cl_Disconnecting*, triggered by *MSG(+OK)* and guarded by the condition $!(nn > 0)$
- From *Cl_Disconnecting* to *Cl_Ready*, triggered by *Cl_Disconnected*

Next, we proceed to the implementation in C++ based on the FSM Library. First, we define symbolic constants specific for this project in a header file, which is typically named *const.h*. The content of this file is the following:

```
#ifndef _CONST_H_
#define _CONST_H_
#include <fsm.h>
const uint8 CH_AUTOMATA_TYPE_ID = 0x00;
const uint8 CL_AUTOMATA_TYPE_ID = 0x01;
const uint8 USER_AUTOMATA_TYPE_ID = 0x02;

const uint8 CH_AUTOMATA_MBX_ID = 0x00;
const uint8 CL_AUTOMATA_MBX_ID = 0x01;
const uint8 USER_AUTOMATA_MBX_ID = 0x02;

// channel messages
const uint16 MSG_Connection_Request = 0x0001;
const uint16 MSG_Sock_Connection_Reject = 0x0002;
```

Implementation

```
const uint16 MSG_Sock_Connection_Accept = 0x0003;
const uint16 MSG_Cl_MSG = 0x0004;
const uint16 MSG_Sock_MSG = 0x0005;
const uint16 MSG_Disconnect_Request = 0x0006;
const uint16 MSG_Sock_Disconnected = 0x0007;
const uint16 MSG_Sock_Disconnecting_Conf = 0x0008;

// pop3 client messages
const uint16 MSG_User_Check_Mail = 0x0009;
const uint16 MSG_Cl_Connection_Reject = 0x000a;
const uint16 MSG_Cl_Connection_Accept = 0x000b;
const uint16 MSG_User_Name_Password = 0x000c;
const uint16 MSG_MSG = 0x000d;
const uint16 MSG_Cl_Disconnected = 0x000f;

// user messages
const uint16 MSG_Set_All = 0x0010;
const uint16 MSG_User_Connected = 0x0011;
const uint16 MSG_User_Connection_Fail = 0x0012;
const uint16 MSG_Mail = 0x0013;
const uint16 MSG_User_Save_Mail = 0x0015;
const uint16 MSG_User_Disconnected = 0x0014;

#define ADRESS "krtlab8"
#define PORT 110

#define TIMER1_ID 1
#define TIMER1_COUNT 10
#define TIMER1_EXPIRED 0x20

#define PARAM_DATA 0x01
#define PARAM_Name 0x02
#define PARAM_Pass 0x03
#endif // _CONST_H_
```

The file *const.h* starts with the definitions of automata types and their private mailbox identifications. The identifications assigned to the classes *ChAuto*, *ClAuto*, and *UserAuto* are CH_AUTOMATA_TYPE_ID, CL_AUTOMATA_TYPE_ID, and USER_AUTOMATA_TYPE_ID, respectively. The identifications of their private mailboxes are CH_AUTOMATA_MBX_ID, CL_AUTOMATA_MBX_ID, and USER_AUTOMATA_MBX_ID, respectively. Next, we define the symbols that correspond to the codes of the messages recognized by the classes *ChAuto*, *ClAuto*, and *UserAuto*, respectively. By convention, these symbols are provided by prefixing the names of the messages from the SDL diagram (Figure 4.19 and Figure 4.20) with the prefix *MSG_*.

At the end of the file *const.h*, we define the domain name and the number of the port, which are used to establish the TCP connection with the e-mail server (symbols *ADDRESS* and *PORT*), channel timer-related constants (symbols *TIMER1_ID*, *TIMER1_COUNT*, and *TIMER1_EXPIRED*), and the identifications of the message parameters (symbols *PARAM_DATA*, *PARAM_Name*, and *PARAM_Pass*).

Next, we write the header file *ClAuto.h*. Its content is the following:

```
#ifndef _Cl_AUTO_H_
#define _Cl_AUTO_H_
#include <NetFSM.h>
#include <fsmsystem.h>
#include "const.h"
class ClAuto : public FiniteStateMachine {
  // for FSM
  StandardMessage StandardMsgCoding;
  MessageInterface *GetMessageInterface(uint32 id);
  void  SetDefaultHeader(uint8 infoCoding);
  void  SetDefaultFSMData();
  void  NoFreeInstances();
  void  Reset();
  uint8  GetMbxId();
  uint8  GetAutomata();
  uint32  GetObject();
  void  ResetData();
  // FSM States
  enum ClStates {
    FSM_Cl_Ready,
    FSM_Cl_Connecting,
    FSM_Cl_Authorizing,
    FSM_Cl_User_Check,
    FSM_Cl_Pass_Check,
    FSM_Cl_Mail_Check,
    FSM_Cl_Receiving,
    FSM_Cl_Deleting,
    FSM_Cl_Disconnecting
  };
public:
  ClAuto();
  ~ClAuto();
  void  Initialize();
  void  FSM_Cl_Ready_User_Check_Mail();
  void  FSM_Cl_Connecting_Cl_Connection_Reject();
  void  FSM_Cl_Connecting_Cl_Connection_Accept();
  void  FSM_Cl_Authorizing_User_Name_Password();
  void  FSM_Cl_User_Check_MSG();
  void  FSM_Cl_Pass_Check_MSG();
  void  FSM_Cl_Mail_Check_MSG();
  void  FSM_Cl_Receiving_MSG();
  void  FSM_Cl_Deleting_MSG();
  void  FSM_Cl_Disconnecting_Cl_Disconnected();
protected:
  int  m_MessageCount;
  char  m_UserName[20];
  char  m_Password[20];
};
#endif /* _Cl_AUTO_H */
```

After listing all necessary header files, we declare the class *ClAuto*, which is derived from the base class *FiniteStateMachine*. The declaration of the class

ClAuto starts with the declaration of field and function members that are mandatory for any class that is derived from the class *FiniteStateMachine* (as explained previously in this chapter). It continues with the declaration of the FSM state names and state transition function prototypes.

By convention, FSM state names are the names from the SDL diagram prefixed with the prefix *FSM_* (e.g., the initial state *Cl_Ready* is named *FSM_Cl_Ready* in the C++ code). The state transition function is named by concatenating the state name and the input message name and by prefixing this composite name with the prefix *FSM_* (e.g., the state transition function performed when the FSM in state *Cl_Ready* receives the message *User_Check_Mail* is named *FSM_Cl_Ready_User_Check_Mail*). As already mentioned, *ClAuto* FSM has nine states and fourteen state transitions.

The reader may be puzzled with the fact that there are fourteen valid FSM state transitions and only ten state transition functions declared in the header file *ClAuto.h*. This circumstance is because some of the state transitions are triggered with the same message type but different message content — e.g., *MSG(+OK)* and *MSG(ERR)* — or they are guarded with the complementary conditions — e.g., $(nn > 0)$ and $!(nn > 0)$. To clearly understand these matters, remember that *FiniteStateMachine* derivatives react to various message types in various FSM states. This is how we calculate the number of state transitions.

If we apply the principle stated above to the class *ClAuto*, we have the situation where all the states react to a single message with the exception of the state *Cl_Connecting*, which reacts to two valid messages, *Cl_Connection_Reject* and *Cl_Connection_Accept*. Because of this, we have $(8 \times 1) + (1 \times 2)$ state transition functions, which resolves to ten state transition functions, as mentioned above.

Finally, we write the class *ClAuto* definition file, named *ClAuto.cpp*. The content of this file is the following:

```
#include <stdio.h>
#include "const.h"
#include "ClAuto.h"
#define StandardMessageCoding 0x00

ClAuto::ClAuto() : FiniteStateMachine(0, 9, 2) {}
ClAuto::~ClAuto() {}

uint8 ClAuto::GetAutomate() {
  return CL_AUTOMATA_TYPE_ID;
}

uint8 ClAuto::GetMbxId() {
  return CL_AUTOMATA_MBX_ID;
}

uint32 ClAuto::GetObject() {
  return GetObjectId();
}
```

```cpp
MessageInterface *ClAuto::GetMessageInterface(uint32 id) {
  return &StandardMsgCoding;
}

void ClAuto::SetDefaultHeader(uint8 infoCoding) {
  SetMsgInfoCoding(infoCoding);
  SetMessageFromData();
}

void ClAuto::SetDefaultFSMData() {
  SetDefaultHeader(StandardMessageCoding);
}

void ClAuto::NoFreeInstances() {
  printf("[%d] ClAuto::NoFreeInstances()\n",   GetObjectId());
}

void ClAuto::Reset() {
  printf("[%d] ClAuto::Reset()\n", GetObjectId());
}

void ClAuto::Initialize() {
  SetState(FSM_Cl_Ready);

  // set message handlers
  InitEventProc(FSM_Cl_Ready, MSG_User_Check_Mail,
(PROC_FUN_PTR)&ClAuto::FSM_Cl_Ready_User_Check_Mail));

  InitEventProc(FSM_Cl_Connecting,   MSG_Cl_Connection_Reject,
(PROC_FUN_PTR)&ClAuto::FSM_Cl_Connecting_Cl_Connection_Reject));

  InitEventProc(FSM_Cl_Connecting,   MSG_Cl_Connection_Accept,
(PROC_FUN_PTR)&ClAuto::FSM_Cl_Connecting_Cl_Connection_Accept));

  InitEventProc(FSM_Cl_Authorizing,   MSG_User_Name_Password,
(PROC_FUN_PTR)&ClAuto::FSM_Cl_Authorizing_User_Name_Password));

  InitEventProc(FSM_Cl_User_Check, MSG_MSG,
    (PROC_FUN_PTR)&ClAuto::FSM_Cl_User_Check_MSG));

  InitEventProc(FSM_Cl_Pass_Check, MSG_MSG,
    (PROC_FUN_PTR)&ClAuto::FSM_Cl_Pass_Check_MSG));

  InitEventProc(FSM_Cl_Mail_Check, MSG_MSG,
    (PROC_FUN_PTR)&ClAuto::FSM_Cl_Mail_Check_MSG));

  InitEventProc(FSM_Cl_Receiving, MSG_MSG,
    (PROC_FUN_PTR)&ClAuto::FSM_Cl_Receiving_MSG));

  InitEventProc(FSM_Cl_Deleting, MSG_MSG,
    (PROC_FUN_PTR)&ClAuto::FSM_Cl_Deleting_MSG));
```

Implementation

```cpp
  InitEventProc(FSM_Cl_Disconnecting,   MSG_Cl_Disconnected,
(PROC_FUN_PTR)&ClAuto::FSM_Cl_Disconnecting_Cl_Disconnected));
}

void ClAuto::FSM_Cl_Ready_User_Check_Mail(){
  PrepareNewMessage(0x00, MSG_Connection_Request);
  SetMsgToAutomata(CH_AUTOMATA_TYPE_ID);
  SetMsgObjectNumberTo(0);
  SendMessage(CH_AUTOMATA_MBX_ID);
  SetState(FSM_Cl_Connecting);
}

void ClAuto::FSM_Cl_Connecting_Cl_Connection_Reject(){
PrepareNewMessage(0x00, MSG_User_Connection_Fail);
SetMsgToAutomata(USER_AUTOMATA_TYPE_ID);
SetMsgObjectNumberTo(0);
SendMessage(USER_AUTOMATA_MBX_ID);
SetState(FSM_Cl_Ready);
}

void ClAuto::FSM_Cl_Connecting_Cl_Connection_Accept(){
  PrepareNewMessage(0x00, MSG_User_Connected);
  SetMsgToAutomata(USER_AUTOMATE_TYPA_ID);
  SetMsgObjectNumberTo(0);
  SendMessage(USER_AUTOMATA_MBX_ID);
  SetState(FSM_Cl_Authorizing);
}

void ClAuto::FSM_Cl_Authorizing_User_Name_Password(){
  char* name = new char[20];
  char* pass = new char[20];
  uint8* buffer = GetParam(PARAM_Name);

  memcpy(m_UserName,buffer+2,buffer[1]);
  m_UserName[buffer[1]] = 0;// terminate string
  buffer = GetParam(PARAM_Pass);
  memcpy(m_Password,buffer+2,buffer[1]);
  m_Password[buffer[1]] = 0;// terminate string
  char l_Command[20] = "user";
  strcpy(l_Command+5,m_UserName);
  strcpy(l_Command+5+strlen(m_UserName),"\r\n");

  PrepareNewMessage(0x00, MSG_Cl_MSG);
  SetMsgToAutomata(CH_AUTOMATA_TYPE_ID);
  SetMsgObjectNumberTo(0);
AddParam(PARAM_DATA,strlen(l_Command),(uint8*)l_Command);
  SendMessage(CH_AUTOMATA_MBX_ID);
  SetState(FSM_Cl_User_Check);
}

void ClAuto::FSM_Cl_User_Check_MSG(){
  char* data = new char[255];
  uint8* buffer = GetParam(PARAM_DATA);
  uint16 size = buffer[1];
```

```
    memcpy(data,buffer + 2,size);
    data[size]=0;
    printf("%s",data);
    if((data[0] == '+')) {
      char l_Command[20] = "pass ";
      strcpy(l_Command+5,m_Password);
      strcpy(l_Command+5+strlen(m_Password),"\r\n");
      PrepareNewMessage(0x00, MSG_Cl_MSG);
      SetMsgToAutomata(CH_AUTOMATA_TYPE_ID);
      SetMsgObjectNumberTo(0);
AddParam(PARAM_DATA,strlen(l_Command),(uint8*)l_Command);
      SendMessage(CH_AUTOMATA_MBX_ID);
      SetState(FSM_Cl_Pass_Check);
      else {
      char l_Command[20] = "quit\r\n";
      PrepareNewMessage(0x00, MSG_Cl_MSG);
      SetMsgToAutomata(CH_AUTOMATA_TYPE_ID);
      SetMsgObjectNumberTo(0);
      AddParam(PARAM_DATA,6,(uint8*)l_Command);
      SendMessage(CH_AUTOMATA_MBX_ID);
      SetState(FSM_Cl_Disconnecting);
    }
}

void ClAuto::FSM_Cl_Pass_Check_MSG(){
  char*   data = new char[255];
  uint8*  buffer = GetParam(PARAM_DATA);
  uint16  size = buffer[1];

  memcpy(data,buffer + 2,size);
  data[size]=0;
  printf("%s",data);
  if((data[0] == '+')) {
    char l_Command[20] = "stat\r\n";
    PrepareNewMessage(0x00, MSG_Cl_MSG);
    SetMsgToAutomata(CH_AUTOMATA_TYPE_ID);
    SetMsgObjectNumberTo(0);
    AddParam(PARAM_DATA,6,(uint8*)l_Command);
    SendMessage(CH_AUTOMATA_MBX_ID);
    SetState(FSM_Cl_Mail_Check);
    else {
    char l_Command[20] = "quit\r\n";
    PrepareNewMessage(0x00, MSG_Cl_MSG);
    SetMsgToAutomata(CH_AUTOMATA_TYPE_ID);
    SetMsgObjectNumberTo(0);
    AddParam(PARAM_DATA,6,(uint8*)l_Command);
    SendMessage(CH_AUTOMATA_MBX_ID);
    SetState(FSM_Cl_Disconnecting);
  }
}

void ClAuto::FSM_Cl_Mail_Check_MSG(){
  char*   data = new char[255];
  uint8*  buffer = GetParam(PARAM_DATA);
  uint16  size = buffer[1];
```

Implementation

```
    memcpy(data,buffer+2,size);
    data[size]=0;
    printf("%s",data);
    int l_nDigit = 1;
    while(buffer[l_nDigit+6] != ' ') l_nDigit++;
    memcpy(data,buffer +6,l_nDigit);
    data[l_nDigit]=0;
    m_MessageCount = atoi(data);

    if((m_MessageCount == 0) {
      char l_Command[20] = "quit\r\n";
      PrepareNewMessage(0x00, MSG_Cl_MSG);
      SetMsgToAutomata(CH_AUTOMATA_TYPE_ID);
      SetMsgObjectNumberTo(0);
      AddParam(PARAM_DATA,6,(uint8*)l_Command);
      SendMessage(CH_AUTOMATA_MBX_ID);
      SetState(FSM_Cl_Disconnecting);
      else {
      char l_Command[20] = "retr ";
      strcpy(l_Command+5,data);
      strcpy(l_Command+5+l_nDigit,"\r\n");
      PrepareNewMessage(0x00, MSG_Cl_MSG);
      SetMsgToAutomata(CH_AUTOMATA_TYPE_ID);
      SetMsgObjectNumberTo(0);

AddParam(PARAM_DATA,5+l_nDigit+2,(uint8*)l_Command);
      SendMessage(CH_AUTOMATA_MBX_ID);
      SetState(FSM_Cl_Receiving);
    }
}

void ClAuto::FSM_Cl_Receiving_MSG(){
  char*   data = new char[255];
  uint8*  buffer = GetParam(PARAM_DATA);
  uint16  size = buffer[1];

  memcpy(data,buffer + 2,size);
  char temp[4];
  memcpy(temp,data,3); temp[3] = 0;
  if((strcmp(temp,"+OK") != 0) {
    PrepareNewMessage(0x00, MSG_Mail);
    SetMsgToAutomata(USER_AUTOMATA_TYPE_ID);
    SetMsgObjectNumberTo(0);
    AddParam(PARAM_DATA,size,(uint8*)data);
    SendMessage(USER_AUTOMATA_MBX_ID);
    if((size < 255) {
      char l_Command[20] = "dele ";
      itoa(m_MessageCount,data,10);
      strcpy(l_Command+5,data);
      strcpy(l_Command+5+strlen(data),"\r\n");
      PrepareNewMessage(0x00, MSG_Cl_MSG);
      SetMsgToAutomata(CH_AUTOMATA_TYPE_ID);
      SetMsgObjectNumberTo(0);

AddParam(PARAM_DATA,5+strlen(data)+2,(uint8*)l_Command);
```

```
      SendMessage(CH_AUTOMATA_MBX_ID);
      SetState(FSM_Cl_Deleting);
    }
  }
}

void ClAuto::FSM_Cl_Deleting_MSG(){
  PrepareNewMessage(0x00, MSG_User_Save_Mail);
  SetMsgToAutomata(USER_AUTOMATA_TYPE_ID);
  SetMsgObjectNumberTo(0);
  SendMessage(USER_AUTOMATA_MBX_ID);
  m_MessageCount—;
  if(m_MessageCount > 0) {
     char data[5];
     char l_Command[20] = "retr ";
     itoa(m_MessageCount,data,10);
     strcpy(l_Command+5,data);
     strcpy(l_Command+5+strlen(data),"\r\n");
     PrepareNewMessage(0x00, MSG_Cl_MSG);
     SetMsgToAutomata(CH_AUTOMATA_TYPE_ID);
     SetMsgObjectNumberTo(0);

AddParam(PARAM_DATA,5+strlen(data)+2,(uint8*)l_Command);
     SendMessage(CH_AUTOMATA_MBX_ID);
     SetState(FSM_Cl_Receiving);
     else {
     char l_Command[20] = "quit\r\n";
     PrepareNewMessage(0x00, MSG_Cl_MSG);
     SetMsgToAutomata(CH_AUTOMATA_TYPE_ID);
     SetMsgObjectNumberTo(0);
     AddParam(PARAM_DATA,6,(uint8*)l_Command);
     SendMessage(CH_AUTOMATA_MBX_ID);
     SetState(FSM_Cl_Disconnecting);
  }
}

void ClAuto::FSM_Cl_Disconnecting_Cl_Disconnected(){
  PrepareNewMessage(0x00, MSG_User_Disconnected);
  SetMsgToAutomata(USER_AUTOMATA_TYPE_ID);
  SetMsgObjectNumberTo(0);
  SendMessage(USER_AUTOMATA_MBX_ID);
  SetState(FSM_Cl_Ready);
}
```

The file *ClAuto.cpp* starts with the list of all necessary header files (*stdio.h*, *const.h*, and *ClAuto.h*), followed by the definition of the symbolic constant *StandardMessageCoding* and the set of mandatory function definitions — class constructor, class destructor, and functions *GetAutomata()*, *GetMbxId()*, *GetObject()*, *GetMessageInterface()*, *SetDefaultHeader()*, *SetDefaultFSMData()*, *NoFreeInstances()*, *Reset()*, and *Initialize()*.

The class constructor *ClAuto()* calls the constructor of the class *FiniteStateMachine* with a list of parameters, which specifies that *ClAuto* FSM has no timers, nine states, and the maximum of two state transitions per state (see

Implementation 233

the FSM Library API specification in Section 6.8, particularly, Section 6.8.11). The class destructor performs no particular operation.

The mandatory functions provide the following functionalities:

- The function *GetAutomata()* returns the *ClAutomata* type identification (the constant *CL_AUTOMATA_TYPE_ID*). See also Section 6.8.24.
- The function *GetMbxId()* returns the associated mailbox identification (the constant *CL_AUTOMATA_MBX_ID*). See also Section 6.8.38.
- The function *GetObject()* returns the object identification (actually, it returns the value returned by the FSM Library function *GetObject Id()*). See also Section 6.8.60.
- The function *GetMessageInterface()* returns the pointer to the message coding object (actually, an instance of the class *StandardMessage*). See also Section 6.8.39.
- The function *SetDefaultHeader()* sets default data in the new message header by calling two FSM Library functions, *SetMsgInfoCoding()* and *SetMessageFromData()*. See also Section 6.8.97, Section 6.8.117, and Section 6.8.108.
- The function *SetDefaultFSMData()* sets the new message header default values by calling the function *SetDefaultHeader()* and specifying the constant *StandardMessageCoding* as its parameter.
- The function *NoFreeInstances()* just prints the information message to the standard output file. See also Section 6.8.78.
- The function *Reset()* also just prints the information message to the standard output file. See also Section 6.8.85.

The most important mandatory function is the function *Initialize()*. It starts by setting the FSM initial state, *Cl_Ready* (denoted with the constant *FSM_Cl_Ready*). It continues by setting the state transition functions (also referred to as message handlers). Each message handler is set by a single call to the FSM Library function *InitEventProc()*. The first parameter of this function is the state name, the second is the input message name, and the third is the address of the corresponding *ClAuto* function member (see also Section 6.8.73).

The set of mandatory functions is followed by the set of state transition functions. As already mentioned, ten such functions are used. Each of these functions processes a single message type in a single state, as follows:

- The function *FSM_Cl_Ready_User_Check_Mail()* processes the message *User_Check_Mail* in the state *Cl_Ready*.
- The function *FSM_Cl_Connecting_Cl_Connection_Reject()* processes the message *Cl_Connection_Reject in the state Cl_Connecting*.

- The function *FSM_Cl_Connecting_Cl_Connection_Accept()* processes the message *Cl_Connection_Accept* in the state *Cl_Connecting*.
- The function *FSM_Cl_Authorizing_User_Name_Password()* processes the message *User_Name_Password* in the state *Cl_Authorizing*.
- The function *FSM_Cl_User_Check_MSG()* processes the message *MSG* in the state *Cl_User_Check*.
- The function *FSM_Cl_Pass_Check_MSG()* processes the message *MSG* in the state *Cl_Pass_Check*.
- The function *FSM_Cl_Mail_Check_MSG()* processes the message *MSG* in the state *Cl_Mail_Check*.
- The function *FSM_Cl_Receiving_MSG()* processes the message *MSG* in the state *Cl_Receiving*.
- The function *FSM_Cl_Deleting_MSG()* processes the message *MSG* in the state *Cl_Deleting*.
- The function *FSM_Cl_Disconnecting_Cl_Disconnected()* processes the message *Cl_Disconnected* in the state *Cl_Disconnecting*.

The function *FSM_Cl_Ready_User_Check_Mail()* is a typical simple state transition function. It first creates a new message by calling the function *PrepareNewMessage()*. (Its first parameter is the message length and the second is the message type; the third parameter is optional and is not used in this example. See also Section 6.8.81.) It then sets the destination FSM type and object identification by calling the function *SetMsgToAutomata()* (its parameter is the FSM type identification; see also Section 6.8.125) and function *SetMsgObjectNumberTo()* (its parameter is the FSM object identification; see also Section 6.8.123), respectively. Next, it sends the new message to the destination mailbox by calling the function *SendMessage()* (its parameter is the mailbox identification; see also Section 6.8.106). Finally, it sets the new FSM state by calling the function *SetState* (its parameter is the state identification; see also Section 6.8.137).

The next two functions, *FSM_Cl_Connecting_Cl_Connection_Reject()* and *FSM_Cl_Connecting_Cl_Connection_Accept()*, are very similar to the one previously described (only the message type and the new state name are different). But the fourth state transition function, *FSM_Cl_Authorizing_User_Name_Password()*, is more complex. It demonstrates well how a state transition function can get a parameter from the current message and how it can add a parameter to the new message. This concrete state transition function gets two parameters (user name and password) from the current message by calling the function *GetParam()* (its parameter is the identification of the parameter type; see also Section 6.8.61). It also adds one parameter (user name) to the new message by calling the function *AddParam()* (its parameters are the message parameter type, length, and pointer; see also Section 6.8.12).

Implementation 235

The fifth state transition function, *FSM_Cl_User_Check_MSG()*, is even more complex because it involves branching depending on the value of the current message parameter. By making a branch, the state transition function actually selects one of two possible paths of the FSM evolution, which yields two different output (new) messages and two different destination FSM states. The sixth state transition function is very similar to the fifth one.

The seventh state transition function, *FSM_Cl_Mail_Check_MSG()*, brings one new important detail. It shows how a state transition function can save some data (in this example, the number of pending e-mail messages, which is stored in the class field member *m_MessageCount*) so that it can be shared or used by other state transitions — in this example, by the ninth state transition function, *FSM_Cl_Deleting_MSG()*.

The rest of the state transition functions do not bring anything essentially new. However, the reader is advised to study them in detail as an additional exercise.

4.5.2 Example 2

The aim of this example is to implement the SIP INVITE client transaction design, which is given in Section 3.10.5 (Chapter 3, Example 5). Briefly, in that section we examined the general collaboration diagram of the SIP Softphone (see Section 2.3.3, Figure 2.16) with the focus on the INVITE client transaction. The result is the general collaboration diagram shown in Figure 3.69. We then made two particular collaboration diagrams and their semantically equivalent sequence diagrams for the cases of successful and unsuccessful SIP session establishment (Figure 3.70, Figure 3.71, Figure 3.72, and Figure 3.73). Finally, we devised the complete dynamic behavior specification in the form of the statechart diagram (Figure 3.74) and semantically equivalent SDL diagram (Figure 3.75, Figure 3.76, Figure 3.77, and Figure 3.78).

We start the implementation of this design by defining the symbolic constants, such as the FSM type names (e.g., the name of the INVITE client FSM type is *InviteClienteTE_FSM*), mailbox names (e.g., the name of the INVITE client mailbox is *InviteClienteTE_FSM_MBX*), names of the FSM Library related message types, timer names (e.g., TIMER_A, TIMER_B, TIMER_D), names of the SIP messages (e.g., INVITE, OPTIONS, CANCEL, ACK, BYE, RESISTER), names of the response codes (e.g., _180_RINGING, _200_OK, _302_MOVED_TEMPORARILY, _401_UNAUTHORIZED, _403_FORBIDDEN, _404_NOT_FOUND), and names of situations (e.g URI_IN_TO_UNRECOGNIZED and NOT_TO_CURRENT_USER). Traditionally, we write definitions of all these constants into the file *constants.h*.

Next, we write the class that represents an SIP message, simply named *Message*. The most important field member of this class is the last (also referred to as the current) SIP message (its type is the C++ type string). Other field members hold the relevant SIP session related information. The function members support SIP message analysis and synthesis (parsing and creation).

Actually, the class *Message* that is used in this example is a simple wrapper around the OpenSIP SIP message parser. (OpenSIP is freely available on the Internet at http://sourceforge.net/projects/opensip/.)

We skip the content of the file *constants.h* and the source code of the class *Message* intentionally to keep this example short enough and easily comprehendible, and we proceed with the introduction of the supplementary class *TALE*. The declaration of the class *TALE* is the following:

```
#ifndef _TALE_FSM_
#define _TALE_FSM_
#include "../kernel/fsm.h"
#include "../message/message.h"
#include "../constants.h"

class TALE : public FiniteStateMachine {
  uint8  MessageCopy[MAX_LENGTH_MESSAGE];
  uint32 IndexTLI;
  BOOL   IndexTLISet;
public:
  void   SetIndexTLI(uint32 newIndexTLI);
  uint32 GetIndexTLI();
  BOOL   IsTransportReliable();
  void   SendMessageToTU();
  void   SendMessageToTPL();
  void   SendErrorMessageToTU();
  void   MakeLocalCopyOfMsg();
  void  SendCopiedMessageToTPL();

public:
  TALE(uint16 numOfTimers, uint16 numOfState, uint16 maxNumOfPrPerSt);
  ~TALE();
};
```

The class *TALE* is a good example of how we can make our implementations more compact. As we can see from the previous example, sending a single message requires a series of FSM Library function calls. For example, forwarding the current message would require a series of calls to the function *CopyMessage()*, *SetMsgToAutomata()*, *SetMsgToGroup()*, *SetMsgObjectNumberTo()*, and function *SendMessage()* — five function calls. In the case of simple designs, we can tolerate repetition of this series of function calls, but in cases of more complex design or platforms with limited resources, this repetition may not be tolerated.

Consider the SIP invite client transaction FSM. It has thirteen state transitions, and most of them require sending a message to either TPL (transport layer) or TU (transaction user). We would need to repeat the same series of function calls about ten times. Consider now the whole SIP Softphone, which supports four types of transactions (invite and non invite, client and server transactions). In such situations, replacing this series of function calls with a single function call, which in its turn performs the original sequence of function calls, makes sense.

Implementation 237

This replacement is exactly the reason why the class *TALE* has been introduced in the first place. This class inherits all field and function members from the class *FiniteStateMachine* from which it is derived. It also adds some new field and function class members. All classes that implement SIP transactions are derived from the class *TALE*. The most important field member of the class *TALE* is the field *MessageCopy*, which holds the copy of the last sent message. Actually, this field is the retransmission buffer (remember that SIP invite client in the state *Calling* must retransmit the message *INVITE* in case the timer A expires).

The two most important function members are the functions *SendMessageToTU()* and *SendMessageToTPL()*. The former sends the current message to TU and the latter to TPL. They are very similar; therefore, it is sufficient to study just one of them. Here is the source code of the former function:

```
void TALE::SendMessageToTU() {
  CopyMessage();
  SetMsgToAutomata(UA_Disp_FSM);
  SetMsgToGroup(INVALID_08);
  SetMsgObjectNumberTo(0);
  SendMessage(UA_Disp_FSM_MBX);
}
```

This is the most elegant way to forward a message in the FSM Library-based implementations. The function *CopyMessage()* copies the current (last received) message to the new (output) message. The symbolic constant *UA_Disp_FSM* is the name of the UA (user agent) FSM type, and the constant *UA_Disp_FSM_MBX* is the name of its mailbox. As we will shortly see, the use of the functions *SendMessageToTU()* and *SendMessageToTPL()* significantly compresses the source code. They make one-to-one mapping of SDL diagrams to C++ code possible.

Next, we proceed to the implementation of the INVITE client transaction FSM. We implement it by writing the class *InviteClientTE*. Note that in Figure 3.69 to Figure 3.73, we used the abbreviation *InClientT* for this name. The declaration of the class *InviteClientTE* is the following:

```
#ifndef _InviteClientTE_FSM_
#define _InviteClientTE_FSM_
#include "TALE.h"

class InviteClientTE : public TALE {
  Message SIPMsg;
  uint32  cseq_number;
  uint32  TimerADuration;

public:
  enum States {
    STATE_INITIAL,
    STATE_CALLING,
    STATE_PROCEEDING,
    STATE_COMPLETED
```

```
};
// state Initial message handlers
void  Evt_Init_INVITE();
// state Calling message handlers
void  Evt_Calng_TIMER_A_EXP();
void  Evt_Calng_RESPONSE_1XX();
void  Evt_Calng_RESPONSE_2XX();
void  Evt_Calng_TIMER_B_EXP();
void  Evt_Calng_RESPONSE_3_6XX();
void  Evt_Calng_TRANSPORT_ERR();
// state Proceeding message handlers
void  Evt_Proc_RESPONSE_1XX();
void  Evt_Proc_RESPONSE_2XX();
void  Evt_Proc_RESPONSE_3_6XX();
// state Completed message handlers
void  Evt_Comptd_TIMER_D_EXP();
void  Evt_Comptd_RESPONSE_3_6XX();
void  Evt_Comptd_TRANSPORT_ERR();
// unexpected messages message handler
void  Event_UNEXPECTED();
// problem specific functions
void  RetransmitInvite();
BOOL  SendAckMessageToTPL();
// FiniteStateMachine abstract functions
StandardMessage StandardMsgCoding;
MessageInterface *GetMessageInterface(uint32 id);
void  SetDefaultHeader(uint8 infoCoding);
void  SetDefaultFSMData();
void  NoFreeInstances();
void  Reset();
uint8 GetMbxId();
uint8 GetAutomate();
uint32 GetObject();
void  ResetData();
public:
```

The class *InviteClientTE* is derived from the class *TALE*. The meaning of its field members is the following:

- The field *SIPMsg* is the SIP message parser (an instance of the class *Message*).
- The field *cseq_number* holds the value of the SIP message header field *CSeq*, which is used to identify and order transactions (see RFC 3261, Subsection 8.1.1.5).
- The field *TimerADuration* contains the current value of the timer A (remember, the value of the timer A is doubled each time it expires).

Next, we enumerate the names of the FSM states. There are altogether four FSM states, *STATE_INITIAL*, *STATE_CALLING*, *STATE_PROCEEDING*, and *STATE_COMPLETED*. A short explanation is needed at this point. According to the original specification (RFC 3261, Figure 5, page 128), the INVITE client

Implementation

transaction FSM also has four explicitly rendered states, namely, *Calling*, *Proceeding*, *Completed*, and *Terminated*. The initial state is omitted in the original specification. In our implementation, we create a pool of *InviteClientTE* objects, which are dynamically allocated on demand by the TU. These objects are never really terminated. Once they play their simple role, they are returned to the pool of free *InviteClientTE* objects, and from there they are dynamically assigned to play the same role again. Therefore, we renamed the state *Terminated* to *Initial*. We also made this state the source of the initial state transition (triggered with the INVITE message from TU), thus making the FSM a never-terminating one.

We then list the state transition function prototypes for each state individually. The naming convention is the same as in the previous example: The name of the state transition function is constructed by concatenating the state name and the message name and by prefixing that name with a certain prefix. The naming convention is applied more freely in this example by shortening the state names. This practice is frequently done to keep the name lengths acceptable (short enough but providing code readability at the same time). Thirteen valid state transitions and their corresponding state transition functions (message handlers) are used. The fourteenth message handler, named *Event_UNEXPECTED()*, handles all unexpected messages in all states.

Finally, we list the function prototypes of the problem-specific functions and mandatory *FiniteStateMachine* abstract functions. These functions — except the function *RetransmitInvite()* — are intentionally skipped in the text that follows to keep the presentation of this example short.

We finish the implementation by writing the class *InviteClientTE* definition file, named *InvClientTE.cpp*. The content of this file is the following:

```
#include <stdio.h>
#include "InvClientTE.h"
#include "../Message/message.h"
#include "timer_values.h"
#define StandardMessageCoding 0x00

InviteClientTE::InviteClientTE() : TALE(10, 10, 10) {}
InviteClientTE::~InviteClientTE() {}

void InviteClientTE::Initialize() {
  SetState(STATE_INITIAL);
  // define timers
  InitTimerBlock(TIMER_A,1,TIMER_A_EXPIRED);
  InitTimerBlock(TIMER_B,1,TIMER_B_EXPIRED);
  InitTimerBlock(TIMER_D,1,TIMER_D_EXPIRED);
  // state STATE_INITIAL message handlers
  InitEventProc(STATE_INITIAL, INVITE,
    (PROC_FUN_PTR)&InviteClientTE::Evt_Init_INVITE);
  // state STATE_CALLING message handlers
InitEventProc(STATE_CALLING, TIMER_A_EXPIRED,
  (PROC_FUN_PTR)&InviteClientTE::Evt_Calng_TIMER_A_EXP);
```

```
InitEventProc(STATE_CALLING, RESPONSE_1XX_T,
  (PRO_FUN_PTR)&InviteClientTE::Evt_Calng_RESPONSE_1XX);

InitEventProc(STATE_CALLING, RESPONSE_2XX_T,
  (PROC_FUN_PTR)&InviteClientTE::Evt_Calng_RESPONSE_2XX);

InitEventProc(STATE_CALLING, TIMER_B_EXPIRED,
  (PROC_FUN_PTR)&InviteClientTE::Evt_Calng_TIMER_B_EXP);

InitEventProc(STATE_CALLING, RESPONSE_3XX_T,
  (PROC_FUN_PTR)&InviteClientTE::Evt_Calng_RESPONSE_3_6XX);

InitEventProc(STATE_CALLING, RESPONSE_4XX_T,
  (PROC_FUN_PTR)&InviteClientTE::Evt_Calng_RESPONSE_3_6XX);

InitEventProc(STATE_CALLING, RESPONSE_5XX_T,
  (PROC_FUN_PTR)&InviteClientTE::Evt_Calng_RESPONSE_3_6XX);

InitEventProc(STATE_CALLING, RESPONSE_6XX_T,
  (PROC_FUN_PTR)&InviteClientTE::Ev_Calng_RESPONSE_3_6XX);

InitEventProc(STATE_CALLING, TRANSPORT_ERR,
  (PROC_FUN_PTR)&InviteClientTE::Evt_Calng_TRANSPORT_ERR);

  // state STATE_PROCEEDING message handlers
InitEventProc(STATE_PROCEEDING, RESPONSE_1XX_T,
  (PROC_FUN_PTR)&InviteClientTE::Evt_Proc_RESPONSE_1XX);

InitEventProc(STATE_PROCEEDING, RESPONSE_2XX_T,
  (PROC_FUN_PTR)&InviteClientTE::Evt_Proc_RESPONSE_2XX);

InitEventProc(STATE_PROCEEDING, RESPONSE_3XX_T,
  (PROC_FUN_PTR)&InviteClientTE::Evt_Proc_RESPONSE_3_6XX);

InitEventProc(STATE_PROCEEDING, RESPONSE_4XX_T,
  (PROC_FUN_PTR)&InviteClientTE::Evt_Proc_RESPONSE_3_6XX);

InitEventProc(STATE_PROCEEDING, RESPONSE_5XX_T,
  (PROC_FUN_PTR)&InviteClientTE::Evt_Proc_RESPONSE_3_6XX);

InitEventProc(STATE_PROCEEDING, RESPONSE_6XX_T,
  (PROC_FUN_PTR)&InviteClientTE::Evt_Proc_RESPONSE_3_6XX);

  // state STATE_COMPLETED message handlers
InitEventProc(STATE_COMPLETED, TIMER_D_EXPIRED,
  (PROC_FUN_PTR)&InviteClientTE::Evt_Comptd_TIMER_D_EXP);

InitEventProc(STATE_COMPLETED, RESPONSE_3XX_T,
  (PROC_FUN_PTR)&InviteClientTE::Evt_Comptd_RESPONSE_3_6XX);

InitEventProc(STATE_COMPLETED, RESPONSE_4XX_T,
  (PROC_FUN_PTR)&InviteClientTE::Evt_Comptd_RESPONSE_3_6XX);

InitEventProc(STATE_COMPLETED, RESPONSE_5XX_T,
  (PROC_FUN_PTR)&InviteClientTE::Evt_Comptd_RESPONSE_3_6XX);
```

Implementation 241

```cpp
  InitEventProc(STATE_COMPLETED, RESPONSE_6XX_T,
    (PROC_FUN_PTR)&InviteClientTE::Evt_Comptd_RESPONSE_3_6XX);

  InitEventProc(STATE_COMPLETED, TRANSPORT_ERR,
    (PROC_FUN_PTR)&InviteClientTE::Evt_Comptd_TRANSPORT_ERR);

  // unexpected messages message handler
  InitUnexpectedEventProc(STATE_INITIAL,
    (PROC_FUN_PTR)&InviteClientTE::Event_UNEXPECTED);

  InitUnexpectedEventProc(STATE_CALLING,
    (PROC_FUN_PTR)&InviteClientTE::Event_UNEXPECTED);

  InitUnexpectedEventProc(STATE_PROCEEDING,
    (PROC_FUN_PTR)&InviteClientTE::Event_UNEXPECTED);

  InitUnexpectedEventProc(STATE_COMPLETED,
    (PROC_FUN_PTR)&InviteClientTE::Event_UNEXPECTED);
}

void InviteClientTE::Evt_Init_INVITE() {
  SendMessageToTPL();
  if (!IsTransportReliable()){
    TimerADuration = GetT1();
    setTimerCount(TIMER_A, TimerADuration);
    StartTimer(TIMER_A);
  }
  setTimerCount(TIMER_B, 64*GetT1());
  StartTimer(TIMER_B);
  MakeLocalCopyOfMsg();
  SetState(STATE_CALLING);
}

void InviteClientTE::Evt_Calng_TIMER_A_EXP(){
  TimerADuration = 2 * TimerADuration;
  setTimerCount(TIMER_A, TimerADuration);
  RestartTimer(TIMER_A);
  RetransmitInvite();
}

void InviteClientTE::Evt_Calng_RESPONSE_1XX(){
  uint16 val;
  StopTimer(TIMER_A);
  StopTimer(TIMER_B);
  SendMessageToTU();
  GetParamWord(INDEX_TLI_PARAM, val);
  SetIndexTLI(val);
  SetState(STATE_PROCEEDING);
}

void InviteClientTE::Evt_Calng_RESPONSE_2XX(){
  StopTimer(TIMER_A);
  StopTimer(TIMER_B);
  SendMessageToTU();
  SetState(STATE_INITIAL);
```

```cpp
}

void InviteClientTE::Evt_Calng_TIMER_B_EXP(){
  StopTimer(TIMER_A);
  SendErrorMessageToTU();
  SetState(STATE_INITIAL);
}

void InviteClientTE::Evt_Calng_TRANSPORT_ERR(){
  StopTimer(TIMER_A);
  StopTimer(TIMER_B);
  SendErrorMessageToTU();
  SetState(STATE_INITIAL);
}

void InviteClientTE::Evt_Calng_RESPONSE_3_6XX(){
  uint16 val;
  StopTimer(TIMER_A);
  StopTimer(TIMER_B);
  SendMessageToTU();
  GetParamWord(INDEX_TLI_PARAM, val);
  SetIndexTLI(val);
  SendAckMessageToTPL();
  if (IsTransportReliable())
    setTimerCount(TIMER_D, ZERO_TIMER_VAL_APPROX);
  else
    setTimerCount(TIMER_D, 64*GetT1());//64T1
  StartTimer(TIMER_D);
  SetState(STATE_COMPLETED);
}

void InviteClientTE::Evt_Proc_RESPONSE_1XX(){
  SendMessageToTU();
}

void InviteClientTE::Evt_Proc_RESPONSE_2XX(){
  SendMessageToTU();
  SetState(STATE_INITIAL);
}

void InviteClientTE::Evt_Proc_RESPONSE_3_6XX(){
  SendMessageToTU();
  SendAckMessageToTPL();
  if (IsTransportReliable())
    setTimerCount(TIMER_D, ZERO_TIMER_VAL_APPROX);
  else
    setTimerCount(TIMER_D, 64*GetT1());//64T1
  StartTimer(TIMER_D);
  SetState(STATE_COMPLETED);
}

void InviteClientTE::Evt_Comptd_TIMER_D_EXP(){
  SetState(STATE_INITIAL);
}
```

Implementation

```
void InviteClientTE::Evt_Comptd_RESPONSE_3_6XX(){
  SendAckMessageToTPL();
}

void InviteClientTE::Evt_Comptd_TRANSPORT_ERR(){
  StopTimer(TIMER_D);
  SendErrorMessageToTU();
  SetState(STATE_INITIAL);
}

void InviteClientTE::Event_UNEXPECTED() {
}

void InviteClientTE::RetransmitInvite(){
  SendCopiedMessageToTPL();
}
```

The mandatory function *Initialize()* starts by setting the FSM initial state *STATE_INITIAL*. It then initializes the timers A, B, and D by calling the FSM Library function *InitTimerBlock()* (its parameters are the timer identification, the timer interval duration, and the identification of the associated message; see also Section 6.8.74). The function *Initialize()* finishes by setting the FSM state transition functions. These functions process various message types in different states, as follows:

- The function *Evt_Init_INVITE()* processes the message *INVITE* in the state *STATE_INITIAL*.
- The function *Evt_Calng_TIMER_A_EXP()* processes the message *TIMER_A_EXPIRED* in the state *STATE_CALLING*.
- The function *Evt_Calng_RESPONSE_1XX()* process the message *RESPONSE_1XX_T* in the state *STATE_CALLING*.
- The function *Evt_Calng_ RESPONSE_2XX()* process the message *RESPONSE_2XX_T* in the state *STATE_CALLING*.
- The function *Evt_Calng_TIMER_B_EXP()* processes the message *TIMER_B_EXPIRED* in the state *STATE_CALLING*.
- The function *Evt_Calng_RESPONSE_3_6XX()* processes the messages *RESPONSE_3XX_T*, *RESPONSE_4XX_T*, *RESPONSE_5XX_T*, and *RESPONSE_6XX_T* in the state *STATE_CALLING*.
- The function *Evt_Calng_TRANSPORT_ERR()* processes the message *TRANSPORT_ERR* in the state *STATE_CALLING*.
- The function *Evt_Proc_RESPONSE_1XX()* processes the message *RESPONSE_1XX_T* in the state *STATE_PROCEEDING*.
- The function *Evt_Proc_RESPONSE_2XX()* processes the message *RESPONSE_2XX_T* in the state *STATE_PROCEEDING*.

- The function *Evt_Proc_RESPONSE_3_6XX()* processes the messages *RESPONSE_3XX_T, RESPONSE_4XX_T, RESPONSE_5XX_T,* and *RESPONSE_6XX_T* in the state *STATE_PROCEEDING*.
- The function *Evt_Comptd_TIMER_D_EXP()* processes the message *TIMER_D_EXPIRED* in the state *STATE_COMPLETED*.
- The function *Evt_Comptd_RESPONSE_3_6XX()* processes the messages *RESPONSE_3XX_T, RESPONSE_4XX_T, RESPONSE_5XX_T,* and *RESPONSE_6XX_T* in the state *STATE_COMPLETED*.
- The function *Evt_Comptd_TRANSPORT_ERR()* processes the message *TRANSPORT_ERR* in the state *STATE_COMPLETED*.
- The function *Event_UNEXPECTED()* processes all unexpected messages in all states.

As we can see from the source code above, the state transition functions (message handlers) are short and easily readable because each program statement is easily traceable back to the original statechart and SDL diagrams. For example, consider the first state transition function *Evt_Init_INVITE()*. The original SDL specification of this state transition starts with the reception of the message *INVITE* (Figure 3.75). This step is provided by the class *FSMSystem*. The next step in the SDL diagram says: "Invite_T to TPL." This step is implemented with a single program statement, namely, the function call to the function *SendMessageToTPL()*.

The next step in the SDL diagram is the question, "Is transport reliable?" We implement it also with a single function call to the function *IsTransportReliable()*. We continue the SDL coding in this manner. If the transport is reliable, the initial value of the timer A is provided by calling the function *GetT1()* — a way to parameterize the software. Next, we set the timer A duration by calling the function *setTimerCount()* — this is the undocumented FSM Library function at the moment, to be included in the next official release — and start the timer A by calling the function *StartTimer()* (the parameter of this function is the timer identification; see also Section 6.8.138).

At the end of this function, we set the duration of the timer B and start it, make the local copy of the last sent message by calling the function *MakeLocalCopy()* — remember that it is needed for the possible retransmission — and set the new state by calling the function *SetState()* (its parameter is the state identification; see also Section 6.8.137).

Next, the state transmission function, *EvtCalng_TIMER_A_EXP()*, performs the reaction to the timer A expiration (see the corresponding SDL specification in Figure 3.75) with only four program statements. The first one doubles the timer A duration, the second sets this new duration, the third restarts the timer A by calling the FSM Library function *RestartTimer()* (see Section 6.8.87), and the fourth retransmits the message *INVITE* by calling the function *RetransmitInvite()*. Also, all other state transition functions are

Implementation

made in this spirit of one-to-one mapping from the original SDL diagram. The reader is advised to study them as an additional exercise.

References

Booch, G., Rumbaugh, J., and Jacobson, I., *The Unified Modeling Language User Guide*, Addison-Wesley, Reading, MA, 1998.

Gamma, E., Helm, R., Johnson, R., and Vlissides, J., *Design Patterns: Elements of Reusable Object-Oriented Software*, Addison-Wesley, Reading, MA, 1995.

5
Test and Verification

The test and verification phase is a phase of communication protocol engineering work that follows the implementation phase. The primary goal of this phase is to verify that the implementation in the higher-level programming language is correct. The implementation is correct if it meets its original requirements, which are modeled in a form of use cases (see Chapter 2).

The correctness of the implementation is checked with the test suite, which is typically designed in TTCN (see Section 3.9). The test suite itself is implemented in a higher-level programming language, e.g., Java or C++. But how do we verify the correctness of the test suite implementation? The answer is that we do not check the correctness of the test suite independently. We always check the correctness of the implementation under test and test suite simultaneously. Theoretically, a bug in a test suite can cover a bug in the implementation; we should be aware of that but such cases seldom happen in practice.

Typical testing activities conducted in the communication protocol engineering test and verification phase are the following:

- Unit testing
- Integration testing
- Conformance testing
- Load testing
- In-field testing
- Formal verification
- Statistical usage testing

The first four types of activities (unit testing, conformance testing, load testing, and in-field testing) are stemming from the traditional software engineering, whereas the last two (formal verification and statistical usage testing) are originating from the Cleanroom engineering. Today, communication protocol engineers tend to complement software engineering with the Cleanroom engineering testing approaches, therefore we cover all the above listed activities in this chapter.

As its name suggests, the unit testing is used for testing individual software units before their integration into the product. Typically, a software unit is a single class written in a separate Java compilation unit or C++ module. This class most commonly implements a simple communication protocol or a part of a more complex communication protocol. In the case of the FSM Library based paradigm, such a unit would be a C++ module that defines the class derived from the class *FiniteStateMachine*.

Unit testing of communication protocols is relatively straightforward. Typically, we construct a set of test cases that check individual FSM state transitions, as well as more complex FSM transactions (series of FSM state transitions). We will use JUnit and CppUinit testing frameworks for unit testing of communication protocols in this book. Details of unit testing are given in Section 5.1 (unit testing) and Section 5.5.1 (Example 1).

The next phase is integration testing. The philosophy of integration testing starts from the fact that some of the units have successfully undergone unit testing and that they are available for further testing, whereas the rest of them are not. For the purpose of integration testing, we introduce replacements for the units that are not available, which are referred to as the imitators (or simulators).

There are two kinds of imitators, namely drivers and stubs. A driver is an active imitator that generates input messages for the real objects (units) under test. A stub is a passive imitator that accepts the output messages generated by the objects under test. Stubs can also send replays that are expected from the objects they are imitating. Of course, we can construct more complex imitators that act as both drivers and stubs. In this book, we will call the collaborations of real objects, *drivers*, and stubs simply *integration test collaborations*.

Generally, communication protocols are well suited for integration testing because families of communication protocols are hierarchically organized in layers with well-defined interfaces. The communication between individual protocols is based on messages, which are traditionally exchanged through the mailboxes (for example, as in implementations based on the FSM Library). Simulating the environment of a real object under test in such a situation is easy. Drivers and stubs simply exchange messages with objects under test. Actually, they act on behalf of the units that will communicate with the units under test in the final product.

Normally, protocol stacks are implemented in the bottom-up fashion, starting from the lowest layer of the protocol stack and building the next layer on top of the previous one. Drivers and stubs in such an approach simulate only a part of the environment, the higher layer of the protocol stack in particular. The example of the simple integration test collaboration is given in Section 5.5.2 (Example 2).

When all software units have undergone unit and integration testing, the final product is integrated and ready for acceptance testing, which comprises conformance testing (also referred to as compliance testing), load testing, and in-field testing. Preliminary acceptance testing can be organized solely by the production organization and conducted on its premises. However, final acceptance testing is organized and conducted by the organization that has the legal authority to issue acceptance certificates.

As suggested by its name, the aim of conformance testing is to prove that the product (implementation) under test conforms to the original requirements. In the area of communication protocol engineering, these requirements would normally be standards issued by the IETF, ISO, ITU-T, ETSI, and similar organizations. The newer standards made by ITU-T and ETSI most frequently include the conformance test suite specification in TTCN.

The conformance testing is a kind of functional testing (also referred to as black box testing). The tester is not interested in the structure of the product and its internal behavior. He only ensures that the external behavior of the product meets the original specification. Typically, this behavior is specified with the set of scenarios described in TTCN. We will return to the subject of conformance testing in Section 5.2.

The load testing typically involves exposing the implementation under test to the conditions of the real exploitation. Conceptually, this means that the implementation under test must service the requests coming from more independent sources simultaneously. While conformance testing focuses on the correctness of services given to the minimal number of request sources, load testing checks the correctness of services driven by the requests coming from independent sources in preferably interleaved fashion.

Normally, load testing is conducted in the simulated environment in the laboratory. Typically, we would construct, purchase, or lease the specialized equipment, referred to as a load generator. A load generator is normally a programmable device that offers a selection of predefined scenarios and their parameters, such as number of request sources, duration of individual communication phases, and so on, as well as definitions of completely new scenarios.

The name *load generator* may be misleading because it suggests that the device generates only the requests — which it does — but it also receives the responses from the implementation under test and checks if it operates correctly. For example, after the connection is successfully established, it sends and receives test tones to check that the connection is really usable. During load testing, we primarily check declared traffic capabilities of the product. A typical requirement would be that the number of lost requests must not exceed the given limit after the given number of requests has been issued in accordance with the given request arrival distribution.

We normally also check the behavior of the implementation under test for both lower and higher rates of request arrivals. With an extremely low rate of requests, we want to check the sustainability of the long-lasting connections, whereas with an extremely high rate we want to make sure that the overload protection mechanisms are in place and that they function correctly. After successful load testing, the implementation under test is integrated into the target network for the in-field testing. The in-field testing is essentially the experimental exploitation of the product for the given interval of time (e.g., three months).

The aim of in-field testing is to detect, locate, and eliminate bugs that are exposed by the real-world scenario (also referred to as a traffic case) that could not be simulated in the laboratory. During this last phase of acceptance testing, log files always prove to be extremely useful. Today, the log files can be collected over the Internet and analyzed remotely. Also, installing software upgrades can be done by uploading new software patches over the Internet.

Detecting bugs through the analysis of the log files can be augmented by adding the program hooks for certain really infrequent traffic cases. Defining state transition preconditions, postconditions, and invariants and checking them at run-time is also extremely useful in detecting bugs during in-field testing, and later during normal system exploitation. Although communication protocol maintenance is an integral part of communication protocol engineering, it is out of scope of this book (see directions for further reading in Section 5.6).

Traditional software engineering comprises a number of development phases, such as requirements, analysis, design, implementation, unit test, integration, integration test, verification, and maintenance. These phases can be cascaded in the case of the waterfall process model or revisited in the case of the spiral-incremental process model. The number of remaining bugs is the main software quality metric. Another important metric used in software engineering is the test coverage (measured as the percentage of tested software paths, variable usages, and so on).

Cleanroom engineering, in contrast to traditional software engineering, is organized as a sequence of the following development activities:

- Formal model development.
- Formal verification of the formal model.
- Handing formal model to the implementation team, which implements it in a higher level programming language.
- Operational profile modeling.
- Automatic test suite generation, which is based on the given operational profile model.
- Statistical usage testing and software reliability estimation. If at least one test case from the automatically generated test suite fails, the implementation under test is thrown away and the complete development cycle is repeated from the very beginning (starting with the formal model development).

The complete treatment of formal modeling and verification is out of the scope of this book (see directions for further reading in Section 5.6). As a means of introduction to the area of formal methods, formal modeling and verification based on theorem proving is covered in Section 5.3. The paradigm described in that section is based on the application of the theorem prover named THEO.

Operational profile modeling, automatic test suite generation, statistical usage testing and reliability software estimation are described in Section 5.4. The paradigm described in that section is based on the application of the software tool, which is named generic test case generator (GTCG).

5.1 Unit Testing

The aim of unit testing is to check the correctness of an individual software unit (Java compilation unit or C/C++ module). A generally accepted belief, especially among proponents of agile methods such as extreme programming, is that unit testing should be conducted by the programmer who is implementing the target software unit because it greatly improves programmer's productivity. In principle, unit tests should be written before, or at least during, the implementation of the target software unit.

Of course the programmer must clearly distinguish two roles, namely implementer and tester roles (the author of extreme programming, Kent Beck, uses the metaphor: "by changing hats" to explain this paradigm). The programmer, as unit tester, concentrates on the unit interface. By thinking about the interface and by writing unit tests, the programmer gets a clearer picture about the services that the target software unit must provide. The programmer should also try to make test cases that cover boundary conditions, as well as situations that would be potentially hard to manage for the target software unit.

The programmer, as unit implementer, concentrates on the implementation of the original unit design. They should forget about unit tests and concentrate on mapping the design to code. This should be a straightforward task if a proper framework (such as the FSM Library) is provided.

Unit testing helps programmers produce software units of better quality in shorter time intervals and this has been proven in practice. First, by creating unit tests, the programmer becomes even more familiar with the implementation at hand. Second, the programmer gets the immediate feedback. If there is a bug, it is easy to detect in the scope of a particular test case. If the test case passes, the programmer gets immediate satisfaction that they have done their job well.

Unit test cases should be executed frequently during the target unit's implementation. As time passes, new test cases are added and old cases are run again. Even if no new test cases are used, we should rerun all existing unit tests every time we add new functionality. Testing that is conducted by running an unchanged test suite to check if the new software functionality has not affected existing functionalities is referred to as regression testing.

Regression testing is the key point of this paradigm. It enables a dramatic increase of productivity because it builds the programmer's confidence that everything is in good order and under control; therefore, the programmer can work more relaxed. Regression testing also encourages experimenting. In situations when alternative paths may be used in the course of implementation, the programmer may try out a way that seems most appropriate. If one or more test cases fail in regression testing that is subsequently conducted, the programmer may decide to reset to the starting point by retrieving the previous version from the installed version control system database.

Unit testing (including regression testing) definitively has a great impact on a programmer's psychology in a positive direction. It is estimated to be the key factor for the increase of the programmer's productivity. The next question is to what extent we should go with the unit testing The answer is not easy. Certainly, any amount of unit testing is better that none. Alternately, an attempt at exhaustive unit testing might be counterproductive.

The right choice is somewhere between these two extremes. We do not need to test trivial things, such as class function members that set or get the value of a certain private field member. Rather, we should concentrate on the boundary conditions and parts of code where it becomes more complex. Although generally unpopular among professionals, copy-paste practice may be tolerated for generating a set of similar test cases.

Three principal preconditions exist for successful unit testing practice:

- A proper unit testing framework must be provided.
- Test cases should not involve any human intervention.
- The implementation under test must not be changed.

A proper unit testing framework must provide three main functions:

- Test case registration: This function enables registering new test cases within the given test suite hierarchy. On each level of the hierarchy, a set of individual test cases may be found, as well as other hierarchically subordinated test suites (very similar to the file system structure).
- Test case execution: This function provides automatic execution of all test cases defined within the given test suite hierarchy. It must not require more than a single push-button to be started. Otherwise, the framework is simply not usable.
- Test case reporting: This function must provide a general report on the outcome of the execution of all test cases, as well as individual reports for all test cases that failed or caused errors.

The second precondition is that test suite execution should not involve any human intervention. This is the essential precondition to make unit testing completely automatic. If we want to eliminate human interventions, we must secure two conditions. First, the input data required by a test case must be defined as symbolic constants in its source code or in other external files. Second, the results of the test case must be automatically checked by a test case itself. The unit testing framework must provide adequate functions for this purpose.

A typical function for checking test case results is the function *assert(condition)*, where *condition* is a Boolean expression that evaluates to either the value *true* or *false*. Test case continues (*pass*) in the former case and breaks (*fail*) in the latter case. If the test case execution successfully reaches the end of the test case, it is considered successful (qualified with the verdict *pass*). Otherwise, it is considered unsuccessful (qualified with the verdict *fail*). If the test case execution breaks because of some error (most typically, an exception such as "divide by zero"), it is qualified with the verdict *error*.

Another typical function for checking test case results is the function *assertEquals(p1,p2)*. This function call is semantically equivalent to the function call *assert(p1==p2)*. This means that if the parameters *p1* and *p2* are equal (of course, they must be comparable), the test case execution continues; otherwise, it breaks. Typically, one of the parameters is a constant and another is a program variable.

Although these two functions are semantically equivalent, the function *assertEquals()* is advantageous when it comes to test case reporting. If the function *assert()* breaks the test case execution, the unit testing framework reports only that the condition evaluated to the value *false*, which is not a very informative report. Alternately, if the function *assertEquals()* breaks, the framework provides the report "expected *C* but was *V*," where *C* is the value of the constant (e.g., *p1*) and *V* is the real value of the variable (e.g., *p2*).

We can further improve the readability of the test case execution reports by using the optional text string parameter of the function *assertEquals()*. Generally, the function call format for this function is *assertEquals(text, condition)*, where *text* is the text string that explains the meaning of this assertion point in more detail. The string *text* is used as a prefix of the test report shown above. For example, if the value of the variable *ch* should be 'A' but it turns out to be 'B' instead, the function call *assertEquals("Check ch:," 'A', ch)* would produce the report, "Check ch: expected 'A' but was 'B'".

Besides the functions *assert()* and *assertEquals()*, unit testing framework typically provides two additional functions for writing test cases, *setUp()* and *tearDown()*. The former sets up the test fixture whereas the latter destroys it. A test fixture is a set of objects that act as samples for testing. Normally, the test fixture comprises the instance of the unit under test (e.g., the instance of the class that is derived from the class *FiniteStateMachine*) and also other supplementary objects, which are required for effective unit testing.

Typically, the unit testing framework offers the base class for writing test cases, which provides the functions *assert()*, *assertEquals()*, *setUp()*, and *tearDown()*. The programmer normally derives his tester class from this base class, fills in *setUp()* and *tearDown()* functions, and starts writing individual test cases. Each function member of the tester class — whose name follows the given naming convention — is a single test case.

Remember that concrete *setUp()* and *tearDown()* implementations are shared by all test cases defined within a single tester class. Actually, these two functions are implemented as null (empty) methods on test cases. The execution of each test case starts with the call to the function *setUp()*, proceeds with the call to the user-defined function that implements a single test case, and ends with the call to the function *tearDown()*. Normally, we put the test case initialization and cleanup code in the functions *setUp()* and *tearDown()*, respectively.

The third unit testing postulate is that the unit under test must not be touched at all. We are only allowed to write new classes that are derived from the base class, which is provided by the unit testing framework. Changing the source code of the unit under test for the purpose of its testing is strictly forbidden, even by adding a simple print statement to the standard output file. Because of that, the only proper way to do the unit testing is to drive the unit under test with various messages, capture its responses, and check the correctness of the unit's external behavior.

This kind of controlled execution of the implementation under test is referred to as the test harness. The key request is that it must be fully automatic. The programmer should provide the mechanisms that support the test harness while he plays the role of the implementer (what we refer to as the *design for testability*). Otherwise, providing a test harness can be a very hard task. For example, consider a simple program that reads its input from the keyboard and writes its output to the monitor by using the operating system services, which cannot be replaced. Because we are not allowed to change the source code of the implementation under test, providing a test harness in this case is hardly achievable.

The example of the unit testing framework is JUnit, an open-source testing framework for unit testing Java programs that was originally developed by Erich Gamma and Kent Beck. Based on this framework, the open-source community came up with CppUnit, a semantically equivalent testing framework for unit testing C++ modules. These frameworks are very simple but powerful enough to enable industrial-strength unit testing of individual software units. Because JUnit and CppUnit are semantically equivalent, we will treat them as two implementations of the same framework.

The framework comprises the interface *Test* and two fundamental classes, the classes *TestSuite* and *TestCase* (Figure 5.1). As shown in the figure, the test suite (an instance of the class *TestSuite*) can contain an arbitrary number of test cases (instances of the class *TestCase*), as well as an arbitrary number of other hierarchically subordinated test suites. This arrangement allows programmers (playing the role of unit testers) to organize test cases into a hierarchy of test suites to their convenience.

Any concrete tester class (such as the class *MyTester* in Figure 5.1) must be derived from the base class *TestCase*, which among others provides the four fundamental functions described above, namely, *setUp()*, *tearDown()*, *assert()*,

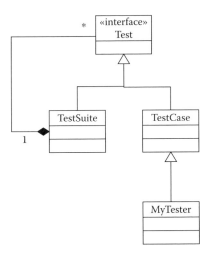

FIGURE 5.1
The structure of the JUnit testing framework.

and *assertEquals()*. By convention, an individual test case is written as the function member of the tester class, whose name starts with the word "test," for example, *test1*, *test2*, and so on.

Next, we illustrate JUnit's usability on a concrete example. In the example that follows, we demonstrate unit testing paradigm for the case where the implementation under test is counter by modulo 2. The particular implementation we are interested in is the one based on the State design pattern. This implementation is presented in Section 4.3.

As already mentioned in Section 4.3, the function *processMsg()*, which processes FSM input (message), prints its results by calling the function member *println()* of the class *MyIO*, rather than by calling the standard I/O function *System.out.println()*. This is a good example of how we can provide the support for the test harness in our design and implementation. Here is the source code of the class *MyIO*:

```
package automata4;
import java.util.*;

public class MyIO {
private static String lastOutput;
  public static String getLastOutput() { return lastOutput; }
  public static void println(String s) {
    lastOutput = s;
    System.out.println(s);
  }
}
```

The field member *lastOutput* is used to store the last output generated by the FSM. The function *getLastOutput()* returns this last output generated by the FSM to its caller. It is used by the test case function to retrieve the last

FSM output to compare it with the expected output (also referred to as the "golden output"). The function *println()* is simple enough — it just stores the output of the FSM and prints it by calling the standard function *System.out.println()*.

Although we do not need it in this example, we can generally use an analogous approach for capturing the FSM inputs also. Instead of calling the standard function *System.in.read()* directly, we can construct and call the function member *read()* of the class *MyIO*. This function would in its own turn read the input by calling the standard input functions and store that input into the corresponding field member of the class *MyIO* (e.g., *lastInput*). The last FSM input would be available through the function member *getLastInput()*.

After providing test harness support, we continue with the definition of the tester class, which is named *Automata4Tester* in this example. The source code of this class is the following:

```
/*
 * Automata4 tester
 *
 */

package automata4;
import junit.framework.*;

public class Automata4Tester extends TestCase {
  protected Automata4 a4;
  public Automata4Tester(String name) {
    super(name);
  }

  protected void setUp() {
    // setup code
    a4 = new Automata4();
  }

  protected void tearDown() {
    // cleanup code
  }

  // test case 1
  public void test1() {
    a4.processMsg('0');
    assertEquals(MyIO.getLastOutput(),"Output 0");
    a4.processMsg('0');
    assertTrue(MyIO.getLastOutput() == "Output 0");
  }

  // test case 2
  public void test2() {
    for(int i=0;i<100;i++) {
      a4.processMsg('0');
      assertEquals(MyIO.getLastOutput(),"Output 0");
```

Test and Verification

```
    }
  }

  // test case 3
  public void test3() {
    a4.processMsg('0');
    assertEquals(MyIO.getLastOutput(),"Output 0");
    a4.processMsg('1');
    assertEquals(MyIO.getLastOutput(),"Output 1");
    a4.processMsg('0');
    assertEquals(MyIO.getLastOutput(),"Output 1");
    a4.processMsg('1');
    assertEquals(MyIO.getLastOutput(),"Output 2");
    a4.processMsg('0');
    assertEquals(MyIO.getLastOutput(),"Output 2");
    a4.processMsg('1');
    assertEquals(MyIO.getLastOutput(),"Output 0");
  }
  // test case 4
  public void test4() {
    a4.processMsg('1');
    assertEquals(MyIO.getLastOutput(),"Output 1");
    a4.processMsg('1');
    assertEquals(MyIO.getLastOutput(),"Output 2");
    a4.processMsg('1');
    assertEquals(MyIO.getLastOutput(),"Output 0");
  }

  // test case 5
  public void test5() {
    for(int i=0;i<1000;i++) {
      test3();
      test4();
    }
  }

  public static TestSuite suite() {
    return new TestSuite(Automata4Tester.class);
  }

  public static void main(String[] args) {
    junit.textui.TestRunner.run(suite());
  }
}
```

The tester class *Automata4Tester* is derived from the class *TestCase*. Its field member *a4* is an instance of the implementation under test, namely, the class *Automata4*. The constructor of the class *Automata4* simply calls the constructor of its super class (the class *TestCase*) and passes its input parameter (*String name*).

The function *setUp()* creates an instance of the implementation under test by instantiating the class *Automata4* and storing its instance into the field member *a4*. The function *tearDown()* is empty in this example because the Java garbage collector takes care of unused objects. The garbage collector destroys the object that is stored in the field member *a4* at the end of the test case.

The function *test1()* is the first test case defined within the tester class *Automata4Tester*. Basically, it tests the FSM state transition from the state *S0* to the state *S0*, which is driven by the input value *0*. It does the same operation twice. Each time it supplies input *0* to the implementation under test (stored in the field member *a4*) by calling its function *processMsg()* and passing it the parameter, '0'.

Assuming that the implementation under test was in its initial state and that it reacted correctly to the given input, its last output should be the text, "Output 0". The test case function *test1()* checks that assumption by calling the function *assertEquals()*. The first real parameter of that function call is the value of the last output, which is returned by the function member *getLastOutput()* of the class *MyIO*, whereas the second parameter is the expected string, "Output 0".

Second, the test case function *test1()* again supplies input *0* to the implementation under test (stored in the field member *a4*) by calling its function *processMsg()* and passing it the parameter, '0'. Assuming that the implementation under test has reacted properly in the first place, it would be in the initial state at the time the second call to the function *processMsg()* happens. Driven with the input '0', it should produce again the output string, "Output 0". The test case function *test1()* checks this assumption again, only this time it does so by calling the function *assert()*. The real parameter of this function call is the condition *MyIO.getLastOutput() == "Output 0"*.

The function *test2()* is the second test case defined within the tester class *Automata4Tester*. This test case is slightly more complex than the previous one. The previous test case checks if the implementation under test reacts correctly when it is driven twice with the same input value '0' in the same current state (*S0*). We did this on purpose — first, to demonstrate the usage of both *assert()* and *assertEquals()*, and second, the implementation under test may not always react correctly if it is driven with a certain input value in the given state, at least not in theory.

This practice may seem paranoid but, in reality, various types of time- and FSM evolution-dependent bugs are hidden at the beginning and become evident only later during the FSM evolution. Returning to the problem at hand, we ask ourselves: Will this FSM react correctly many times, for example, 100 times? With JUnit at our disposal, we can easily construct a test case that resolves such dilemmas.

This is exactly what the test case function *test2()* does. It does so by executing the body of the *for* loop 100 times. Inside the body of the loop, it drives the implementation under test with input value '0' by calling its function *processMsg()*. After each of these calls, it checks if the last output was the string "Output 0" by calling the function *assertEquals()*.

The function *test3()* is the third test case defined within the tester class *Automata4Tester*. This is a typical FSM-related test case, characterized with the complete coverage of the FSM state transition graph. The flow of the state transitions checked by this test case is the following:

- From S0 to S0, driven with the input 0 (expected output 0)
- From S0 to S1, driven with the input 1 (expected output 1)
- From S1 to S1, driven with the input 0 (expected output 1)
- From S1 to S2, driven with the input 1 (expected output 2)
- From S2 to S2, driven with the input 0 (expected output 2)
- From S2 to S0, driven with the input 1 (expected output 0)

The function *test4()* is the fourth test case defined within the tester class *Automata4Tester*. This is another typical FSM-related test case, characterized by its progressive nature. The counter is always driven with the input "1" so that its content is incremented every time. This test case does not provide the full state transition graph coverage, but it is valid and we can think of many partial graph coverage test cases. The flow of the state transitions checked by this test case is the following:

- From S0 to S1, driven with the input 1 (expected output 1)
- From S1 to S2, driven with the input 1 (expected output 2)
- From S2 to S0, driven with the input 1 (expected output 0)

The function *test5()* is the fifth, and the last, test case defined within the tester class *Automata4Tester*. It is a fairly simple, yet rather intensive, test case that is based on the combination of the previous two test cases. The test case function *test5()* repeats the body of the *for* loop 1,000 times. Inside the body of the loop, it just calls the functions *test3()* and *test4()* in succession.

The function *suite()* returns the test suite, which it creates by calling the constructor of the class *TestSuite*. The real parameter of this function call is the name of the implementation under test class file (*Automata4Tester.class*). The constructor of the class *TestSuite* finds all the functions whose names start with the word "test" defined within the class *Automata4Tester* and automatically adds them to the test suite it creates.

The function *main()* runs the test suite defined by the previous function *suite()*. It does that by calling the function *run()* of the class *TestRunner*, which is an integral part of the JUnit testing framework. The real parameter of this function call is the test suite that is created by the function *suite()*. This test suite contains all test cases defined within the class *Automata4Tester*.

In the case of more complex implementations, we may decide to create more tester classes rather than define all test cases within a single tester class, such as the class *Automata4Tester*. In such a situation, we would need to create a hierarchy of test suites and the overall tester class that would automatically run all test cases in all test suites. The source code of such a tester class is the following:

```
/*
 * Tester
 *
 */

package automata4;
import junit.framework.*;

/*
 * TestSuite that runs all test suites
 *
 */

public class AllTests {
  public static void main (String[] args) {
    junit.textui.TestRunner.run(suite());
  }
  public static TestSuite suite() {
    TestSuite suite = new TestSuite("All Tests");
    suite.addTest(Automata4Tester.suite());
    // add other test suites here
    return suite;
  }
}
```

The class *AllTests* comprises two function members, namely, the functions *suite()* and *main()*. The former function creates and returns the test suite that is in the root of the test suite hierarchy. This means that it contains all other hierarchically subordinated test suites. The latter function executes the root test suite, i.e., it executes all test suites that were added to it.

The function *suite()* creates the root test suite simply by calling the constructor of the class *TestSuite*. The real parameter of this function call is the name of that test suite (the string "All Tests"). It then adds the test suite that contains the test cases defined within the tester class *Automata4Tester* to the root test suite. It does this by calling the function member *addTests()* of the root test suite object *suite*. Generally, in the case when we have multiple tester classes, we would repeat the call to the function *addTests()* for each tester class.

The function *main()* runs the test suite defined by the previous function *suite()*. It does this by calling the function member *run()* of the class *TestRunner*. The real parameter of this function call is the test suite created by the function member *suite()* of the class *AllTests*. This test suite contains a single hierarchically subordinated test suite, which in turn contains all test cases defined within the class *Automata4Tester*.

We start the automatic execution of all test cases defined within the class *Automata4Tester* by running the file *Automata4Tester.class*. Similarly, we start the automatic execution of all test cases defined within all tester classes (in this simple example we have just one of them, the class *Automata4Tester*) by running the file *AllTests.class*. In both cases, we should get the same result. Each test case function will print its own outputs to the standard output file.

At the end, the test runner will print out the final report, which should look like this:

```
Time: 1,783
OK (5 tests)
Press any key to continue...
```

The number 1783 corresponds to the number of seconds that were needed to execute all test cases, whereas the number 5 in parenthesis corresponds to the total number of test cases that were executed.

5.2 Conformance Testing

As already mentioned at the beginning of this chapter, conformance testing is the first step of acceptance testing (followed by the load testing and in-field testing). The aim of conformance testing is to check the functional correctness of external behavior of the implementation under test without checking its inner workings. Essentially, conformance testing is functional testing that is based on the "black box" approach.

The main goal of conformance testing is to separately check the correctness of each individual function of the implementation under test (IUT). The sample test case for a simple SIP softphone (IUT) is: "Initiate session setup. Check if IUT sends the message INVITE to the outbound proxy server (imitated by the testing framework). Make the testing framework replay with the message 404 (not found). Check if IUT replays with the message ACK" (see sequence diagram in Figure 5.2). We are intentionally making test cases as simple as possible so we can easily interpret their outcomes. Of course, some of the test cases are inevitably complex and we cannot do anything about this, but we should never make them more complex than they need to be.

More precisely, we do not try to check more functions simultaneously by interleaving the corresponding scenarios. For example, consider the SIP proxy server as the implementation under test. In the case of conformance testing, we are interested only if it can support a single session establishment at a time. Normally, we would not be interested in checking if it can support multiple session establishments simultaneously. Actually, that is exactly the purpose of the load testing.

When it comes to specifying official conformance test suites for real-world protocols (like SIP), this is a really serious business conducted by the international standardization institutions, such as IEEE, ISO, IETF, ITU-T, ETSI, and others. The results are rather voluminous specifications that most frequently use TTCN language (see Section 3.9, Tree and Tabular Combined Notation). The most recent version of TTCN at the time of this writing is the TTCN-3, which enables both tabular and program formats of specifications.

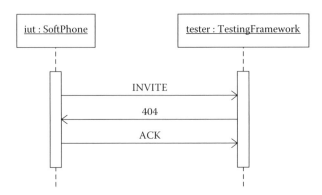

FIGURE 5.2
An example of the conformance testing test case.

For a better understanding of the scope of conformance testing, consider the documents currently available from ETSI (you can download them from the Internet; see www.etsi.org) that are related to conformance testing of SIP (IETF RFC 3261). These documents are the following:

- Conformance test specification for SIP, Part 1: Protocol implementation conformance statement proforma (ETSI TS 102 027-1)
- Conformance test specification for SIP, Part 2: Test suite structure and test purposes (ETSI TS 102 027-2)
- Conformance test specification for SIP, Part 3: Abstract test suite and partial protocol implementation of extra information for testing (ETSI TS 102 027-3)

The first document is the proforma to be completed by the vendor of the implementation to claim implementation capabilities. The guidance for completing the proforma is given is its Section 5. This document is used both during static conformance review and during the test suite parameterization phase of conformance testing.

The second document describes the test suite structure and the purposes of individual test cases. This document was used as the test plan before the test suite was written in the TTCN-3 language. Now it is used as the reference document for understanding the abstract test suite, which is given in the third document.

The third document specifies the abstract test suite to be used for SIP conformance testing. Actually, it is comprised of two files, the archive (ZIP file) that contains SIP test suite in TTCN-3 program format and the SIP test suite overview file (PDF file). The SIP test suite in TTCN-3 program format can be executed using a commercially available TTCN-3 tool.

The SIP conformance test suite specification by ETSI (the three documents listed above) considers four types of implementations under test. The implementations are the following (see IETF RFC 3261 for their definitions):

- User agent that behaves as client or server
- Registrar
- Proxy server (both outbound and simple proxy server)
- Redirect server

The present version of the specification considers the following three types of sessions:

- Sessions that are established using a proxy server
- Sessions that are established directly (without proxy)
- Sessions that are established using the redirect server

The way the SIP conformance test suite is structured is a good example of typical conformance test suite structuring. All test cases are classified into the following four main groups (which correspond to the main SIP functionalities):

- Registration
- Call control
- Querying for capabilities
- Messaging

The test cases in the main groups are further classified according to the role that should be checked. The roles for the main group *registration* are the *registrant* and the *registrar*. The roles for the main group *call control* are *originating endpoint, terminating endpoint, proxy,* and *redirect server*. The roles for the main group *querying for capabilities* are *originating endpoint, terminating endpoint,* and *proxy.* The roles for the main group *messaging* are *registrant, registrar, originating endpoint, terminating endpoint, proxy,* and *redirect server*.

Some of the role subgroups are further divided into functional subgroups. For example, the role subgroup *originating endpoint* of the main group *call control* is divided into three functional subgroups, namely, *call establishment, call release,* and *session modification*. Finally, functional subgroups of test cases can be divided into three test groups: *valid behavior* (V), *invalid behavior* (I), and *inopportune behavior* (O).

Notice that official conformance testing can be conducted only by the authorized organizations (national certification centers, telecom operators, and so on) that use special tools that themselves were certified for such a usage. These tools are professional equipment, most frequently referred to as testers, e.g., a SIP tester. A tester typically comprises the framework that supports test suite administration, execution (most frequently based on interpretation), and associated reporting. Such a framework is referred to as the testing framework.

The testers may be rather sophisticated. Most of them support most of — if not all — the state-of-the-art protocols. Alternately, almost unique testers are also used that support ultramodern protocols that have not become part of the main stream protocols. Both of these types of testers can be rather expensive. Most frequently, competent and efficient operating of protocol testers requires special training.

Because of that, most of the small- and even middle-scale organizations involved in protocol development can not afford purchasing testers and employing full-time employees (confusingly enough, also called testers) for the purpose of conformance testing. Rather, they rent the equipment or the person who can operate it for the purpose of the unofficial and preliminary conformance testing at the client location. The goals of this preliminary conformance testing are to reduce the overall cost and to minimize the risk of failing the official conformance testing.

Some organizations use open source test suites to reduce the cost of the preliminary conformance testing. An example of such a test suite is the SIP Forum Basic UA Test Suite created by Nils Ohlmeier, freely available on the Internet at www.sipfoundry.org/downloads/ (in accordance with the GPL license). This test suite is comprised of the following two parts:

- SIP Forum Testing Framework (SFTF)
- Basic UA tests

SFTF provides regular functions of test suite administration (e.g., adding new test cases, simply referred to as the tests), test suite execution control (executing all tests, selected groups of tests, or individual tests), and test suite execution reporting (both by printouts in the interactive window and in the log files with five possible levels of logging details). The testing framework contains the logic required to execute the test, parse incoming messages, and create replies.

The second part (listed above) is simply a subdirectory that contains all basic user agent tests (i.e., test cases). The tests and SFTF itself are written in Python. The goal of these tests is not to provide the complete conformance testing of SIP implementations, as the ETSI specification does. Rather, the goal is to check the well-known SIP interoperability problems, which frequently occur in immature SIP User Agent (UA) implementations, such as the simple SIP softphone.

Additionally, these tests can discover the implementation under test behavior that conforms to the original SIP specification but is considered a suboptimal implementation solution. Such cases are reported as *warnings* (W). The developer should consider revising the implementation in the case of warnings to make it more robust.

Many tests in this test suite are adopted from the IETF's SIP torture tests Internet draft (available on the Internet under the name *draft-ietf-sipping-torture-tests-02*). The rest of the tests are the contributions from the SIP Forum members. Original IETF SIP torture tests focus on areas that have caused problems in the past or have particularly unfavorable characteristics if handled improperly. Some of them test only the parser and others test both the parser and the application above it. Some use valid and some use invalid SIP messages to check the target functionality.

The SIP Forum tests are classified into the following eight test groups: protocol tortures (26 tests), authentication (4 tests), registration (1 test), dialog and transaction processing (19 tests), DNS (2 tests), NAT capabilities (2 tests), services (2 tests), and warnings about obsolete features (5 tests). All tests are defined in one spreadsheet (XLS file). The test attributes (spreadsheet columns) are the following: number, title, tested device, expected behavior, typical failures, notes, call flow, source (the corresponding section in RFC 3261), and comment.

For example, the test number 201 entitled "A Short Tortuous Request" tests the SIP user agent server behavior. The expected behavior is, "Server considers the request valid and generates a proper response". The call flow is illustrated with the sequence diagram shown in Figure 5.3.

5.3 Formal Verification Based on Theorem Proving

This section covers the formal verification of communication protocols based on automated theorem proving. The reader will learn how to use automated theorem proving for formal verification of both communication protocol specification and its implementation. Normally, the communication protocol

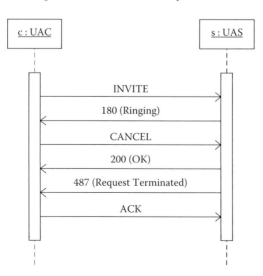

FIGURE 5.3
An example of the SIP protocol torture test.

is modeled as the finite state machine. Basic knowledge of predicate calculus (first-order logic) is assumed for easy and complete understanding of this section.

The outline of this section is the following:

- Axiomatic specification of finite state machines
- Theoretic specification of test cases
- Formal verification of the specification
- Directions for generating test cases
- Formal verification of the implementation
- Software development process based on the formal verification
- A realistic example

The axiomatic specification of the finite state machine is the model of the FSM in the predicate calculus. This model is the set of well-formulated formulas. The first well-formulated formula in the model is optional and it defines the initial state of the FSM. Its general format is the following:

```
State(INITIAL).
```

State is a predicate and *INITIAL* is the name (label) of the FSM initial state. The names *State* and *INITIAL* are noninterpretative user-defined names (like names of the user-defined functions and constants in the higher-level programming languages). For brevity, in this section we use the name S instead of *State* and we label finite state machine states with numbers (0, 1, 2...) rather than with symbolic names.

The fact that this first well-formulated formula is optional requires a short comment. In most of the formal FSM descriptions, such as UML activity diagrams and statecharts, the specification of the FSM initial state is mandatory. Here, it is not. If we always want to examine the FSM evolution beginning from the same state, we will define it as the FSM initial state in the FSM axiomatic specification. Alternately, sometimes it is possible and preferable to examine the FSM evolution beginning from different FSM states. In that case, we do not define the FSM initial state in the FSM axiomatic specification and we define it on the left-hand side of the concluding well-formulated formula instead.

The rest of the well-formulated formulas in the FSM axiomatic specification are obligatory. Each of the mandatory well-formulated formulas models a single FSM state transition (also referred to as a FSM branch). The format of the well-formulated formula that models time invariant FSM state transition from the state X to the state Y triggered with the input T and generating the output R is the following:

$$\{State(X) \& Input(T)\} => \{State(Y) \& Output(R)\}$$

State, *Input*, and *Output* are predicates. *X, Y, T,* and *R* are constants that label the source FSM state, the destination FSM state, the particular FSM input, and the particular FSM output, respectively. Most frequently, we use abbreviated names *I* and *S* instead of *Input* and *Output*, respectively. In the case that the state transition generates more, say *N*, output signals (messages), the corresponding well-formulated formula has the following format:

$$\{State(X)\&Input(T)\} => \{State(Y)\&Output(R\_1)\& \\ Output(R\_2)\&…\&Output(R\_N)\}.$$

where *R_1, R_2…R_N* are the labels of particular output signals.

Next, we introduce the concept of control predicates. As their name suggests, the control predicates are used to control the FSM activity. A global control predicate is used to enable or disable the complete FSM activity. Usually we name it *A(N_I)*, where *A* stands for *Automata* and *N_I* labels the particular FSM.

Besides the global control predicate, state transition control predicates also exist, one for each FSM state transition. A state transition control predicate enables or disables the associated state transition. We typically name it *T(M_I)*, where *T* stands for *Transition* and *M_I* labels the particular FSM state transition. The state transition well-formulated formula that includes control predicates has the following format:

$$\{Automata(I)\&Transition(J)\&State(X)\&Input(T)\} => \{State(Y)\&Output(R)\}$$

I is the label of the particular FSM and *J* is the label of the particular state transition modeled with this formula. If we include both *Automata(I)* and *Transition(J)*, the state transition is enabled. If we skip *Automata(I)*, the FSM (i.e., all its state transitions) are disabled. If we skip *Transition(J)*, this individual state transition is disabled. This concludes the presentation of the axiomatic specification of a single FSM.

A theoretical test case for a single FSM is the theorem about the particular FSM evolution path, which states that for a given series of inputs $(I_1, I_2…I_n)$, FSM performs a series of state transitions $(S_1, S_2…S_n)$, which will produce a series of particular output values $(O_1, O_2…O_n)$. The corresponding well-formulated formula has the following format:

$$\{Automata(N)\&Transition(M)\&Input(I_1)\&…\&Input(I_n)\} => \\ \{Output(O_1)\&…\&Output(O_n)\&State(S_1)\&…\&State(S_n)\}$$

Most frequently, we only want to check that FSM produces the expected series of outputs and that at the end it reaches the expected final state S_n. The corresponding theorem has a very similar but simpler format:

$$\{Automata(N)\&Transition(M)\&Input(I_1)\&…\&Input(I_n)\} =>$$

$$\{Output(O_1)\&…\&Output(O_n)\&State(S_n)\}$$

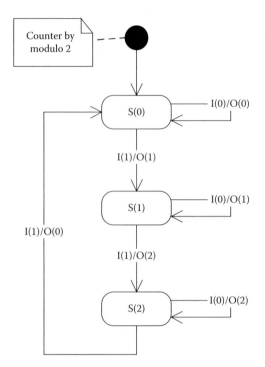

FIGURE 5.4
The counter by modulo two statechart.

Before proceeding to modeling the groups of communicating FSMs, let us look at a simple example. The following shows the axiomatic specification of the counter by modulo 2 (see the statechart diagram in Figure 5.4) and a sample theorem about its expected behavior. The FSM axiomatic specification is the following:

S(0)

{A(0)&T(0)&S(0)&I(0)} => {S(0)&O(0)}

{A(0)&T(1)&S(0)&I(1)} => {S(1)&O(1)}

{A(0)&T(2)&S(1)&I(0)} => {S(1)&O(1)}

{A(0)&T(3)&S(1)&I(1)} => {S(2)&O(2)}

{A(0)&T(4)&S(2)&I(0)} => {S(2)&O(2)}

{A(0)&T(5)&S(2)&I(1)} => {S(0)&O(0)}

The first well-formulated formula defines the state S(0) as the FSM initial state. Next, six well-formulated formulas define six FSM state transitions — from the state S(0) to S(0), from S(0) to S(1), from S(1) to S(1), from S(1) to S(2), from S(2) to S(2), and from S(2) to S(0), respectively. A(0) is the global control predicate. T(0), T(1)...T(5) are the individual state transition control predicates. The sample theorem is the following:

{A(0)&T(0)&I(0)&T(1)&I(1)} => {O(0)&O(1)&S(1)}

It may be interpreted as follows: The FSM is globally enabled by including the general control predicate A(0) on the left-hand side of the concluding well-defined formula. The first FSM state transition is enabled by including the state transition predicate T(0). The FSM is stimulated with the input I(0), which should result in the output O(0). The second FSM state transition is enabled by including the state transition control predicate T(1). The FSM is stimulated with the input I(1) and the FSM should generate O(1) at its output. Finally, the FSM should reach the state S(1).

We can prove this theorem with the automated theorem prover THEO developed by Monty Newborn (Newborn, 2001). To do that, we must write the theorem to a text file, compile it using the program Compile (*cc.exe*), and prove it by running the program THEO (*teo.exe*). The final result looks like this:

```
Predicates: S A T I O
Functions: 0 1 . 2 3 4 5 :
EQ:
ESAF:
ESAP:
  0 <BC: 19 NC: 6 AC: 3 U: 0>
  1 {T0 N1 R1 F0 C9 H0 h0 U11} *
.Proof Found!
```

Of course, realistic finite state machines never operate in isolation. Rather, they normally operate in groups of cooperating finite state machines. For example, according to ITU-T the system consists of functional blocks interconnected with communication channels (see Section 3.7, SDL). Each functional block comprises finite state machines (processes) interconnected with signaling paths (routes). A communication channel may comprise one or more signaling paths. Finite state machines communicate by exchanging signals (events, messages) over signaling paths.

We can use such a kind of traditional system decomposition to our convenience, but it is not required. As the opposite extreme, we can have a chaotic system in which each FSM talks to all other FSMs (like stations in wireless networks). We can even connect more FSMs in signaling networks with all kinds of topologies, such as start, bus, or a network that connects an arbitrary number of FSMs. The means to model all these abstractions in the first-order logic are predicates and their compositions.

To start, we can introduce the notation *Signal(SIG_N)* that represents the act of signaling the particular signal, where *Signal* is a predicate and *SIG_N* is the label of a particular signal. We then can introduce the notation *SignalOverPath(SIG_N,PATH_M)* that represents the act of signaling the particular signal over the particular signaling path, and so on. The well-formulated formulas that model state transitions do not change much. For example, the state transition from the state *X* to the state *Y* is triggered with the signal *P* and generates the signal *Q*, and looks like this:

{*State(X)&Signal(P)*} => {*State(Y)&Signal(Q)*}

In the formula above, *Signal(P)* is received and *Signal(Q)* is sent out of any signaling path, channel, or network. In the case where the former signal is transferred over path *M* and the latter signal is sent over the path *N*, the formula would look like this:

{*State(X)&SignalOverPath(P,M)*} => {*State(Y)&SignalOverPath(Q,N)*}

After introducing the concept of signaling between finite state machines in a group of cooperating FSMs, we can proceed to the axiomatic specification of the group of FSMs. As shown above, each FSM in a group is specified with a set of well-formulated formulas (one optional for the initial state and one mandatory for each individual state transition). Consequently, the specification of a group of FSMs is the union of sets of well-formulated formulas for individual FSMs that constitute that group.

The theoretical test case for the group of FSMs is just a generalization of the theoretical test case for the individual FSM. The left-hand side of the corresponding well-formulated formula consists of control predicates, if any, and staring signals whereas the right-hand side of the formula lists the resulting signals and final states of individual FSMs. The format of the typical theorem about the evolution of the group of FSMs is the following (assume the system with two FSMs):

{*Signal(A)*} => {*Signal(B)&Signal(C)&Signal(D)&State(X)&State(Y)*}

In the sample theorem above, *Signal(A)* triggers the evolution of the system. As the result of the evolution, the system generates three signals: *Signal(B)*, *Signal(C)*, and *Signal(D)*. At the end of the evolution, the FSMs reach their final states, namely, *State(X)* and *State(Y)*.

We now illustrate the concepts introduced above by the means of a simple example. Consider a simple system with three FSMs (see their statechart diagrams in Figure 5.5). The first FSM waits for the signal *E(0)* in its state *S(0)*. After receiving that signal, it sends the signal *E(10)* and goes to the state *S(1)*, where it waits for the signal *E(1)*. Once it receives the signal *E(1)*, it sends the signal *E(20)* and goes to the state *S(2)*. The second and the third FSMs are very much alike. The former waits for the signal *E(10)* and after

Test and Verification

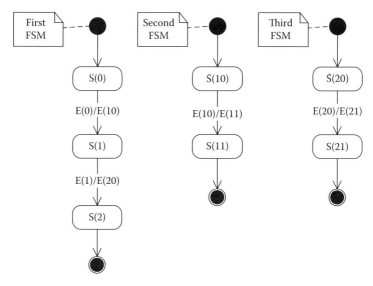

FIGURE 5.5
The statecharts of three communicating FSMs.

receiving that signal, it sends the signal *E(11)*. The latter waits for *E(20)* and sends *E(21)*.

Next, we construct the theorem about the expected behavior of this simple system. This theorem says that if we supply signals *E(0)* and *E(1)* to this system, the first FSM will start evolving and will generate the signals *E(10)* and *E(20)*. These two signals will trigger the second and the third FSMs, which will in their turn generate signals *E(11)* and *E(21)*, respectively. Finally, these FSMs will reach final states *S(2)*, *S(11)*, and *S(21)*, respectively.

The axiomatic specification of this simple system and the theorem explained above are specified in the following sequence of well-formulated formulas:

```
; Simple system with 3 FSMs
; Axiomatic spec. of the first FSM
S(0).
{S(0)&E(0)} => {S(1)&E(10)}.
{S(1)&E(1)} => {S(2)&E(20)}.

; Axiomatic spec. of the second FSM
S(10).
{S(10)&E(10)} => {S(11)&E(11)}.

; Axiomatic spec. of the third FSM
S(20).
{S(20)&E(20)} => {S(21)&E(21)}.

; Theorem
conclusion
{E(0)&E(1)} => {S(2)&S(11)&S(21)&E(10)&E(20)&E(11)&E(21)}.
```

To automatically prove this theorem, we run Compile and THEO once again. The final result looks like this:

```
Predicates: S E
Functions: 0 1 10 2 20 11 21 : .
EQ:
ESAF:
ESAP:
  0 <BC: 14 NC: 3 AC: 3 U: 0>
  1 {T0 N1 R1 F0 C1 H0 h0 U14} *
.Proof Found!
```

Next, we introduce the concept of a theoretical log file. As already mentioned, a theoretical test case is a theorem about an FSM's expected behavior. It defines starting (input) signals on its left-hand side and a series of expected output signals and traversed FSM states (including the final ones that we are most interested in) on its right-hand side. We refer to the right-hand side of the theoretical test case as the theoretical log file.

A strong similarity exists between the theoretical and the real log files. The real log file is the result of the system execution in real time. It represents a particular path of the system evolution. The theoretical log file is the result of the virtual (speculative) system execution. It shows the expected outcomes, such as generated signals and traversed states (including the final states).

However, one principal difference between the two of them is that the logs in the real log file usually have a time stamp. The value of the time stamp is usually unique (with the exception of the logs in multiprocessor systems). Alternately, logs in the theoretical log files are individual predicates that correspond to signals and states, and they do not have any time stamp at all.

Furthermore, we can write logs in the theoretical log file in any order, because the operator "&" is a commutative one. The easiest way to think about it is that the theoretical test case is true forever. Hence, it really does not matter in which order we name the logs. Another way to think about it is that it all have happened at the same moment of time. Therefore, all logs have the same "time stamp," which may be omitted because it does not provide any meaningful information, and then again the order of logs does not matter.

Actually, when we look at the FSM axiomatic specification and to the theoretical test case more closely, we notice that no explicit notion of time exists at all. The only notion of time present there is the implicit one, and it is made through the control predicates. Although the absence of the explicit notion of time may seem confusing and disadvantageous, it is the main source of power of proving theorems.

To understand why, imagine that we made a system that reacts in certain ways when it receives two different messages but we are not sure what will happen if these two signals arrive at exactly the same time. If the probability of this event is very low, it can take a long period of time before the event happens and we face a system failure. With the theorem-proving approach, we check such situations immediately. Imagine the enormous amounts of test time that are saved this way.

Another powerful characteristic of this approach is that each theoretical test case actually represents a family of test cases. For example, let us return to the counter by modulo 2. Consider the theorem:

$$\{A(0)\&T(0)\&I(0)\} => \{O(0)\&S(0)\}$$

Because in first-order logic, $I(0) <=> I(0)\&I(0)$, we can rewrite the theorem as follows:

$$\{A(0)\&T(0)\&I(0)\&I(0)\&I(0)\&I(0)\&I(0)\} => \{O(0)\&S(0)\}$$

We may interpret this theorem as follows: If we apply the same signal $I(0)$ many times (even up to infinity), we will always get the signal $O(0)$ at the FSM output and it will remain in the state $S(0)$. Therefore, by proving individual theoretical test cases, most frequently we are actually checking the families of test cases. This concludes the presentation of the axiomatic specification and theoretical test cases related to FSMs.

Now let us see how we can use this in communication protocol engineering. We start with the formal verification of the specification. The concept is rather simple, although it can prove to be difficult to realize in practice. Ideally, two independent teams must be present (or at least a person who is "changing hats"), namely, the design and testing teams. The former writes the axiomatic specification of the family of communication protocols that is modeled as a group of FSMs. The latter writes and proves the theoretical test cases.

If a theoretical test case fails (the proof of the theorem cannot be found), at least one error is generated in either axiomatic specification or in the theorem. It may be the case that two or even more errors occur in both of them. Most frequently, the errors are trivial oversights made by theorem writers because they are not so familiar with the system at hand. If not, the errors are typically caused by rather nontrivial oversights in the system design.

Finding these errors is not a trivial task at all. Typically, we would try to shorten the theorem or the axiomatic specification and see what happens. Of course, with an automated theorem prover such as THEO at our disposal, this is much easier than doing it by hand. Control predicates may help, also — with them, we can sequence the events to our convenience. The need for them is typically a clue that we have synchronization problems.

We can also use an automated theorem prover for automatic test case generation. To do that, we assume that axiomatic specification of the system is errorless. We start by selecting one of the possible input signals on the left-hand side of the theorem. We then check various output signals at the right-hand side of the theorem by trying to prove the theorem. If the proof is found, our assumption was correct and we keep that signal at the right-hand side. If not, we continue by checking other signals.

Of course, some input signals can just cause internal state transitions and no signals at the output of the system. The right-hand side will remain empty in that case. By continuing this process, we can generate theoretical test cases of arbitrary length:

$$\{I(A)\&I(B)\&I(C)\} => \{O(X)\&O(Y)\&O(Z)\}.$$

Similarly, we can make guesses about transient or final states of the system, for example:

$$\{I(A)\&I(B)\&I(C)\} => \{O(X)\&O(Y)\&O(Z)\&S(P)\&S(Q)\}$$

The real benefit of such automatically generated test cases is that they can be translated into executable test cases and used for automatic testing of the system implementation. Generating test cases in the previously described fashion is not very efficient, and neither it is well coordinated. We can generate test cases more cleverly by respecting the structure of the FSM axiomatic specification rather then viewing it as a black box. Actually, the FSM axiomatic specification introduced in this section is yet another means of modeling the FSM state transition graph.

Generating test cases by traversing the FSM state transition graph is possible with the goal to achieve its complete coverage. Three possible types of FSM state transition coverage exist, namely, node, branch (arc), and path coverage. That the path coverage cannot be achieved if the graph is cyclic is well known. Alternately, branch coverage subsumes node coverage and, because of that, seems to be the best selection.

Sometimes we may have the opposite problem. The test suite (a set of test cases) may already be available, such as the SIP conformance test suite available from ETSI in TTCN-3 language (see Section 5.3). In such a situation, we can use a tool to translate TTCN-3 test cases into theorems, and then we can use the automated theorem prover to formally verify conformance of the system axiomatic specification with the standard.

Yet another application of the automated theorem prover is the formal verification of the system implementation. To do this, we assume that a conformance test suite is already available and use the reverse engineering tool to extract the axiomatic specification of the system from the implementation source code and, optionally, from log files if some are available. The reverse engineering tool normally relies on conventions that govern the structure of the source code and log files.

For example, the reverse engineering tool for the FSM Library-based implementations relies on the specification of the FSM Library API (see Section 6.8). This tool simply searches the source code for the specific library functions and their real parameters to retrieve the well-formulated formulas that constitute system axiomatic specification. More precisely, the tool extracts the elements of the left-hand side of the state transition well-formulated formula by searching for library functions *InitEventProc()* and *InitUnexpectedEventProc()*.

The real parameters of the function *InitEventProc()* are the source state, the triggering signal (event, message), and the state transition function. The first two parameters (state and signal) are exactly the elements of the left-hand side of the corresponding well-formulated formula. The real parameters of the function *InitUnexpectedEventProc()* are the source state and the state transition function. The state is the first element of the left-hand side of the well-formulated formula. The second element is any signal that is not valid for the given state.

The reverse engineering tool proceeds by examining an individual state transition function. It creates one well-formulated formula (they all have the same left-hand side) for each state transition function execution path. For example, a state transition function with a simple sequence of statements yields a single formula, whereas a state transition function that has a switch with three cases yields three formulas.

The right-hand side of the state transition well-formulated formula is constructed by the analysis of the state transition function. The tool first searches for the functions *PrepareNewMessage()* and *SendMessage()* to extract symbolic names of the signals that are generated by that execution path of the state transition function. It then searches for the function *SetState()*, whose real parameter is the name of the destination state. If this function is not found, the tool assumes that the FSM state should not be changed and copies the state name from the left-hand side to the right-hand side of the formula.

This procedure is repeated for all state transition functions. Finally, the tool provides the complete axiomatic specification of the system in ASCII format, which is readable by the automated theorem prover. We then use already available test cases to formally verify the system implementation source code.

Although most frequently we assume that the tools and other components we use are bug-free (in this particular case, these tools are the reverse engineering tool, compiler, linker, loader, and operating system), sometimes they are not. No matter how low the probability of such a failure is, it can happen and when it does, it compromises the formal verification of the source code. In such a case, we can use the reverse engineering tool that extracts the axiomatic system specification from log files. The example of the particular log file that was created by the FSM Library-based implementation is given in Section 5.5.1. Principally, the axiomatic specification that is provided from the log file is usually incomplete (except when it contains traces of all possible system execution paths), but even as such, it is sufficient to locate and eliminate the problem at hand.

When it comes to the application of formal verification methods, software development processes can be classified into three different categories. The Cleanroom engineering is a typical representative of the first category. It uses formal verification methods to formally verify the system design. The second category uses formal methods to formally verify the system implementation whereas the third uses it to formally verify both the system design and implementation.

We will end this section with a more realistic example — the axiomatic specification of the FSM that implements both ITU-T Q.71 FE1 and FE5 call control functional entities (see Figure 3.38, Section 3.7.1) and a sample theoretical test case. The former functional entity models the functionality of the calling party (also referred to as the subscriber A) whereas the latter models the functionality of the called party (also referred to as the subscriber B). The following is the axiomatic specification of the FSM, named FE1FE5 (ITU-T Q.71 FE1 and FE5 merged together):

```
;
;   FE1FE5 definition
;
;   Initial state definition:
S(FE1FE5_ON_HOOK).

{S(FE1FE5_ON_HOOK)&E(r3_DisconnectReqInd)} =>
{S(FE1FE5_ON_HOOK)&E(r3_DisconnectRespConf)}.

{S(FE1FE5_ON_HOOK)&E(r3_SetupReqInd)} =>
{S(FE1FE5_WAIT_OFF_HOOK)&E(r3_ReportReqInd)}.

{S(FE1FE5_ACTIV)&E(r3_SetupReqInd)} =>
{S(FE1FE5_ACTIV)&E(r3_DisconnectReqInd)}.

{S(FE1FE5_ACTIV)&E(r3_DisconnectReqInd)} =>
{S(FE1FE5_WAIT_ON_HOOK)&E(r3_DisconnectRespConf)}.

{S(FE1FE5_ACTIV)&E(User_ON_HOOK)} =>
{S(FE1FE5_ON_HOOK?)&E(r3_DisconnectReqInd)}.

{S(FE1FE5_WAIT_ON_HOOK)&E(User_ON_HOOK)} =>
{S(FE1FE5_ON_HOOK)}.

{S(FE1FE5_WAIT_ON_HOOK)&E(r3_DisconnectReqInd)} =>
{S(FE1FE5_WAIT_ON_HOOK)&E(r3_DisconnectRespConf)}.

{S(FE1FE5_WAIT_ON_HOOK)&E(r3_SetupReqInd)} =>
{S(FE1FE5_WAIT_ON_HOOK)&E(r3_DisconnectReqInd)}.

{S(FE1FE5_WAIT_OFF_HOOK)&E(User_OFF_HOOK)} =>
{S(FE1FE5_ACTIV)&E(r3_SetupRespConf)}.

{S(FE1FE5_WAIT_OFF_HOOK)&E(r3_DisconnectReqInd)} =>
{S(FE1FE5_ON_HOOK)&E(r3_DisconnectRespConf)}.

{S(FE1FE5_WAIT_OFF_HOOK)&E(r3_SetupReqInd)} =>
{S(FE1FE5_WAIT_OFF_HOOK)&E(r3_DisconnectReqInd)}.

conclusion
;   {S(FE1FE5_ON_HOOK)&E(User_OFF_HOOK)} =>
```

Test and Verification

```
; {S(FE1FE5_UNKNOWN_FE2)&E(r1_SetupReqInd)}.

; {S(FE1FE5_UNKNOWN_FE2)&E(User_ON_HOOK)} =>
; {S(FE1FE5_DISCONNECTING_FE2)}.

{S(FE1FE5_ON_HOOK)&E(User_OFF_HOOK)&E(User_ON_HOOK)} =>
{S(FE1FE5_DISCONNECTING_FE2)&E(r1_SetupReqInd)}.
```

Actually, this file contains three theorems (starting after the keyword *conclusion*). The first two are commented out (the semicolon character ";" at the beginning of the line means that the line is a comment) leaving only the third open as a subject to prove by the automated theorem prover. The first commented theorem claims that if the FSM FE1FE5 is stimulated with the input signal *User_OFF_HOOK* in its initial state *FE1FE5_ON_HOOK*, it will generate the output signal *r1_SetupReqInd* and move to the state *FE1FE5_UNKNOWN_FE2*. The second commented theorem claims that if the FSM FE1FE5 is further stimulated with the signal *User_ON_HOOK* in the state *FE1FE5_UNKNOWN_FE2*, it will just move to the state *FE1FE5_DISCONNECTING_FE2*.

Finally, the third theorem — which is actually the subject of automated theorem proving — is a simple composition of the previous two theorems. It states that if the FSM FE1FE5 is stimulated by the sequence of the input signals *User_OFF_HOOK* and *User_ON_HOOK* in its initial state *FE1FE5_ON_HOOK*, it will generate the output signal *r1_SetupReqInd* and finish in the state *FE1FE5_DISCONNECTING_FE2*. To automatically prove this theorem, we run Compile and THEO once again. The final result looks like this:

```
Predicates: S E
Functions: FE1FE5_ON_HOOK User_OFF_HOOK r1_SetupReqInd User_ON_HOOK
FE1FE5_DISCONNECTING_FE2 . r1_DisconnectRespConf FE1FE5_UNKNOWN_FE2
r1_DisconnectReqInd User_DIGIT r1_ProceedingReqInd
FE1FE5_WAIT_FOR_DIGITS r1_ADDL_AddrReqInd r3_DisconnectReqInd
FE1FE5_WAIT_ON_HOOK r1_SetupRespConf FE1FE5_ACTIV r1_ReportReqInd
r3_DisconnectRespConf r3_SetupReqInd FE1FE5_WAIT_OFF_HOOK
r3_ReportReqInd FE1FE5_ON_HOOK? r3_SetupRespConf :
EQ:
ESAF:
ESAP:
0 <BC: 56 NC: 4 AC: 4 U: 0>
1 {T1 N1 R1 F0 C49 H1 h0 U8} *
.Proof Found!
```

5.4 Statistical Usage Testing

Statistical usage testing, also referred to as *statistical testing* or *behavioral testing*, is today the main industry standard for quality assessment of embedded systems. As its name suggests, the goal of statistical usage testing is to test the product under conditions that it is expected to face in its real exploitation. The description of these conditions is given with the set of product's operational profiles. Two key ideas are behind the concept of statistical usage testing.

The first addresses the focus of testing whereas the second addresses the quality of the final product. We start with the genesis of the first of these two ideas. That any nontrivial product requires a vast amount of test cases for its verification should be obvious by now. The order of this amount can very easily go up to hundreds of thousands of test cases or even more. Because some of the product working modes (also referred as states) are more frequently used than others, selecting the number of associated test cases accordingly makes sense, especially if we want to limit the size of the test suite.

This reasoning led to the concept of the operational profile. Remember that the motivation for its introduction was to respect the usage frequencies of individual operational states. Actually, because product state transitions are triggered by the corresponding events (signals, messages), the state usage frequencies are equal to the frequencies of these events. Furthermore, if we want to make our considerations independent of the total number of usages (tests), introducing the probabilities of events is convenient. (In this context, we define the probability as the number of real occurrences of the event divided by the total number of its possible occurrences.)

Mathematically, the operational profile is a Markov process. It can be modeled as a special kind of graph whose vertices are product states and whose arcs are state transitions triggered with the corresponding events of the given probability. The operational profile is essentially a finite state machine with given probabilities of its state transitions. Of course, the sum of probabilities of all outgoing state transitions for a single state must be equal to 1 (100%).

The second idea behind the concept of statistical usage testing is to use the product reliability as the main measure of its quality. The genesis of this idea is that traditional software engineering measures of product quality are the *number of remaining bugs* and the *test coverage* of the implementation under test that was achieved through its testing. However, achieving good results with respect to these two measures is not sufficient for assuring the high quality of the product.

For example, consider the following paradox. Imagine a software product that has a single bug that causes a system crash every time the software is started. Although the product has the excellent value of the metric *number of remaining bugs* (only 1 bug remaining), it is completely unreliable and therefore practically unusable. In real life, we are not interested in how good the product is with respect to number of remaining bugs and test coverage. Rather, we are primarily interested in its reliability.

Of course, we cannot measure the product reliability directly, but we can estimate this from the number of test cases that it has successfully passed. More precisely, in real engineering practice we have the opposite problem.

We want to calculate the number of test cases needed for the desired product reliability and for the given level of risk we are ready to accept. We can do this by solving the following equation:

$$B = R^N$$

where
  B is an upper bound on the probability that the model assertions are erroneous
  R is a lower bound on the estimate of product reliability
  N is the number of random test cases that the product must successfully pass

For example, achieving even moderate reliability of $R = 0.999$ with $B = 0.007$ would require the successful pass of $N = 5{,}000$ random test cases. Similarly, achieving $R = 0.9999$ with $B = 0.007$ requires $N = 50{,}000$ random test cases, and achieving $R = 0.99999$ with $B = 0.007$ requires $N = 500{,}000$ random test cases. Alternatively, we can run a smaller number of test cases on more product samples in parallel. For example, instead of running $N = 500{,}000$ random test cases on a single sample, we can run $N = 50{,}000$ random test cases on 10 product samples simultaneously.

By considering these examples, we can deduce two conclusions. The first is that conducting statistical usage testing of the final product may require a significant amount of time. The order of magnitude of this amount is calendar weeks or even months, depending on the characteristics of the concrete product. The second conclusion is that we definitely need tools that automatically generate and execute test suites of that size. We simply cannot do this by hand.

An example of the automated working environment for generating statistical test suites is described by Popovic (Popovic and Velikic, 2005). This working environment consists of two parts, namely, the front-end and the back-end (Figure 5.6). The front-end is the Generic Modeling Environment (GME) developed at the Institute for Software Integrated Systems at Vanderbilt University. GME is a configurable toolkit for creating domain-specific modeling and program synthesis environments.

Generally, we configure GME by creating metamodels that specify the *modeling language*, and therefore the *modeling paradigm*, of the application domain. Once we create a metamodel, we must interpret and register it by GME to create a new working environment for making domain-specific models. We normally use such working environments for building domain-specific models and for storing them in a model database. The domain-specific models are essentially graphs, and we render them by dragging and dropping the graphical symbols on the working sheet that is maintained by the GME graphical user interface (GUI). The symbols in GME have their attributes, preferences, and properties.

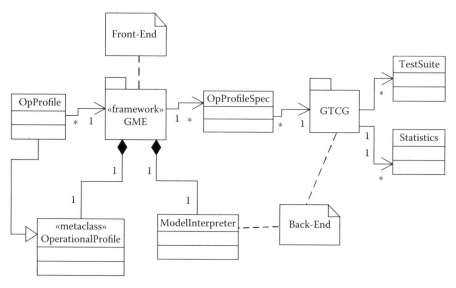

FIGURE 5.6
The working environment for generating statistical test suites.

The particular metamodel that specifies the language (and the paradigm) for modeling operational profiles is represented with the metaclass *OperationalProfile* in Figure 5.6. Each concrete operational profile model (represented with the class *OpProfile* in Figure 5.6) is created by using the operational profile modeling paradigm (the class *OpProfile* is derived from the class *OperationalProfile*). Creating operational profile models by using this paradigm is quite easy.

The modeling language for rendering operational profile models has a single symbol, *State*. This symbol has a single attribute, which is the name of the state. Normally, we just drag and drop the state symbol icon to the working sheet, click on the name field, and type in its name. Each of the state symbols we place on the working sheet represents a single working state (mode) of the product that we want to test.

Rendering state transitions requires a little more work. To render a state transition, we select a connecting tool (symbolized by the operator "+"), click on the source state, and click on the destination state. When the state transition is in place, we enter the particular data for its attributes. A state transition has the following three attributes:

- *EventClass*: specifies the class of events that trigger the state transition
- *Output*: specifies the expected output of the state transition
- *Probability*: specifies the probability of the state transition (in percent)

The most frequently used format of the attribute *EventClass* definition is the following:

```
E(a,b,c…);->a := A1/A2/…; b := B1/B2/…; c := C1/C2/…
```

The event class definition above consists of two parts. The first one is on the left-hand side of the substring "->" and is referred to as the event class. The event class $E(a,b,c…);$ is a string with an arbitrary number of parameters (substrings), labeled here as *a*, *b*, *c*, and so on. The second part of the definition is on the right-hand side of the substring "->". It provides definitions of possible replacements (which are also strings) for each event class parameter. As indicated above, the parameter *a* may be replaced with the string A_1 or A_2 and so on.

A particular event (also referred to as the *constant event*) is an event class without parameters. We may also think about it as the event class with a single member. Particular events are generated from the event class by substituting each event class parameter with the randomly selected replacement from the list of possible replacements. All replacements have equal selection probabilities. Examples of particular events for the event class definition given above are $E(A_1,B_1,C_1…)$, $E(A_1,B_1,C_2…)$, $E(A_1, B_2, C_1…)$, $E(A_2, B_1, C_1…)$, and so on.

The event class format shown above is feasible as far as the number of the possible values of event class parameters is relatively small. But when the number of the possible values is large, writing them explicitly becomes impractical, if not impossible. For example, consider the integer parameter whose possible values are from the interval [0,10000). Writing all 10,000 of its possible values would be really annoying. To make it easier for the user, the working environment supports the following two intrinsic functions:

- *randInt<i,j>* randomly selects an integer number from the interval [*i,j*)
- *randFloat<x,y>* randomly selects a float number from the interval [*x,y*)

When we place and name all state symbols, interconnect them with state transitions, and enter the data for attributes of all state transitions, the operational profile model is finished and we can store it in a file (or a database). This is exactly the main purpose of the working environment front-end (Figure 5.6). Of course, later we may modify the model by adding or deleting states or state transitions, as well as by changing the data for attributes of state transitions, and store it again. All these manipulations are supported by the GME's GUI.

The working environment back-end consists of two parts. The first is the operational profile model interpreter (represented by the class *ModelInterpreter* in Figure 5.6), which is registered to GME. The second part of the back-end is a separate program written in Java, which is named Generic Test Case

Generator (GTCG). The main task of the model interpreter is to transform the operational profile model to the operational profile specification, a simple text file of the well-defined format (represented with the class *OpProfileSpec* in Figure 5.6.). Alternately, the main task of GTCG is to automatically generate the test suite to be used for statistical usage testing and the corresponding statistical report (represented with the classes *TestSuite* and *Statistics* in Figure 5.6).

The operational profile model interpreter is a Java package that is registered to GME with the program JavaCompRegister. The package comprises the following three classes:

- *OPBONComponent:* the interface between GME and the model interpreter
- *OPState*: the state interpreter
- *OPTransition*: the state transition interpreter

The model interpreter behaves similarly to traditional plug-in components of GUIs. We activate it by a click on the corresponding model interpreter icon. As the result of this activation, GME calls the model interpreter interface function *invokeEx*, which in its turn creates temporary container objects for state names, event classes, state transition probabilities, event class definitions, and next state definitions.

Next, the model interpretation is performed by traversing the multigraph architecture of the model in focus. While visiting individual states and state transitions, GME calls the function *traverseChildren* of the class *OPState* and *OPTransition*, respectively. These two functions effectively interpret the model by reading the data of the attributes and filling the above mentioned container objects. At the end of the interpretation, the content of these container objects is saved into the operational profile specification file named *opspec.txt*.

The automatic test case generator GTCG uses the following input items:

- The operational profile data from the file *opspec.txt*.
- The initial operational profile state. Most frequently, the initial state is fixed, but sometimes it may be selectable.
- The number of test cases to be generated. This item determines the size of the test suite. As mentioned earlier, it depends on the product reliability we want to guarantee.
- The test case length, defined as the number of test steps in a test case. A test step is the particular event that is randomly selected from the given event class.

The operational profile specification file *opspec.txt* consists of the following four parts:

- Part I defines the number of states (M) and the number of event classes (N).
- Part II is a matrix of state transition probabilities. The matrix element P_{ij} defines the probability of the event class number j in the operational profile state number i.
- Part III is a matrix of event class definitions. The matrix element E_{ij} defines the event class number j in the operational profile state number i. Most frequently, E_{ij} is the same in all states ($E_{i1} = E_{i2} = \ldots E_{iM}$).
- Part IV is a matrix of next states. The matrix element T_{ij} defines the next state number (index) for the event class number j in the operational profile state number i.

GTCG provides the following two files at its output:

- *testcases.txt*: contains the test suite to be used for statistical usage testing
- *statistics.txt*: contains the corresponding statistical report, which is the important measure of the generated test suite quality

The file *testcases.txt* contains the series of test cases. Each test case starts with its number followed by the column character ':' (e.g., 0:, 1:, 2:). The next line contains the test bed setup command *TestBox.initialize()*, which essentially initializes the hardware connected to product inputs and outputs for the purpose of automatic testing. The test bed setup command is followed by the series of lines that contain particular events randomly selected from the associated event classes (the number of these lines is determined by the given test case length). The event class itself is selected randomly from the distribution defined by the operational profile data (*opspec.txt*, Part II).

The file *statistics.txt* consists of two parts. The first part contains a series of lines, one per operational profile state. Each of these lines indicates the number of occurrences of the corresponding operational profile state (c_i), the discrepancy between the observed and expected frequency of state occurrence (d_i), and the significance level (SL_i). The significance level is actually the probability that the discrepancies as large as those observed would occur with random variation. The second part of the statistical report shows the mean value of the discrepancy and the mean value of the significance value.

The detailed explanation of the statistical measures mentioned above is outside the scope of this book but can be found elsewhere (e.g., Woit, 1994). Practically, it is enough to remember the following guides:

- A significance level greater than or equal to 20% is considered large. This result means that the test suite is of sufficient quality and we may use it for statistical usage testing.

- A significance level less than or equal to 1% is considered small. This result means that the test suite quality is poor and it should not be used for statistical usage testing.

The statistical usage testing methodology governs the usage of tools that create the working environment. The methodology subsumes the following steps:

- Make the operational profile model of the product (implementation under test).
- Interpret the model.
- Determine the desired level of reliability.
- Calculate the required size of the test suite (the number of test cases).
- Generate the test suite.
- Check the test suite quality. If the quality is not acceptable, return to the previous step.
- Execute the test suite. If all test cases successfully pass, the final verdict is *pass*. In that case, we can claim that the product reliability is at least at the level of the desired reliability. If at least one test case fails, the final verdict is *fail* and the product is considered not usable, at least not at the desired level of reliability.

This methodology can be used for testing both parts of products and complete products. We will illustrate such applications by the following two examples. The implementation under test in the first example is the SIP invite client transaction. We start with modeling its operational profile in accordance with the methodology outlined above (Figure 5.7).

The operational profile shown in Figure 5.7 has five working states, namely, *Initial, Calling, Proceeding, Completed,* and *Terminated*. At the same time, it has nine event classes that are intentionally labeled with names that resemble the original specification (see RFC 3261, Figure 5). The definitions of the event classes (not shown in Figure 5.7) are the following:

- The event class labeled *INVITE* is defined as *INVITE* (this class has a single member).
- The event class labeled *300–699* is defined as `M3->M3:=randInt<300,700>;`
- The event class labeled *TA* is defined as TA (original RFC 3261 label: Timer A fires).
- The event class labeled *1XX* is defined as `M1->M1:=randInt<100,200>;`
- The event class labeled *TB* or *TransportERR* is defined as `E->E:=TB/TransportERR;`

Test and Verification

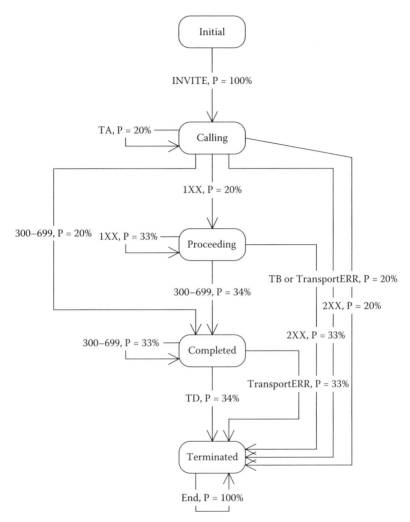

FIGURE 5.7
The SIP INVITE client transaction operational profile.

- The event class labeled *2XX* is defined as `M2->M2:=randInt<200,300>;`
- The event class labeled *TD* is defined as TD (original RFC 3261 label: Timer D fires).
- The event class labeled *TransportERR* is defined as TransportERR (constant event).
- The event class labeled *End* is defined as End (added because the sum of outgoing state transition probabilities for each state must be equal to 100%).

The probabilities of individual state transitions are shown in Figure 5.7. Note that outgoing state transition probabilities add up to 100% for each state (an essential request for a Markov process). Generally, we set the state transition probabilities according to what we expect the product will face in its real exploitation. Of course, we should use statistical data available for some similar product or the previous version of the same product whenever we can.

Next, we start the model interpreter, which transforms the model into the operational profile specification file *opspec.txt*. When writing GME model interpreters, we should make no assumptions about the order in which the model is traversed. For example, to assume that individual states and state transitions are going to be visited in the same order in which they were originally entered would be a mistake because this is not going to happen. The best assumption we can make in this respect is to assume the completely random visiting order.

Based on this assumption, the model interpreter simply assigns identifications to states and state transitions according to the order they are visited. The particular assignment of identifications to operational profile states in this example is the following:

- The state *Terminated* is assigned the identification 0.
- The state *Calling* is assigned the identification 1.
- The state *Proceeding* is assigned the identification 2.
- The state *Completed* is assigned the identification 3.
- The state *Initial* is assigned the identification 4.

The particular assignment of identifications to operational profile event classes is the following:

- The event class *E* is assigned the identification 0.
- The event class *M1* is assigned the identification 1.
- The event class *TA* is assigned the identification 2.
- The event class *INVITE* is assigned the identification 3.
- The event class *TransportERR* is assigned the identification 4.
- The event class *M2* is assigned the identification 5.
- The event class *M3* is assigned the identification 6.
- The event class *TD* is assigned the identification 7.
- The event class *End* is assigned the identification 8.

The content of the file *opspec.txt* is the following:

```
5    9
0.0   0.0    0.0   0.0   0.0   0.0   0.0   0.0   1.0
0.2   0.2    0.2   0.0   0.0   0.2   0.2   0.0   0.0
0.0   0.33   0.0   0.0   0.0   0.33  0.34  0.0   0.0
0.0   0.0    0.0   0.0   0.33  0.0   0.33  0.34  0.0
0.0   0.0    0.0   1.0   0.0   0.0   0.0   0.0   0.0
null null null null null null null null End
E->E:=TB/TransportERR;  M1->M1:=randInt<100,200>; TA null null
  M2->M2:=randInt<200,300>; M3->M3:=randInt<300,700>; null null
null M1->M1:=randInt<100,200>; null null null
  M2->M2:=randInt<200,300>; M3->M3:=randInt<300,700>; null null
null null null null TransportERR null M3->M3:=randInt<300,700>;
  TD null
null null null INVITE null null null null null
0   0   0   0   0   0   0   0   0
0   2   1   0   0   0   3   0   0
0   2   0   0   0   0   3   0   0
0   0   0   0   0   0   3   0   0
0   0   0   1   0   0   0   0   0
```

Note: The specifications of event classes for the states 1, 2, and 3 (*Calling*, *Proceeding*, and *Completed*) were too long to fit into a single line. Therefore, definitions of event classes for each of these states spans across two lines (the second starts at the next level of indentation).

Next, we activate GTCG with the script that specifies the starting state identification 4 (*Initial*), the number of test cases that is equal to 1,000, and the test case length that is equal to 4 (this means 4 steps, i.e., particular events, per test case). Selection of this particular test case length requires a short comment. This value is exactly the length of the shortest path across all five states starting from the state *Initial* (path *Initial-Calling-Proceeding-Completed-Terminated*, with five states and four state transitions). Of course, other paths of length 4 are possible and will be generated.

As already mentioned, the GTCG creates two output files, *testcases.txt* and *statistics.txt*. According to the methodology outlined above, we first check the quality of the generated test suite by inspecting the file *statistics.txt*. Its content is the following:

```
Calculating statistics

i=0   ci=1104           di=0.0                  SLi=1.0
i=1   ci=1237           di=2.470493128536783    SLi=0.0
i=2   ci=291            di=0.7498208280500565   SLi=0.7014229616104999
i=3   ci=368            di=0.1864198248469353   SLi=0.910066579962014
i=4   ci=1000           di=0.0                  SLi=1.0
Mean  d=0.6813467562867549
Mean  SL=0.7222979083145027
```

The average significance level *SL* is equal to 72% (0.72). Because this number is greater than the required 20%, we conclude that the quality of the generated test suite is sufficient and that we can use it for the statistical usage testing.

Next, we look more closely to a couple of test cases from the beginning of the file *testcases.txt* to get a better feeling of the nature of statistical test cases. The relevant comments are interleaved with the test cases:

```
0:
TestBox.initialize();
INVITE
443
TransportERR
End
```

Test case number 0: After the initial *INVITE*, GTCG randomly selects the event class labeled *300–699* and the particular event 443 from that class. This action causes the state transition to the state *Completed* (Figure 5.7). Next, GTCG randomly selects the event *TransportERR*, thus causing the state transition to the state *Terminated*. *End* is the only possible event in that state.

```
1:
TestBox.initialize();
INVITE
TA
586
TD
```

Test case number 1: After the initial *INVITE*, GTCG randomly selects the event class *TA* (Timer A fires). The current state remains the state *Calling* (Figure 5.7). Next, GTCG randomly selects the event *586*, thus causing the state transition to the state *Completed*. Finally, GTCG randomly selects the event *TD* (Timer D fires), which causes the state transition to the state *Terminated*.

```
2:
TestBox.initialize();
INVITE
190
267
End
```

Test case number 2: After the initial *INVITE*, GTCG randomly selects the event class *1XX* and the particular event *190*. This causes the state transition to the state *Proceeding* (Figure 5.7). Next, GTCG randomly selects the event *267*, thus causing the state transition to the state *Terminated*. The next event must be the event *End*.

```
3:
TestBox.initialize();
INVITE
494
TD
End
```

Test case number 3: After the initial *INVITE*, GTCG randomly selects the event class *300–699* and the particular event *494*. This causes the state transition to the state *Completed* (Figure 5.7). Next, GTCG randomly selects the event *TD*, thus causing the state transition to the state *Terminated*. The next event must be the event *End*.

In the short descriptions of the generated test cases given above, we used the construct, "GTCG randomly selects the event class X and the particular event Y," for brevity. One should remember that the selection of the event class is always in accordance with the given operational profile probability distribution whereas the selection of the particular event from the given class is really random.

The previous example shows how we can use statistical usage testing for testing a part of the product. As already mentioned, we can employ statistical usage testing for testing the whole products, too. The next example shows such an application — statistical usage testing of the simple SIP softphone.

The operational profile of the SIP softphone is shown in Figure 5.8. It has 8 states and 13 event classes. The states are *Connecting*, *Terminating*, *Disconnecting*, *Connected*, *Calling*, *Initial*, *Proceeding*, and *Ringing* (listed here in the ascending order of their identification). The event classes are *RELEASE*, *200*, *ACK*, *180*, *ERR*, *END*, *ANSWER*, *100*, *INVITE*, *SETUP*, *BYE*, *TH*, and *TB* (also listed in the ascending order of their identification).

All event classes have just one member and their definition is equal to the label shown in Figure 5.8 with the exception of the event class that is labeled *ERR*, which is defined as follows:

```
M3->M3:=randInt<300,381>/randInt<400,494>/randInt<500,514>/randInt<600,607>;
```

This definition is a good example of how we can specify a random value that may be selected from more disjoint intervals of values. Next, we generate 1,000 test cases with five test steps each. The content of the file *statistics.txt* is the following:

```
Calculating statistics

i=0   ci=360         di=0.625                 SLi=0.4686783191616166
i=1   ci=1564        di=0.0                   SLi=1.0
i=2   ci=244         di=0.0                   SLi=1.0
i=3   ci=546         di=1.6483516483516483    SLi=0.21453651135488572
i=4   ci=496         di=0.5843413978494628    SLi=0.7503695231083775
i=5   ci=1000        di=0.064                 SLi=0.8248262531456066
i=6   ci=286         di=3.0879953379953404    SLi=0.21451818049555796
i=7   ci=504         di=0.4897959183673477    SLi=0.4966702889206116
Mean  d=0.8124355378204748
Mean  SL=0.6211998845233321
```

Because the average significance level is 62% (greater than 20%), we can conclude that the test suite quality is acceptable. A couple of typical test cases are taken from the file *testcases.txt* and shown here without comments (the reader should study them for their own exercise):

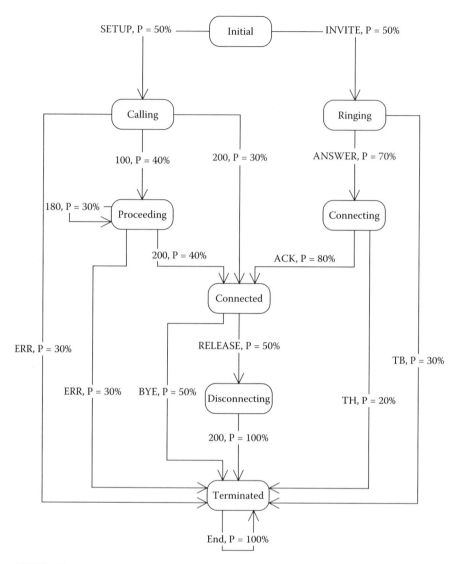

FIGURE 5.8
The SIP softphone operational profile.

```
15:
TestBox.initialize();
SETUP
100
180
200
BYE

16:
TestBox.initialize();
INVITE
```

```
ANSWER
ACK
BYE
END

17:
TestBox.initialize();
SETUP
100
200
BYE
END

18:
TestBox.initialize();
INVITE
ANSWER
ACK
RELEASE
200
```

5.5 Examples

This section includes two examples and two related problems. The first example demonstrates unit testing of the FSM Library-based implementations. The second example illustrates integration testing of the FSM Library-based products.

5.5.1 Example 1

This example demonstrates unit testing of the SIP invite client transaction implementation, which is described in Section 4.5.2 (Example 2). The SIP invite client transaction implementation is based on the requirements and analysis made in Section 2.3.3 (Figure 2.16) and the design presented in Section 3.10.5 (Example 5).

Because the implementation under test (SIP invite client transaction) is implemented in C++, we use CppUnit implementation of the unit testing framework, introduced in Section 5.1. In this simple example, we will construct just one test case to keep it short enough. Also, we will skip some SIP message-specific message handling, which is really not essential for this example.

We start this example by constructing two classes: *ExampleTestCase* and *ExampleMessageFactory*. The former is the tester class, which comprises one sample test case, whereas the latter is the supplementary class, which provides the functions for message management. The content of the class *ExampleTestCase* declaration file, named *ExampleTestCase.h,* is the following:

```
#ifndef CPP_UNIT_EXAMPLETESTCASE_H
#define CPP_UNIT_EXAMPLETESTCASE_H
// CppUnit helper macros
#include <cppunit/extensions/HelperMacros.h>
// Problem specific headers
#include "../kernel/fsmsystem.h"
#include "../kernel/logfile.h"
#include "../NewSIP/InvClientTE.h"
#include "ExampleMessageFactory.h"
/*
 * A sample test case
 *
 */
class ExampleTestCase : public CPPUNIT_NS::TestFixture {
  CPPUNIT_TEST_SUITE(ExampleTestCase);
  CPPUNIT_TEST(example);
  CPPUNIT_TEST_SUITE_END();

protected:
  FSMSystemWithTCP *pSys;
  LogFile *lf;
  InviteClientTE* pInviteCltTE[NUMBER_OF_TES];
  ExampleMessageFactory* pEMF;
  uint8 *msg;
  uint16 msgcode;
public:
  void setUp();
protected:
  void example();
};
#endif
```

The declaration file above includes the CppUnit helper macros header file (*HelperMacros.h*) and the problem-specific header files (*fsmsystem.h*, *logfile.h*, *InvClientTE.h*, and *ExampleMessageFactory.h*). The class *ExampleTestCase* is derived from the class that is defined by the macro instruction *CPPUNIT_NS::TestFixture*. The definition of the test suite starts with the macro instruction *CPPUNIT_TEST_SUITE()* and ends with the macro instruction *CPPUNIT_TEST_SUITE_END()*. The parameter of the former macro instruction is the name of the test suite (*ExampleTestCase*, in this example).

Generally, we use the macro instruction *CPPUNIT_TEST()* to define individual test cases inside the body of the test suite definition. The parameter of this macro instruction is the name of the test case function that is defined within the tester class and that we want to add to the test suite. In this particular example, we add a single test case function, named *example()*, with a single macro instruction, *CPPUNIT_TEST()*, whose real parameter is the string "example".

Next, we define the test case fixture. In this example, it comprises the following:

Test and Verification

- The pointer to the instance of the class *FSMSystemWithTCP* (see Section 6.8.9)
- The pointer to the instance of the class *LogFile* (which is the interface to the log file)
- The array of pointers to the instances of the class *InviteClientTE* (which is actually the implementation under test)
- The pointer to the instance of the class *ExampleMessageFactory* (which is the supplementary tester class)
- The pointer to the message
- The code of the message

At the end of this file we declare the function *setUp()* and the test case function *example()*. The content of the class *ExampleTestCase* definition file, named *ExampleTestCase.cpp*, is the following:

```
#include "ExampleTestCase.h"
#include "../kernel/fsmsystem.h"
#include "../kernel/logfile.h"
#include "../NewSIP/InvClientTE.h"
#include "ExampleMessageFactory.h"

CPPUNIT_TES_SUITE_REGISTRATION(ExampleTestCase);
void ExampleTestCase::setUp() {
  pSys = new FSMSystemWithTCP(11,11);
  pEMF = new ExampleMessageFactory();
  for (int i = 0; i < NUMBER_OF_TES; i++){
    pInviteCltTE[i] = new InviteClientTE();
  }

  uint8 buffClassNo = 4;
  uint32 buffsCount[4] = {50, 50, 50, 50};
  uint32 buffsLength[4] = {1025, 1025, 1025, 1025};
  pSys->InitKernel(buffClassNo, buffsCount, buffsLength, 1);

  lf = new LogFile("log.log", "log.ini");
  LogAutomateNew::SetLogInterface(lf);

  pSys->Add(pInviteCltTE[0], InviteClientTE_FSM, 10, true);
  for (i = 1; i < NUMBER_OF_TES; i++){
    pSys->Add(pInviteCltTE[i], InviteClientTE_FSM);
  }
}

void ExampleTestCase::example() {
  msg = pEMF->MakeInviteToTALMsg();
  pInviteCltTE[0]->Process(msg);
  msgcode = pEMF->GetMsgCodeFromMBX(TLI_Test_FSM_MBX);
  CPPUNIT_ASSERT_EQUAL(msgcode,(uint16)INVITE);

  msg = pEMF->Make1XXToTAL();
  pInviteCltTE[0]->Process(msg);
```

```
    msgcode = pEMF->GetMsgCodeFromMBX(UA_Disp_FSM_MBX);
    CPPUNIT_ASSERT_EQUAL(msgcode,(uint16)RESPONSE_1XX);

    msg = pEMF->Make2XXToTAL();
    pInviteCltTE[0]->Process(msg);
    msgcode = pEMF->GetMsgCodeFromMBX(UA_Disp_FSM_MBX);
    CPPUNIT_ASSERT_EQUAL(msgcode,(uint16)RESPONSE_2XX);
}
```

At the beginning of this file, we register the test suite with the macro instruction *CPPUNIT_TEST_SUITE_REGISTRATION()*. The real parameter of this macro instruction is the name of the test suite. Next, we define the function *setup()* and the test case function *example()*.

The function *setup()* starts by creating an instance of the class *FSMSystemWithTCP*, an instance of the class *ExampleMessageFactory*, and the given number (*NUMBER_OF_TES*) of instances of the implementation under test (the class *InviteClientTE*). After that, it defines the types of buffers to be used by the FSM Library kernel, initializes the kernel by calling the function *InitKernel()* (see Section 6.8.4), creates the log file by calling the function *LogFile()*, and sets the log interface by calling the function *SetLogInterface()* (see Section 6.8.105). At the end, it adds the given number (*NUMBER_OF_TES*) of instances of the implementation under test to the FSM system by calling its function *Add()* (see Section 6.8.2 and Section 6.8.3).

The function *example()* performs the test case by checking state transitions of the implementation under test in the following three steps:

- Check the state transition form the state *STATE_IDLE* (see Section 4.5.2) to the state *STATE_CALLING*, driven by the message *INVITE*
- Check the state transition from the state *STATE_CALLING* to the state *STATE_PROCEEDING*, driven by the message *1XX*
- Check the state transition from the state *STATE_PROCEEDING* to the state *STATE_INITIAL*, driven by the message *2XX*

Each of these three steps consists of the following four substeps:

- Create the message (*INVITE*, *1XX*, or *2XX*).
- Send the message to the implementation under test by calling its function member *Process()* (see Section 6.8.82).
- Get the message code of the output message by calling the function member *GetMsgCodeFromMBX()* of the class *ExampleMessageFactory*. The output message is retrieved from the destination FSM Library mailbox. The destination mailbox is either the mailbox of the transport layer (TPL) or the mailbox of the transaction user (TU).
- Check the retrieved message code against the expected one (message code of the message *INVITE*, *1XX*, or *2XX*) by calling the macro *CPPUNIT_ASSERT_EQUAL()*.

Test and Verification

The particular substeps of the first step are the following:

- Create the message *INVITE* by calling the function member *MakeInviteToTALMsg()* of the class *ExampleMessageFactory*
- Send the message to the implementation under test
- Get the message code of the message that is retrieved from the TPL mailbox
- Check it against the code of the message *INVITE*

The particular substeps of the second step are the following:

- Create the message *1XX* by calling the function member *Make1XXToTAL()* of the class *ExampleMessageFactory*
- Send the message to the implementation under test
- Get the message code of the message that is retrieved from the TU mailbox
- Check the message code against the code of the message *1XX*

The particular substeps of the third step are the following:

- Create the message *2XX* by calling the function member *Make2XXToTAL()* of the class *ExampleMessageFactory*
- Send the message to the implementation under test
- Get the message code of the message that is retrieved from the TU mailbox
- Check the message code against the code of the message *2XX*

Next, we construct the supplementary class *ExampleMessageFactory*. The content of its declaration file, named *ExampleMessageFactory.h*, is the following:

```
#ifndef _ExampleMessageFactory_FSM_
#define _ExampleMessageFactory_FSM_
#include "../constants.h"
#include "../kernel/fsm.h"
#include "../message/message.h"

class ExampleMessageFactory : public FiniteStateMachine {
  int cseq_number;
  Message SIPMsg;
  sip_t *mes;
  stringresponseBody;
public:
  uint8* MakeInviteToTALMsg();
  uint16 GetMsgCodeFromMBX(uint8 mbx);
  uint8* Make1XXToTAL();
```

```
  uint8*  Make2XXToTAL();

  // FiniteStateMachine abstract functions
  StandardMessage StandardMsgCoding;
  MessageInterface *GetMessageInterface(uint32 id);
  void  SetDefaultHeader(uint8 infoCoding);
  void  SetDefaultFSMData();
  void  NoFreeInstances();
  void  Reset();
  uint8  GetMbxId();
  uint8  GetAutomate();
  uint32  GetObject();
  void  ResetData();
public:
  ExampleMessageFactory();
  ~ExampleMessageFactory();
  void Initialize();
};
#endif
```

The content of the class *ExampleMessageFactory* definition file, named *ExampleMessageFactory.cpp*, is the following (the parts that are not essential for this example are omitted to keep the example short):

```
#include "ExampleMessageFactory.h"
#include "../parser/smsgtypes.h"
#include "../parser/smsg.h"
#define SipMessageCoding 0x00
extern char* IPString(unsigned int addr, char* buf, int len);

ExampleMessageFactory::ExampleMessageFactory() : FiniteStateMachine(16, 2, 3) {}

ExampleMessageFactory::~ExampleMessageFactory() {}

void ExampleMessageFactory::Initialize() {}

uint8* ExampleMessageFactory::MakeInviteToTALMsg(){
  char  temp[10];
  char  szHostName[255];
  hostent*  HostData;
  uint8*  recmsg;
  uint8*  msg;
  ...
  PrepareNewMessage(0x00,INVITE);
  SetMsgToAutomate(InviteClientTE_FSM);
  SetMsgToGroup(INVALID_08);
  SetMsgObjectNumberTo(0);
  AddParam(SIP_RAW_MESSAGE, SIPMsg.getLastMessage().length(),
    (uint8*) SIPMsg.getLastMessage().c_str());
  AddParamDWord(SIP_PARSED_MESSAGE, (unsigned long)  mes);
  SendMessage(InviteClientTE_FSM_MBX);
  msg = GetMsg(InviteClientTE_FSM_MBX);
  return msg;
}
```

Test and Verification 297

```
uint16 ExampleMessageFactory::GetMsgCodeFromMBX(uint8 mbx) {
  uint8*  msg;
  uint16  msgCode;
  msg = GetMsg(mbx);
  msgCode = GetUint16((uint8*)(msg+MEC_CODE));
  return msgCode;
}

uint8* ExampleMessageFactory::Make1XXToTAL(){
  uint8* msg;
  ...
  PrepareNewMessage(0x00,RESPONSE_1XX_T);
  SetMsgToAutomate(TAL_Disp_FSM);
  SetMsgToGroup(INVALID_08);
  SetMsgObjectNumberTo(0);
  AddParamDWord(SIP_PARSED_MESSAGE, (unsigned long)  mes);
  SendMessage(InviteClientTE_FSM_MBX);
  msg = GetMsg(InviteClientTE_FSM_MBX);
  return msg;
}

uint8* ExampleMessageFactory::Make2XXToTAL(){
  uint8* msg;
  SIPMsg.makeResponse("200","OK",responseBody,0);
  PrepareNewMessage(0x00,RESPONSE_2XX_T);
  SetMsgToAutomate(TAL_Disp_FSM);
  SetMsgToGroup(INVALID_08);
  SetMsgObjectNumberTo(0);
  AddParamDWord(SIP_PARSED_MESSAGE, (unsigned long)  mes);
  SendMessage(InviteClientTE_FSM_MBX);
  msg = GetMsg(InviteClientTE_FSM_MBX);
  return msg;
}
...
```

The main reason we must introduce the supplementary class *ExampleMessageFactory* is because most of the functions defined in the FSM Library API are protected, which means that they cannot be used in the tester class directly. Alternately, as defined at the moment, CppUnit does not allow us to use multiple-inheritance when we are defining tester classes. Rather, a tester class may be derived only from the class that is defined by the macro instruction *CPPUNIT_NS::TestFixture*.

The source code from the file *ExampleMessageFactory.cpp* should be obvious by now. The only detail that deserves a short explanation is the method by which we create messages. We use typical snippets of code, which start with the *PrepareNewMessage()* function call and are followed with the series of *SetXX()* and *AddParamXX()* function calls. The way we end these code snippets may seem odd. First, we send the new message by calling the function *SendMessage()* and immediately after that, we read that message from the same destination mailbox by calling the function *GetMsg()*. Although it may seem odd, this is the most effective method of creating the complete message in the format that is expected by the function *Process()*.

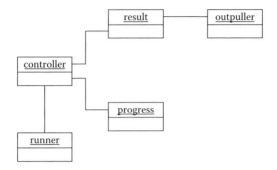

FIGURE 5.9
The collaboration of objects necessary for the automatic execution of the CppUnit test suite.

Finally, we write the main module, named *Main.cpp*. This module creates the collaboration of objects necessary to automatically execute the test suite and report the results of its execution (Figure 5.9). The function *main()* performs the following steps:

- Create the event manager and the test controller
- Add a listener that collects test results
- Add a listener that prints dots as test cases are executed (one dot per test case)
- Add the top suite to the test runner
- Print the test results in a compiler-compatible format

The source code of the module *Main.cpp* follows:

```
#include <cppunit/BriefTestProgressListener.h>
#include <cppunit/CompilerOutputter.h>
#include <cppunit/extensions/TestFactoryRegistry.h>
#include <cppunit/TestResult.h>
#include <cppunit/TestResultCollector.h>
#include <cppunit/TestRunner.h>

int main(int argc,char* argv[]) {
  CPPUNIT_NS::TestResult controller;
  CPPUNIT_NS::TestResultCollector result;
  controller.addListener(&result);
  CPPUNIT_NS::BriefTestProgressListener progress;
  controller.addListener(&progress);
  CPPUNIT_NS::TestRunner runner;
    runner.addTest(CPPUNIT_NS::TestFactoryRegistry::getRegistry().makeTest());
  runner.run(controller);
  CPPUNIT_NS::CompilerOutputter   outputter(&result,std::cerr);
  outputter.write();
  return result.wasSuccessful() ? 0 : 1;
}
```

Test and Verification

As the result of the automatic test suite execution, we get the following report on the monitor:

```
ExampleTestCase::example : OK
OK(1)
Press any key to continue...
```

Additionally, we will get the log file with the following content:

```
Fri Sep 16 19:32:50 2005
Msg To:   UNKNOWN (0x02),  Automate ID: 0x00000000
MsgFrom:  UNKNOWN (0x0f),  Automate ID: 0xcdcdcdcd
Received Msg: (0x0000),  Length: 502  Coding type: 0
0f cd 02 ff | 00 00 cd cd | cd cd 00 00 | 00 00 cd cd | cd cd 00 f6 |
...
Start Timer:  (2)
State: 0 -> 1
-----------------------------------------------------
Fri Sep 16 19:32:50 2005
Msg To:   UNKNOWN (0x02),  Automate ID: 0x00000000
MsgFrom:  UNKNOWN (0x0f),  Automate ID: 0xcdcdcdcd
Received Msg: (0x0029),  Length: 9  Coding type: 0
0f cd 06 ff | 29 00 cd cd | cd cd 00 00 | 00 00 cd cd | cd cd 00 09 | 00 01 00 04
00 | 50 9c 4c 00 | 00
Stop Timer:  (2)
State: 1 -> 2
-----------------------------------------------------
Fri Sep 16 19:32:50 2005
Msg To:   UNKNOWN (0x02),  Automate ID: 0x00000000
MsgFrom:  UNKNOWN (0x0f),  Automate ID: 0xcdcdcdcd
Received Msg: (0x002a),  Length: 9  Coding type: 0
0f cd 06 ff | 2a 00 cd cd | cd cd 00 00 | 00 00 cd cd | cd cd 00 09 | 00 01 00 04
00 | 50 9c 4c 00 | 00
State: 2 -> 0
-----------------------------------------------------
```

Each record of the log file indicates date and time, message source and destination, message type, message length, message coding type, the content of the message (in hexadecimal code), timer operations, and state transition information (e.g., "0 -> 1" means a transition from the state S_0 to the state S_1). By looking at this particular log file, we see that the implementation under test behaves as expected. But normally we do not look at the log file if all test cases pass. The real value of the log file is that it is of great help in localizing bugs if a test case fails. Additionally, we could use the log file to check the internal operation of the implementation under test automatically by the tester class. We skipped that step to keep the example simple enough.

5.5.2 Example 2

This example illustrates one of the steps in the integration testing of the SIP-based softphone. Imagine that the SIP invite client transaction and the

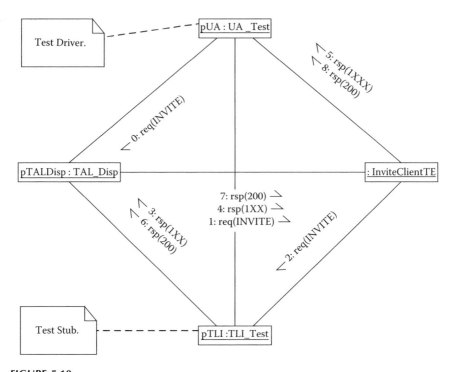

FIGURE 5.10
The example of the integration testing collaboration.

transaction layer dispatcher have undergone complete unit testing. The next normal step would be to integrate them into the final product. Furthermore, imagine that TU and TPL are not yet developed. The only thing we can do is to replace TU and TPL with their imitator classes, named *UA_Test* and *TLI_Test* (TLI stands for Transport Layer Interface), respectively (see the collaboration diagram in Figure 5.10).

The aim of this simple example is to check one particular interaction, illustrated with the collaboration diagram in Figure 5.10. To achieve that goal, we construct the class *UA_Test* that acts as a simple test driver and the class *TLI_Test* that acts as a simple test stub. Both classes are derived from the class *FiniteStateMachine*. The former class has a single state and a single state transition, whereas the latter has two states and two state transitions.

The class *UA_Test* declaration file, named *UA_Test.h*, has the following content:

```
#ifndef _UA_Test_FSM_
#define _UA_Test_FSM_
#include "../constants.h"
#include "../kernel/fsm.h"
#include "../message/message.h"

class UA_Test : public FiniteStateMachine {
```

Test and Verification

```
    int cseq_number;
    Message SIPMsg;
    void SendInviteToTAL();
public:
    enum States { STATE_INITIAL };
    void Evt_Init_TIMER_TINV_EXP();
    void Event_UNEXPECTED();
    // FiniteStateMachine abstract functions
    StandardMessage StandardMsgCoding;
    MessageInterface *GetMessageInterface(uint32 id);
    void SetDefaultHeader(uint8 infoCoding);
    void SetDefaultFSMData();
    void NoFreeInstances();
    void Reset();
    uint8 GetMbxId();
    uint8 GetAutomate();
    uint32 GetObject();
    void ResetData();
public:
    UA_Test();
    ~UA_Test();
    void Initialize();
};
#endif
```

As mentioned above, the class *UA_Test* has a single state, named *STATE_INITIAL*, and a single state transition function, named *Evt_Init_TIMER_TINV_EXP()*. The class *UA_Test* definition file, named *UA_Test.cpp*, has the following content (parts that are not essential are omitted):

```
#include "UA_Test.h"
#include "../parser/smsgtypes.h"
#include "../parser/smsg.h"
#define SipMessageCoding 0x00
extern char* IPString(unsigned int addr, char* buf, int len);

UA_Test::UA_Test() : FiniteStateMachine(16, 2, 3) {}
UA_Test::~UA_Test() {}

void UA_Test::Initialize() {
    SetState(STATE_INITIAL);
    InitTimerBlock(TIMER_TINV,1,TIMER_TINV_EXPIRED);
    InitEventProc(STATE_INITIAL,TIMER_TINV_EXPIRED,
        (PROC_FUN_PTR)&UA_Test::Evt_Init_TIME_TINV_EXP);
    InitUnexpectedEventProc(STATE_INITIAL,
        (PROC_FUN_PTR)&UA_Test::Event_UNEXPECTED);
    StartTimer(TIMER_TINV);
}

void UA_Test::Evt_Ini_TIMER_TINV_EXP() {
    SendInviteToTAL();
}
```

```
void UA_Test::SendInviteToTAL(){
  char  temp[10];
  char  szHostName[255];
  hostent* HostData;
  uint8*  recmsg;
  sip_t  *mes;
  ...
  PrepareNewMessage(0x00,INVITE);
  SetMsgToAutomate(TAL_Disp_FSM);
  SetMsgToGroup(INVALID_08);
  SetMsgObjectNumberTo(0);
  AddParam((SIP_RAW_MESSAGE,    SIPMsg.getLastMessage().length(),
    (uint8*) SIPMsg.getLastMessage().c_str());
  AddParamDWord((SIP_PARSED_MESSAGE, (unsigned long)    mes);
  SendMessage(TAL_Disp_FSM_MBX);
}
...
```

The function *Initialize()* sets the FSM initial state, initializes the timer *TIMER_TINV* to a 1-sec delay, sets the state transition functions, and starts the timer *TIMER_TINV*. When the timer expires, the state transition function *Evt_Init_TIMER_TINV_EXP()* is called. This function sends the *INVITE* message to the transaction layer dispatcher (*TAL_Disp*) by calling the function *SendInviteToTAL()*, which is very similar to the one given in the Example 1 (see Section 5.5.1). Further on, the *INVITE* message is routed toward the test stub class *TLI_Test*.

The class *TLI_Test* declaration file, named *TLI_Test.h*, has the following content (parts that are not essential are omitted):

```
#ifndef _TLI_Test_FSM_
#define _TLI_Test_FSM_
#include "../constants.h"
#include "../kernel/fsm.h"
#include "../message/message.h"

class TLI_Test : public FiniteStateMachine {
  ...
  Message SIPMsg;
  sip_t *mes;
  // Message management functions
  void Send1XXToTAL();
  void Send2XXToTAL();
public:
  enum States {
    STATE_INITIAL,
    STATE_1XX_SENT
  };
  void Evt_Init_INVITE_T();
  void Evt_1XXSent_TIMER_T2XX_EXP();
  void Event_UNEXPECTED();
// FiniteStateMachine abstract functions
  ...
public:
```

Test and Verification

```
  TLI_Test();
  ~TLI_Test();
  void Initialize();
};
#endif
```

As mentioned above, the class *TLI_Test* has two states, named *STATE_INITIAL* and *STATE_1XX_SENT*, and two state transition functions named *Evt_Init_INVITE_T()* and *Evt_1XXSent_TIMER_T2XX_EXP()*. The class *TLI_Test* definition file, named *TLI_Test.cpp*, has the following content (parts that are not essential are omitted):

```
#include "TLI_Test.h"
#define SipMessageCoding 0x00
extern char* IPString(unsigned int addr, char* buf, int len);
TLI_Test::TLI_Test() : FiniteStateMachine(16, 2, 3) {}
TLI_Test::~TLI_Test() {}

void TLI_Test::Initialize() {
  char szHostName[255];
  hostent* HostData;
  SetState(STATE_INITIAL);
  InitTimerBlock(TIMER_T2XX,2,TIMER_T2XX_EXPIRED);
  InitEventProc(STATE_INITIAL,INVITE,
    (PROC_FUN_PTR)&TLI_Test::Evt_Init_INVITE_T);
  InitEventProc(STATE_1XX_SENT,TIMER_T2XX_EXPIRED,

(PROC_FUN_PTR)&TLI_Test::Evt_1XXSent_TIME_T2XX_EXP);
  InitUnexpectedEventProc(STATE_INITIAL,
    (PROC_FUN_PTR)&TLI_Test::Event_UNEXPECTED);
  // Problem specific part
  ...
}

void TLI_Test::Evt_Init_INVITE_T() {
  Send1XXToTAL();
  StartTimer(TIMER_T2XX);
  SetState(STATE_1XX_SENT);
}

void TLI_Test::Evt_1XXSent_TIMER_T2XX_EXP() {
  Send2XXToTAL();
}

void TLI_Test::Send1XXToTAL(){
  uint8* recmsg;
  recmsg = GetParam(SIP_RAW_MESSAGE);
  ...
  SIPMsg.makeResponse("100","Trying",responseBody,0);
  PrepareNewMessage(0x00,RESPONSE_1XX_T);
  SetMsgToAutomate(TAL_Disp_FSM);
  SetMsgToGroup(INVALID_08);
  SetMsgObjectNumberTo(0);
  AddParamDWord((SIP_PARSED_MESSAGE, (unsigned long)  mes);
```

```
  SendMessage(TAL_Disp_FSM_MBX);
}

void TLI_Test::Send2XXToTAL(){
  SIPMsg.makeResponse("200","OK",responseBody,0);
  PrepareNewMessage(0x00,RESPONSE_2XX_T);
  SetMsgToAutomate(TAL_Disp_FSM);
  SetMsgToGroup(INVALID_08);
  SetMsgObjectNumberTo(0);
  AddParamDWord((SIP_PARSED_MESSAGE, (unsigned long)   mes);
  SendMessage(TAL_Disp_FSM_MBX);
}
...
```

The function *Initialize()* sets the initial state, initializes the timer *TIMER_T2XX* to a 2-sec delay, sets the state transition functions, and finishes with some problem-specific initializations. The state transition function *Evt_Init_INVITE_T()*, triggered with the reception of the message *INVITE*, sends the preliminary response *100 (Trying)* by calling the function *Send1XXToTAL()*, starts the timer *TIMER_T2XX*, and changes its state to *STATE_1XX_SENT*. The state transition function *Evt_1XXSent_TIMER_T2XX_EXP()*, triggered with the expiration of the timer *TIMER_T2XX*, sends the final response *200 (OK)* by calling the function *Send2XXToTAL()*.

The content of the main module, named *test_main.cpp*, is the following (parts that are not essential are omitted):

```
#include <conio.h>
#include "kernel/fsmsystem.h"
#include "kernel/logfile.h"
#include "NewSIP/TAL_Disp.h"
#include "Test/UA_Test.h"
#include "Test/TLI_Test.h"
#include "NewSIP/InvClientTE.h"
  FSMSystemWithTCP *pSys;
  LogFile *lf;
  TAL_Disp* pTALDisp;
  TLI_Test* pTLI;
  UA_Test* pUA;
  InviteClientTE* pInviteCltTE[NUMBER_OF_TES];
  DWORD thread_id;
  HANDLE thread_handle;
  ...
DWORD WINAPI SystemThread(void *data){
  FSMSystem *sysAutomate = (FSMSystem *)data;
  sysAutomate->Start();
  return 0;
}
int init(){
  pSys = new FSMSystemWithTCP(11,11);
  pTALDisp = new TAL_Disp();
  pTLI = new TLI_Test();
```

```
  pUA = new UA_Test();
  for (int i = 0; i < NUMBER_OF_TES; i++){
    pInviteCltTE[i]= new InviteClientTE();
  }
  uint8 buffClassNo = 4;
  uint32 buffsCount[4] = { 50, 50, 50, 50 };
  uint32 buffsLength[4] = { 1025, 1025, 1025, 1025};
  pSys->InitKernel(buffClassNo, buffsCount,    buffsLength, 1);
    lf = new LogFile("log.log", "log.ini");
  LogAutomateNew::SetLogInterface(lf);
    pSys->Add(pTALDisp, TAL_Disp_FSM, 1, false);
    pSys->Add(pInviteCltTE[0], InviteClientTE_FSM, 10,    true);
    pSys->Add(pTLI, TLI_Test_FSM, 1, false);
    pSys->Add(pUA, UA_Test_FSM, 1, false);
    for (i = 1; i < NUMBER_OF_TES; i++){
      pSys->Add(pInviteCltTE[i], InviteClientTE_FSM);
    }
    thread_handle = CreateThread(NULL, 0, SystemThread,   pSys,
      THREAD_PRIORITY_ABOVE_NORMAL, &thread_id);
    return 1;
}
...
void main (void){
  parser_init();
  init();
while(!kbhit());
  exit_app();
}
```

As the result of the execution of the main module, we get the log file with nine records that correspond to the messages that are exchanged between implementations under test (transaction layer dispatcher and SIP invite client transaction) and test driver (*UA_Test*) and test stub (*TLI_Test*). This file is very similar to the one given in Example 1 (see Section 5.5.1) but three times longer, hence not included here.

Test automation of integration tests based on log files is possible for simple collaborations like the one shown in this example, although it may be cumbersome. However, if we must deal with more complex collaborations that evolve concurrently, this approach is hardly applicable. Using log files in such situations would normally require human intervention for checking the results of the integration tests. Generally, we should try to use the style of unit testing based on automatic checking of results (see Section 5.1), even for the integration of the parts of the system.

5.6 Further Reading

The reader can find more information related to this chapter in the references. The first reference (Berard et al., 2001) contains a comprehensive coverage of the state-of-the-art model-checking techniques and tools. The second (Newborn, 2001) provides detailed information on the theorem prover THEO used in Section 5.3. The third (Popovic et al., 2001) provides a software maintenance case study in the area of communication protocol engineering. The fourth (Popovic and Velikic, 2005) contains more information on the generic test case generator used in Section 5.5. Finally, the fifth reference (Woit, 1994; Chapter 3 and Section 3.1.1, in particular) provides more information on the reliability estimation model used in Section 5.5.

References

Berard, B., Bidoit, M., Finkel, A., Laroussinie, F., Petit, A., Petrucci, L., Schnoebelen, Ph., and McKenzie, P., *Systems and Software Verification: Model-Checking Techniques and Tools*, Springer-Verlag, Berlin, 2001.

Newborn, M., *Automated Theorem Proving*, Springer-Verlag, New York, 2001.

Popovic, M., Atlagic., B., and Kovacevic, V., "Case study: A maintenance practice used with real-time telecommunication software," *Journal of Software Maintenance and Evolution: Research and Practice*, John Wiley & Sons, West Sussex, No. 13, pp. 97, 2001.

Popovic, M. and Velikic, I., "A generic model-based test case generator," *Proc. IEEE International Conference and Workshop on Engineering of Computer Based Systems*, Greenbelt, MD, April 4–7, 2005.

Woit, D.M., "Operational profile specification, test case generation, and reliability estimation for modules," Ph.D. thesis, Queens University Kingstone, Ontario, Canada, February 1994.

6
FSM Library

The purpose of this chapter is to familiarize the reader with an example of a real-world library for making families of communication protocols. Although it is not perfect, it is in use and evolving. The main argument against it may be that there are too few C++ classes with too many function members. Alternately, this disadvantage is a tradeoff for a rather simple API, which is quite easy to learn and use.

6.1 Introduction

The FSM library described in this book is created to be used as a working environment for the implementation of groups of communication protocols. The programmer has two basic classes at his or her disposal, namely, *FSMSystem* and *FiniteStateMachine*. The class *FSMSystem* models a platform for a group of communication processes (otherwise called finite state machines or automata). An instance of this class interconnects individual communication processes by handling all of the resources needed for the operation of individual finite state machines.

The class *FiniteStateMachine* models a generic communication process (i.e., communication protocol). Each individual communication protocol is represented by an instance of this class. The implementation of a particular communication protocol is narrowed down to writing state-transition functions in C++. The transition function comprises procedures that process the message received in a given FSM state. This processing results in a transition to a new FSM state and the optional generation of the corresponding outgoing messages. All state transition functions must be defined for all finite state machines registered to a single FSM system (an instance of the class *FSMSystem*). Additionally, all the FSM system run-time elements must be initialized properly before it can be successfully started.

The relationship between the classes *FSMSystem* and *FiniteStateMachine* is a symbiosis — one cannot operate without the other. The FSM system clearly represents just an infrastructure, or an unused platform. In reality, an FSM system is always used so that at least a couple of finite state machines are registered to it, together representing a group of finite state machines. Because of that and to achieve simplicity and brevity, we frequently use the term "FSM system" as a synonym for the group of automata, assuming that some individual automata are actually registered to it, and vice versa. Although an instance of the class *FiniteStateMachine* cannot operate on its own, we simply refer to it as a "finite state machine."

6.2 Basic FSM System Components

The FSMSystem library is written in C++ using an object-oriented approach. The basic components are written as C++ classes that provide functionality of both individual finite state machines and a group of finite state machines. These classes are the following:

- *FiniteStateMachine*
- *FSMSystem*

A class can inherit the functionality of a single finite state machine by specializing the base class *FiniteStateMachine*. The programmer implements this class by writing the real functions for those declared as virtual, by adding new problem-specific functions (e.g., state transition functions), and by optionally overriding the inherited functions to redefine the functionality of the base class.

A class can inherit the functionality of a group of finite state machines by specializing the class *FSMSystem*. Normally, this class is simply instantiated as an oracle of a group of finite state machines.

6.2.1 Class *FSMSystem*

An instance of the class *FSMSystem* is an object representing a finite state machine system, i.e., a group of finite state machines (a group of automata). The protected attributes of this class represent the resources available for all the automata included in a group of automata. The basic task of this class is the initialization and management of FSMs, buffers (memory zones), messages, and timers. During a normal lifecycle of an instance of the class *FSMSystem*, its user typically performs the following steps or operations:

- Create FSM system
- Initialize FSM system
- Start FSM system

- Stop FSM system

In the list above, the idiom "FSM system" represents an instance of the class *FSMSystem*.

6.2.1.1 FSM System Initialization

The initialization of the FSM system consists of the following steps:

- Create the FSM system — see the constructor *FSMSystem()*
- Create and initialize individual finite state machines — see the constructor *FiniteStateMachine()*
- Add individual finite state machines to the FSM system
- Initialize the FSM system
- Start FSM system logging

The constructor *FSMSystem()* requires two parameters:

- The number of types of finite state machines
- The number of mailboxes

Individual instances of the class *FiniteStateMachine* can be added to the FSM system by using one of two the possible functions:

```
void Add(ptrFiniteStateMachine object,  // Automata instance address
  uint8 automataType,   // Automata type
  uint32 numOfObjects,  // Number of instances
  bool useFreeList = false);  // List of free automata

void Add(ptrFiniteStateMachine object,  // Automata instance address
  uint8 automataType);  // Automata type
```

The first of the overloaded functions above is used to add the first finite state machine of each type. The other instances of the same type are added using the second function.

The initialization of the FSM system kernel is performed by calling the following function:

```
void InitKernel(uint8 buffClassNo,  // Number of different types
  uint32 *buffersCount,   // Number of buffers per type
  uint32 *buffersLength,  // Buffer lengths per type
  uint8 numOfMbxs=0,  // Number of mailboxes
  TimerResolutionEnum timerRes = Timer1s);  // Timer  resolution in ms
```

The parameters of the function *InitKernel* specify the number of buffer types, the numbers of the instances of different types, their sizes, the number of mailboxes to be used by the automata in a group, and the basic timer resolution. The default number of mailboxes is 0. The default basic timer resolution is 1 sec (just as an example, it can be much smaller, e.g., 10 ms).

The FSM system logging functionality provides message content recording in a sequence resulting from the evolution of the FSM system. These messages are recorded automatically into a file created at the FSM system startup. The file *log.ini* is optional and is used to define textual titles (names) of the messages exchanged among the finite state machines included in the corresponding FSM system. If *log.ini* file is defined, the message binary codes are substituted by the corresponding message names, thus making the log files human readable. On Windows® machines, the *log.ini* file must be placed in the system folder (*c:\winnt* or *c:\windows*). The format of this file is the following:

```
[AUTOMATA]
1=AUTOMATA1_FSM
2=AUTOMATA2_FSM
SequenceNumber=AUTOMATA_TYPE
[MESSAGES]
0=0xe000,MSG_1,0
1=0xe002,MSG_2,0
SequenceNumber=MSG_CODE,TEXT_TITLE,0
```

A typical example is the following:

```
#define NO_BUFFERS   3
#define NO_AUTOMATA_1  5
#define NO_AUTOMATA_2  9
...

// Definition of buffers: three types, where number of buffers per type
// is 50, 30, and 20, and their lengths are 128, 256, and 512 bytes,
// respectively.
uint8  buffClassNo = NO_BUFFERS;
uint32 buffersCount[NO_BUFFERS]  = {50,30,20};
uint32 buffersLength[NO_BUFFERS] = {128,256,512};

// Create FSM system that has two automata types and uses
// two mailboxes (one mailbox per each automata type)
FSMSystem *fsmSystem = new FSMSystem(2,2);

// Create individual automata
Automata1 *automata1 = new Automata1[NO_AUTOMATA_1];
Automata2 *automata2 = new Automata2[NO_AUTOMATA_2];

// Add individual automata to FSM system and implicitly initialize each
// automata instance by calling its function Initialize(). This call is
// made from the function Add.
fsmSystem->Add(&automata1[0],AUTOMATA1_FSM,NO_AUTOMATA_1,false);
for((i=1; i<NO_AUTOMATA_1; i++))
  fsmSystem->Add(&automata1[i],AUTOMATA1_FSM);

fsmSystem->Add(&automata2[0],AUTOMATA2_FSM,NO_AUTOMATA_2,true);
```

FSM Library

```
for((i=1; i<NO_AUTOMATA_2; i++))
  fsmSystem->Add((&automata2[i],AUTOMATA2_FSM);

// Initialize kernel
fsmSystem->InitKernel(buffClassNo,buffersCount,buffersLength,2);

// Create and set logging system (log file name, message definition file)
lf = new LogFile("log.log", "log.ini");
LogAutomataNew::SetLogInterface(lf);
...
```

The example above starts with the definition of the number of buffer types. In this example, three buffer types are defined (i.e., small, medium, and large buffers) by setting the symbolic constant *NO_BUFFERS* value to 3. Next, we define the number of instances of two automata types by setting the values of symbolic constants *NO_AUTOMATA_1* to 5 and *NO_AUTOMATA_2* to 9. This means that five instances of the first automata type and nine instances of the second automata type will exist in the group of automata we are going to create.

Next, the program paragraph defines the number of buffers, as well as their size, for each buffer type. Fifty small buffers of size 128 bytes, thirty medium buffers of size 256 bytes, and twenty large buffers of size 512 bytes would be used. The number of buffer types is stored in the variable *buffClassNo*. The number of buffers of each type and their lengths are stored in the arrays *buffersCount* and *buffersLength*.

We then create the FSM system by calling the constructor of the class *FSMSystem*. This constructor has two parameters: the number of automata types and the number of mailboxes to be used by the system for its own purposes. Next, we create two groups of automata of two different types. In the program, these groups are represented as arrays of instances of classes, namely, the classes *Automata1* and *Automata2*. In this example, we assume that these classes have already been defined by extending the base class *FiniteStateMachine*.

After creating two groups of automata of different types, all the automata are added to the already created FSM system. The first instance of each automata type is added by calling the overloaded function *Add* with the first type of signature, which specifies the instance address, the instance type, the total number of instances of this type, and the indicator specifying if a list of free automata of this type exists or not. The rest of the instances are added by calling the overloaded function *Add* with the second type of signature, specifying just the instance address and its type.

The first automata type in this example does not have a list of free automata, whereas the second type does have a list of free automata. This means that the instance of the second automata type can be viewed as a pool of resources of the same type. They may be dynamically allocated to be engaged in a certain communication scenario. When a programmer decides to use this opportunity, he must provide the function *NoFreeInstance*, which is called when the dynamic allocation request cannot be satisfied, because no more free automata instances of that type are found.

The FSM system is initialized by simply calling its function *InitKernel*. The parameters of this function specify the number of buffer types, the number of buffers of each type, their sizes, and the number of mailboxes to be used for FSMs. Normally, we use one mailbox per each automata type. This is not a restriction imposed by the class *FSMSystem*, it is simply a convention. Other arrangements are also allowed; for example, we can create more mailboxes for messages of different priorities, or we can create additional mailboxes dedicated to communication between the given groups of automata types. Most generally, we can use mailboxes just as queues of any kinds of messages. Because the last parameter of the function *InitKernel* is omitted, the timer resolution is set to its default value (1 sec, in this example).

At the end of this example, we create and set the logging system by calling its constructor *LogFile* and the function *SetLogInterface*, respectively. The parameters of the constructor specify the name of the log file (*log.log*) as well as the name of the file containing the textual names of the messages (*log.ini*). The parameter of the function *SetLogInterface* specifies the logging system interface, which generally is a file. In this example, the disk file is named *log.log* but it could be any file, including special files representing devices handled by the corresponding device drivers, such as */dev/lpt* or */dev/com1*.

6.2.1.2 FSM System Startup

The FSM system is started by calling its function *Start*. Most frequently, this function is called by the thread assigned to the FSM system. Here is an example:

```
DWORD WINAPI FsmSystemThreadFunc((void* param)){
  try {
    fsmSystem->Start();
  }
  catch(...){
    OutputDebugString('Exception — terminating FSM system\n');
    return 0;
  }
  OutputDebugString('FSM system terminated\n');
  return 0;
}
...

// Somewhere in the main function
DWORD fsmSystemThreadId;
CreateThread(NULL,0,FsmSystemThreadFunc,0,0,fsmSystemThreadId);
...
```

In the example above, we start the FSM system by calling its function *Start* from the thread function *FsmSystemThreadFunction*. We assume that thread

FSM Library 313

has already been created and that its identification is stored in the variable *fsmSystemThreadId*.

6.2.2 Class *FiniteStateMachine*

All the automata added to the FSM system are implemented by extending the base class *FiniteStateMachine*. This class defines a set of virtual functions that must be defined by the programmer. These functions are the following:

```
MessageInterface *GetMessageInterface(uint32 id);
void SetDefaultHeader(uint8 infoCoding);
uint8 GetMbxId();
uint8 GetAutomata();
void SetDefaultFSMData();
void NoFreeInstances();
void Initialize();
```

The following example illustrates the most frequently used definitions of *FiniteStateMachine* functions. The detailed description of all the functions is given in Section 6.8 describing API functions.

```
// This function returns the message interface for the given interface ID.
// It is assumed that standardMsgCoding is defined as:
// StandardMessage standardMsgCoding;
MessageInterface *Automata::GetMessageInterface(uint32 id){
  switch(id){
    case 0x00:
    return &standardMsgCoding;

    // Other definitions
    // case 0x01:
    // case 0x02:
  }
  throw TErrorObject(__LINE__,__FILE__,0x01010400);
}

// This function fills in the message header.
void Automata::SetDefaultHeader(uint8 infoCoding){
  SetMsgInfoCoding(infoCoding);
  SetMessageFromData();
}

// This function defines the mailbox number (ID) to be used as default
// by the automata of the type defined by this class.
uint8 Automata::GetMbxId(){
  return AUTOMATA_MB_ID;
}

// This function returns the number (ID) which identifies the automata
// type defined by this class.
uint8 Automata::GetAutomate(){
  return AUTOMATA_TYPE_ID;
```

}

```
// This function sets the values of the instance attributes.
void Automata::SetDefaultFSMData(){
  attribut1 = VALUE_1;
  attribut2 = VALUE_2;
}

// This function is called if there are no more free automata of this
// type. It may be used if the instances of this class have been added to
// the FSM system with the parameter useFreeList set to value true.
void Automata::NoFreeInstances(){
  // The activity if there are no free automata of this   type.
}

// This function defines state transition functions and timers to be used
// by the automata of this type. It is called by the function Add, which
// is used to add an automata instance to the given FSM system.
// It is assumed that state transition functions are declared and defined
// elsewhere.
void Automata::Initialize(){
  // Here we place a series of initializations:
  // InitEventProc(uint8 state, uint16 event,  PROC_FUN_PTR fun);
  // InitUnexpectedEventProc(uint8 state, PROC_FUN_PTR   fun);
  // InitTimerBlock(uint16 timerId, uint32 timerCount,  uint16 signalId);

  InitEventProc(IDLE, MSG_SEND, (PROC_FUN_PTR)     &Automata::Idle_MsgSend);
  InitEventProc(IDLE, MSG_RCV,  (PROC_FUN_PTR)     &Automata::Idle_MsgReceive);

  InitEventProc(SEND, MSG_NEW,  (PROC_FUN_PTR)     &Automata::Send_MsgNew);
  InitEventProc(SEND, MSG_END,  (PROC_FUN_PTR)     &Automata::Send_MsgEnd);
  InitEventProc(IDLE, T200_CODE,(PROC_FUN_PTR)     &Automata::T200Expired);

  InitUnexpectedEventProc(IDLE, (PROC_FUN_PTR)     &Automata::Idle_Unexpected);
  InitUnexpectedEventProc(SEND, (PROC_FUN_PTR)     &Automata::Send_Unexpected);

  InitTimerBlock(T200,T200_VALUE,T200_CODE);
}
```

In the example above, we would like to create the class *Automata* that models one type of finite state machines (automata). The definition of the class comprises the definitions of its function members. The function member *GetMessageInterface* returns the object that embodies the coding of the messages to be used by the instances of the class *Automata*. In this example, it is an instance of the class *StandardMessage*.

The member function *SetDefaultHeader* is used to automatically fill in the message header defaults. Normally, these are the data about the automata instance that has created the message to send to some other automata instance. In this example, it uses the function *SetMsgInfoCoding* to specify the type of the coding to be applied. It also uses the function *SetMessageFromData* to specify the type of the originating automata instance, the identification of the group to which the automata instance belongs, and the identification of the originating automata instance.

The member function *GetMbxId* returns the identification of the mailbox used by the automata instance of this type. In this example, it is the value of the symbolic constant *AUTOMATA_MBX_ID*. The member function *GetAutomata* returns the identification of the automata type. It is the value of the symbolic constant *AUTOMATA_TYPE_ID*. The member function *SetDefaultFSMData* is used by the automata instance to set its specific data before it commences its normal operation. In this example, *attribute1* is set to the value *VALUE_1* and *attribute2* is set to the value *VALUE_2*.

The member function *NoFreeInstances* can be used to specify the action to be performed if no more free automata instances of this type are found, e.g., to make a small system restart, allocate some additional automata instances, and so on. This mechanism is available to the programmer if the instances of automata have been added (function *Add*) to the FSM system with the parameter *useFreeList* set to the value *true*.

The member function *Initialize* is used to define automata state transition functions and timers (referred to as timer blocks throughout the FSM library documentation) to be used by the automata. The FSM library distinguishes two types of events, expected and unexpected, and allows the programmer to specify the corresponding event handlers, which are just specialized C++ functions. These handlers are defined by calling the registration functions, namely, the function *InitEventProc* for the expected events and the function *InitUnexpectedEventProc* for the unexpected events. The parameters of both of these functions specify the state code, the event (message) code, and the pointer to the event handler.

In this example, we have defined seven automata state transition functions altogether, five of them triggered by the expected events and two triggered by the unexpected events. The part of the automata shown in the example has two states, *IDLE* and *SEND*. The expected events in the state *IDLE* are *MSG_SEND*, *MSG_RCV*, and *T200_CODE*. The corresponding event handlers are *Idle_MsgSend*, *Idle_MsgReceive*, and *T200Expired*, respectively. Two legible events exist in the state *SEND*, *MSG_NEW* and *MSG_END*. The corresponding handlers are *Send_MsgNew* and *Send_MsgEnd*. The unexpected event handler for the state *IDLE* is *Idle_Unexpected* whereas for the state *SEND* it is *Send_Unexpected*. The corresponding state transition table is shown in Table 6.1.

TABLE 6.1

Example of a State Transition Table

	MSG_RCV	MSG_SEND	T200_CODE	MSG_NEW	MSG_END	?
Idle	Idle_MsgReceive	Idle_MsgSend	T200Expired			Idle_Unexpected
Send				Send_MsgNew	Send_MsgEnd	Send_Unexpected

The timers are initialized by calling the function *InitTimerBlock*. The parameters of this function specify the unique timer identification, its duration (as the number of basic timer resolution units), and the code of the message sent when the timer expires. In the example above, these are the symbolic constants *T200*, *T200_VALUE*, and *T200_CODE*.

To sum, automata states and attributes are defined in accordance with the problem at hand. The state transition function, referred to as the event handler, is called upon the reception of a given message in a given state, as defined by the function *Initialize*. Each event handler is defined as a class member function responsible for handling a given event.

The timers to be used by the automata are defined also by the function *Initialize*. This is done by calling the function *InitTimerBlock*, which in turn creates the internal kernel timer block (essentially a program object) and fills in its identification, duration, and the corresponding timer message code.

6.3 Time Management

In the previous section, automata timers are initialized during the FSM system startup by the function *Initialize*. The automata type that uses timers in its regular operation manages them through the corresponding FSM library API functions, which maintain the internal kernel object behind the scenes. The API functions are the following:

```
void InitTimerBlock(uint16 tmrId,uint32 count,uint16 signalId);
void StartTimer(uint16 tmrId);
void StopTimer(uint16 tmrId);
void RestartTimer(uint16 tmrId)
bool IsTimerRunning(uint16 tmrId);
```

The function *InitTimerBlock* is used to define (initialize) the timer. Its parameters specify the unique timer identification, its duration as a multiple of the basic timer resolution unit, and the code of the message sent to the automata mailbox when the timer expires. This is explained in the previous section. Notice that each timer has the unique identification *tmrId* used as a parameter of all the API functions to identify the timer.

Each API function represents a primitive timer operation. The function *StartTimer* is used to start the timer, the function *StopTimer* stops the timer, the function *RestartTimer* restarts the timer, and the function *IsTimerRunning* is used to check if the timer is running or not.

The following example illustrates the usage of these primitives:

```
if(!IsTimerRunning(T200)){
  StartTimer(T200);
}
```

FSM Library

```
else
  StopTimer(T200);
...
```

A normal timer life cycle has the following phases:

- Define, i.e., initialize the timer
- Use the timer by alternative application of the following primitives:
  - *Start* (applicable if the timer is not running, hence either newly defined or previously stopped)
  - *Stop* (applicable if the timer is running)
  - *Restart* (logically equivalent to *Stop* plus *Start*)
  - *IsTimerRunning* (returns true if it does, otherwise returns false)

6.4 Memory Management

Because the main application of the FSM library is in real-time systems, efficient memory allocation must be provided. The FSM library does not rely on the hosting operating system because some of the operating systems suffer from memory fragmentation problem. Furthermore, in some applications on bare machines, the operating system may not even be available. Because of that, memory management is one of the main functions of the FSM library.

The working memory is partitioned into certain zones referred to as **buffers**. The programmer defines the number of different buffer types, the numbers of buffers of each type, and their sizes. The programmer specifies this data as parameters of the function *InitKernel* (see Section 6.8.4) and the FSM library kernel in its turn creates them as its own internal objects.

The buffers are most frequently used indirectly through message management (message create, send, receive, and similar operations) and timer operations (timer definition and usage operations). Besides this indirect buffer usage, the buffers can be managed directly if needed through the following API functions:

```
uint8  *GetBuffer(uint32 length);
void    RetBuffer(uint8 *buff);
bool    IsBufferSmall(uint8 *buff,uint32 length);
uint32  GetBufferLength(uint8 *buff);
```

The programmer requests a buffer by calling the function *GetBuffer*. The parameter of this function is the minimal size of the desired buffer. All the buffers provided by the kernel must be returned to it by calling the function *RetBuffer*. Untidy memory management can cause buffer loss, commonly referred to as **memory leak**, which may cause irregular kernel operation and a system crash.

Besides the memory allocation (*malloc*) and *free* primitives, two additional primitives provide the information about the buffer already allocated to the finite state machine. The function *IsBufferSmall* checks if the buffer size is smaller than the value of its parameter. If yes, it returns *true*, otherwise, it returns *false*. Another function, named *GetBufferLength*, returns the buffer size in octets (bytes).

The following example illustrates the usage of the buffer management primitives:

```
// We define two buffer types, small and large.
// There are ten small buffers and fifteen large buffers.
// The small buffer size is 128 bytes. The large buffer size is
// 256 bytes.
uint8   buffClassNo = 2;
uint32  buffersCount[2]  = {10,15};
uint32  buffersLength[2] = {128,256};
...

// Kernel initialization (noMBX is irrelevant in this example)
fsmSystem->InitKernel(buffClassNo,buffersCount,buffersLength,noMBX);
...

uint32 bufferLength;
uint8 *pointer = GetBuffer(100);
if((IsBufferSmall(pointer,129)){
  RetBuffer(pointer);
  pointer = GetBuffer(129);
}
if((pointer != NULL))
  bufferLength = GetBufferLength(pointer);
...
```

In the example above, we first define two buffer types — small and large — by calling the function *InitKernel*. Its fourth parameter (*noMBX*, the number of the mailboxes) is not relevant for this example. The rest of the program illustrates the usage of FSM library's buffer management functions. First, the program asks for a buffer not smaller than 100 bytes, then it checks if this buffer is smaller than 129 bytes. If yes, it returns the allocated buffer and requests a new one not smaller than 129 bytes (in this example, it will get one large buffer of size 256 bytes). At the end, the program checks if the pointer is defined, which also means that it points to a certain buffer. If it is defined, the program asks for its size by calling the function *GetBufferLength*.

6.5 Message Management

The main communication among individual automata included in the FSM system is achieved through the messages exchanged through the mailboxes typically assigned to the individual automata. The message sent from the originating automata instance towards the destination automata instance is placed temporarily in the mailbox assigned to the destination automata instance. There it waits to be taken over and subsequently processed by the destination automata instance (process).

As already mentioned, a **mailbox** is a message queue that can contain messages for any automata type, thus it does not need to be assigned to some particular automata type. In contrast to a typical paradigm, it can be used as a general message queue shared by more destination automata. Essentially, in such a paradigm the source automata instance can put the message in any mailbox hosted by the FSM system and it will eventually be delivered to its proper destination.

This message routing and delivery is performed automatically by the FSM system and is hidden from the automata, which are just service users. The FSM system has an abstraction of the mailbox from which it takes messages, one at a time (mailbox abstraction provides buffering functionality by employing the FIFO memory type). Upon the reception of each individual message, the FSM system consults the message header to determine the destination automata instance and passes the message to it. The destination automata instance looks up the message code and, based on the current automata state, calls the appropriate automata state transition function.

Message reception is completely transparent for the programmer writing the program code for the finite state machine. The above mechanism is absolutely hidden from him. The programmer must simply accept that the message reception and its classification are done automatically by the system. He just writes the message processing functions that are called automatically by the system upon the reception of the corresponding message.

The API functions can be partitioned into two groups:

- The functions that work with the received message
- The functions that work with the new message that must be prepared and sent

The functions in the first group are used to provide the information about the originating automata instance. The source of this information is the message header and the values of the message parameters. The functions in the second group provide primitives needed to make and send a message:

- Buffer allocation (indirect call to *GetBuffer* primitive)
- Filling the message header with the data about the originating automata instance
- Adding the message parameters and setting them to the given values
- Sending the message to the mailbox assigned to the destination automata instance

The messages may be sent only from a finite state machine or a FSM system. Note that during normal system operation, a FSM system does not send any messages. In this context, a finite state machine is an instance of the class *FiniteStateMachine*, or a class derived from it, and an FSM system is an instance of the class *FSMSystem*.

Example 1:

```
// Get parameter of type PARAM_1 from the received message.
// The size of PARAM_1 is WORD.
WORD word;
GetParamWord(PARAM_1,word);

// Get parameter of arbitrary size. Maximum size for StandardMessage is
// 256 bytes. If that is not sufficient, a programmer must derive a new
// class and redefine its functions.
uint8 *pointer;
uint8 text[300];
uint8 msgLength;

pointer = GetParam(TEXT);
if(pointer != NULL){
  // StandardMessage format: bytes 1 and 2 contain   parameter name,
  // byte 3 contains parameter length in bytes,
  // byte 4 and further contain the parameter itself.
  memcpy(text,pointer+3,*((pointer+2)));

  // Make a string by placing null at the end of  character array.
  memset(text+(*((pointer+2))),0x00,1);
}
```

The example above shows how the programmer can get a parameter from the *current* message. A current message is the last message received by the automata instance, i.e., it is the last message taken from the mailbox and assigned to the automata instance for processing. The parameter size is WORD (2 bytes). First, the programmer declares the variable *word* in which he wants to store the parameter value.

The message can contain many parameters, therefore the programmer must specify the unique identifier of the parameter he wants to get. In this example, the identifier is the value of the symbolic constant *PARAM_1*. Finally, a copy of the desired parameter is provided by calling the API function *GetParam*. The first parameter of this function is the parameter identifier (*PARAM_1*) and the second is the variable (*word*) in which the desired parameter is to be copied.

The second part of the example above demonstrates how the programmer may handle textual parameters of arbitrary size. The *StandardMessage* format prescribes that the first 2 bytes of such a parameter are reserved for the parameter name, the next byte is used for the parameter length (in bytes), and the rest of the bytes in the parameter represent its value. The example shows

FSM Library

how a copy of such a parameter can be provided and how a null terminated string can be constructed by adding the NULL character at its end.

Example 2:

```
...
// PrepareNewMessage parameters: buffer size and message type.
PrepareNewMessage(0xAA,MSG_NAME);

// Fill in the message header:
// destination automata type, its ID, and optionally its group ID.
SetMsgToAutomata(AUTOMATA_TYPE);
SetMsgObjectNumberTo(automataId);
SetMsgToGroup(INVALID_08);

// Add parameters: see also other AddParam functions.
AddParamByte(PARAM_1,byte);
AddParamWord(PARAM_2,word);
AddParam(PARAM_3,parameterLength,parameterPointer);

// Send message to the specified mailbox.
SendMessage(AUTOMATA_MBX_ID);
```

The example above shows a common way to construct and send a message. The first step is to call the function *PrepareNewMessage*. The parameters of this function specify the expected buffer size (*0xAA* in this example) and the message name, which also specifies the message type (*MSG_NAME*).

Next, we fill in the message header by calling the following functions:

- *SetMsgToAutomata*: set the destination automata instance type (*AUTOMATA_TYPE*)
- *SetMsgObjectNumberTo*: set the destination automata instance identification (*automataId*)
- *SetMsgToGroup*: set the automata instance group identification (*INVALID_08*)

We then add three message parameters by calling the members of the *AddParam* family of functions. The first function shown in the example is *AddParamByte*. Its parameters specify the unique parameter identifier (*PARAM_1*) and the variable containing the value of the parameter to be copied to the corresponding field of the message (*byte*). The second function is *AddParamWord*. Similarly, its parameters specify the parameter identification (*PARAM_2*) and the variable holding its value (*word*). The last function is *AddParam*. The parameters of this function specify the parameter identification (*PARAM_3*), its length (*parameterLength*), and a pointer to it (*parameterPointer*).

At the end of the example above, we send the message by calling the function *SendMessage*. The parameter of this function specifies the destination mailbox identification (*AUTOMATA_MBX_ID*).

Example 3:
```
// Send a message from the FSM system.
uint8 *msg = GetBuffer(messageInfoLength+MSG_HEADER_LENGTH);

// infoBuffer must be properly formatted.
memcpy(msg+MSG_HEADER_LENGTH,infoBuffer,infoBufferLength);

SetMsgFromAutomata(AUTOMATA_TYPE_FROM_ID,msg);
SetMsgFromGroup(INVALID_08,msg);
SetMsgObjectNumberFrom(automataFromId,msg);

SetMsgToAutomata(AUTOMATA_TYPE_TO_ID,msg);
SetMsgToGroup(INVALID_08,msg);
SetMsgObjectNumberTo(automataToId,msg);

SetMsgInfoCoding(0,msg);// 0 = StandardMessage
SetMsgCode(MSG_FROM_SYSTEM_AUTOMATA,msg);
SetMsgInfoLength(infoBufferLength,msg);
SendMessage(AUTOMATA_TO_MBX_ID,msg);
...
```

The example above shows how a message can be created and sent within the FSM system. This process is done through the following steps:

- Allocate a buffer by calling the function *GetBuffer*.
- Copy the information payload.
- Fill in the data about the originating automata instance by calling the function *SetMsgFromAutomata* fill in the originating automata instance type identification (AUTOMATA_TYPE_FROM_ID); by calling the function *SetMsgFromGroup*, fill in the originating automata instance group identification (*INVALID_08*); and by calling the function *SetMsgObjectNumberFrom*, fill in the automata instance identification (*automataFromId*).
- Fill in the data about the destination automata instance. The function *SetMsgToAutomata* sets the destination automata instance type identification (*AUTOMATA_TYPE_TO_ID*), the function *SetMsgToGroup* sets the destination automata instance group identification (*INVALID_08*), and the function *SetMsgObjectNumberTo* sets the destination automata instance identification (*automataToId*).
- Finalize the message. The function *SetMsgInfoCoding* sets the type of coding (*StandardMessage*), the function *SetMsgCode* sets the message code (*MSG_FROM_SYSTEM_AUTOMATA*), and the function *SetMsgInfoLength* sets the payload length (*infoBufferLength*).
- Send message by calling the function *SendMessage* with the second type of the signature. The parameters of this function specify the destination mailbox identification (*AUTOMATA_TO_MBX_ID*) and the pointer to the message to be sent (*msg*).

6.6 TCP/IP Support

One of the primary design goals of creating the FSM library was to support the design of scalable applications based on distributed processing. The FSM library enables both single-processor and multiprocessor applications. In the former case, all groups of automata execute in a single processor. They share processor resources, such as its processing unit, operating memory, flash, and so on. The automata communicate over the mailboxes placed in the common operating memory.

In the latter case, various groups of automata are deployed on more processors, which can be logically viewed as a multiprocessor system. The groups of automata execute on different processors in parallel and use the mailboxes physically located in separate operating memories. The FSM library transparently uses the network infrastructure to pass messages among the communicating automata. Most frequently, the communication infrastructure is TCP/IP technology.

In both cases, the communicating automata are unaware of the real physical infrastructure because the physical details are hidden from them. This is accomplished by providing a unique API. An individual automata instance manages just its timers, buffers, and messages (new and current, i.e., last received). The rest is handled by the FSM library kernel behind the scenes. This means that the FSM library inherently provides implicit support for TCP/IP. For example, if an automata instance wishes to send a message to some other automata instance physically located on a different machine, it just prepares the message and calls the API function *SendMessage*. The class *FSMSystem* takes care of transporting the message over the TCP/IP network and placing it in the local mailbox assigned to the destination automata.

As far as individual automata based on the FSM library need to communicate only among themselves, implicit TCP/IP support is sufficient. The need to communicate with other program components that are not based on the FSM library and that use TCP/IP sockets directly leads to the requirement for explicit TCP/IP support. To fulfill that requirement, the FSM library also provides explicit (in addition to implicit) TCP/IP support in a form of traditional TCP/IP socket abstraction. Of course, the automata instance that uses these additional API features must be aware and capable of handling details of TCP/IP communication (IP addresses and port numbers).

Explicit TCP/IP support is provided by two additional classes, namely, *FSMSystemWithTCP* and *NetFSM*. These two classes enable the FSM library-based automata to directly communicate over the TCP/IP protocol stack with other FSM library based automata or with other TCP/IP program components, e.g., Web server, SIP client. As their names suggest, the class *FSMSystemWithTCP* is used instead of the class *FSMSystem*, and the class *NetFSM* is a logical counterpart of the class *FiniteStateMachine*.

6.6.1 Class *FSMSystemWithTCP*

The class *FSMSystemWithTCP* is derived from the class *FSMSystem* by extending it with support for communication over the TCP/IP family of protocols. It inherits the basic functionality of the base class, which has been described previously (see Section 6.2.1 describing the class *FSMSystem*).

In contrast to single-processor applications, distributed applications comprise parts (i.e., groups of automata) that are started independently. Because of this, two groups of automata executing on different processors must establish a TCP/IP connection at their startup. The connection establishing procedure is symmetric: This means that either side of the party — or both — must start their local TCP servers by calling the function *InitTCPServer*. The opposite side establishes the connection by calling the function *establishConnection*.

Example:
```
// In processor 1 (server)
//
// Initialize kernel.
fsmSystem1->InitKernel(buffClassNo,buffersCount,buffersLength,2);

// Initialize TCP/IP server on port number 5000.
// NetFSM_Automata1 is derived from NetFSM.
fsmSystem1->InitTCPServer(5000,NetFSM_Automata1);

// In processor 2 (client)
//
// Set server TCP/IP parameters (port, IP address).
// Establish the connection.
fsmSystem2.setPort(5000);
fsmSystem2.setIP("192.168.77.77");
fsmSystem2.establishConnection();
...
```

This example shows the code excerpts for the TCP/IP server and client machines, named processor 1 and processor 2. At startup, the server initializes the FSM library kernel by calling the function *InitKernel* (its parameters are the number of buffer types, their count, length, and the number of the mailboxes to be used). Next, it calls the function *InitTCPServer* to start the TCP/IP server. We assumed in this example that the class *NetFSM_Automata1* is derived from the class *NetFSM*.

Alternately, the client sets the TCP port number (5000) by calling the function *setPort* and the IP address of the TCP server (192.168.77.77) by calling the function *setIP*, and establishes the connection with the server by calling the function *establishConnection*.

6.6.2 Class *NetFSM*

The class *NetFSM* is derived from the base class *FiniteStateMachine* by extending its basic functionality with support for the communication over TCP/IP infrastructure. The inherited basic functionality has been described previously (see Section 6.2.2 describing the class *FiniteStateMachine*). The basic functionality is extended with the abstraction enabling TCP/IP communication by adding three new function members. The new functions are the following:

```
virtual void convertFSMToNetMessage()=0;
virtual uint16 convertNetToFSMMessage()=0;
virtual uint8 getProtocolInfoCoding()=0;
```

These functions are used to convert the internal message format (abbreviated as *FSM*) into external, or network message format (abbreviated as *Net*), and vice versa. Normally, automata executing in the same processor exchange internal messages coded in internal message format. However, this message format is not suitable for transmission over the network. Most commonly, the message must be serialized, i.e., transformed from data object and structure form into the external message in accordance with a given external message format. This is a series of bits, sometimes grouped in octets or words, that are transmitted over the communication line.

The functions listed above are virtual functions and therefore the programmer must define them while he writes a class that is derived from the class *NetFSM*. The message format conversion functions naturally read a message from some input buffer, convert it into a requested format, and write the output to an output buffer.

The function *convertFSMToNetMessage* is not intended to be used directly by the communicating automata but rather to be called internally by the FSM library kernel to convert an internal message into the external one before it can be sent over the network. Therefore, the input of this function is the internal message and its output is the corresponding output message. The parameters of this function specify the pointer to the internal message *fsmMessageS*, its length *fsmMessageLength*, the pointer to the output, the external message *protocolMessageS*, and its length *sendMsgLength*. The programmer must specify the mapping algorithm by writing this function.

Symmetrically, the function *convertNetToFSM* is intended to be used by the FSM library kernel to convert an external message received over the network into an internal message representation, which must be delivered to the local mailbox and processed further by the corresponding local automata. The input of this function is the external message and the output is the internal message. The parameters of this function specify the pointer to the external message *protocolMessageR*, its length *receivedMessageLength*, the pointer to the output, internal message *fsmMessageR*, and its length *fsmMessageRLength*.

The function *getProtocolInfoCoding* returns the code of the type of external information coding. An instance of the class *NetFSM*, referred to as *net automata*, initiates the transmission of the message across the TCP/IP network by calling the function *sentToTCP*. This function may throw an exception in the case of an error, e.g., when net automata wants to send a message after the TCP connection has been closed.

Example:
```
// PrepareNewMessage parameters: buffer size and message type
PrepareNewMessage(0xAA,MESSAGE_NAME);

// Fill in message header:
// destination automata type, its ID, and its group ID (if relevant)
SetMsgToAutomata(AUTOMATA_TYPE);
SetMsgObjectNumberTo(automataId);
SetMsgToGroup(INVALID_08);
// Add parameters.
AddParamByte(PARAM_1,byte);
AddParamWord(PARAM_2,word);
AddParam(PARAM_3,parameterLength,parameterPointer);

// Send message to local mailbox:
// SendMessage(AUTOMATA_MBX_ID);
// or send it over TCP/IP network:
sendToTCP();
```

The example above demonstrates how automata can prepare a message and send it over a TCP/IP network. The message is prepared like any other message. The function *PrepareNewMessage* is used to allocate a buffer for the message and to specify a message name. A series of already described functions is then used to fill in the message header and add the message parameters (see the second example in Section 6.5 describing the message management). At the end, instead of sending the message to the local mailbox by calling the function *SendMessage*, the message is sent over the TCP/IP network by calling the function *sendToTCP*.

A net finite state machine receives the messages equally as simple automata (instances of the class *FiniteStateMachine*) do, just by reading its local mailbox.

6.7 Global Constants, Types, and Functions

The file *kernelConsts.h* defines the global constants, the types, and the functions used by the FSM library kernel. The constants and their values are the following:

```
MSG_FROM_AUTOMATA = 0;  // Source automata ID (BYTE)
MSG_FROM_GROUP    = 1;  // Source automata group ID (BYTE)
MSG_TO_AUTOMATA   = 2;  // Destination automata ID (BYTE)
```

FSM Library

```
MSG_TO_GROUP = 3; // Destination automata group ID (BYTE)
MSG_CODE = 4; // Message code(WORD)
MSG_OBJECT_ID_FROM = 6; // Source automata instance ID (DWORD)
MSG_OBJECT_ID_TO = 10; // Destination automata ID (DWORD)
CALL_ID = 14; // Call (process) ID
MSG_INFO_CODING = 18; // Info coding type, 0 = StandardMessage
MSG_LENGTH = 19; // Message payload length
MSG_INFO = 21; // Message payload offset
MSG_HEADER_END = MSG_INFO; // End of message header

INVALID_08 = 0xff; // Mask for 8 bits
INVALID_16 = 0xfff; // Mask for 16 bits
INVALID_32 = 0xffffffff; // Mask for 32 bits
```

The global data types are the following:

```
int8, uint8 // BYTE
int16, uint16 // WORD
int32, uint32 // DWORD
```

The utility functions provided for the load-store manipulation with various data types are the following:

```
void SetUint16(uint8 *addr,uint16 value);
void SetUint32(uint8 *addr,uint32 value);
uint16 GetUint16(uint8 *addr);
uint32 GetUint32(uint8 *addr);
```

The utility functions are provided to avoid cast operators in C/C++ programs because some microcontrollers do not allow word or double-word memory access to odd memory addresses.

6.8 API Functions

The FSM library API functions are grouped into the following eight groups:

- *FSMSystem* constructor (Table 6.2)
- *FSMSystem* member functions (Table 6.3)
- *FSMSystemWithTCP* constructor (Table 6.4)
- *FSMSystemWithTCP* member functions (Table 6.5)
- *FiniteStateMachine* constructor (Table 6.6)
- *FiniteStateMachine* member functions (Table 6.7)
- *NetFSM* constructor (Table 6.8)
- *NetFSM* member functions (Table 6.9)

TABLE 6.2

FSMSystem Constructor Summary

FSMSystem(uint8 numOfAutomata, uint8 numberOfMbx)

    The constructor initializes the object that represents the FSM system along with the data structures needed for its proper operation.

TABLE 6.3

FSMSystem Member Functions Summary

Type	Member Function
Void	Add (ptrFiniteStateMachine object, uint8 automataType, uint32 numOfObjects, bool useFreeList=false)
	This function adds the first instance of each automata type to the FSM system.
Void	Add(ptrFiniteStateMachine object, uint8 automataType)
	This function adds all the automata instances of the given type to the FSM system, except for the first instance.
Void	InitKernel(uint8 buffClassNo, uint32 *buffersCount, uint32 *buffersLength, uint8 numOfMbxs=0, TimerResolutionEnum timerRes=Timer1s)
	This function initializes the elements of the kernel responsible for time, buffer, and message management.
Void	Remove(uint8 automataType)
	This function removes all the instances of the given automata type from the FSM system.
ptrFiniteStateMachine	Remove(uint8 automataType, uint32 object)
	This function removes the given instance of the given automata type.
Virtual void	Start()
	This function starts the FSM system.
Void	StopSystem()
	This function stops the FSM system.

TABLE 6.4

FSMSystemWithTCP Constructor Summary

FSMSystemWithTCP(uint8 numOfAutomata, uint8 numberOfMbx)

    The constructor initializes the object that represents the FSM system supporting communication over TCP/IP network along with the data structures needed for its proper operation.

FSM Library

TABLE 6.5

FSMSystemWithTCP Member Functions Summary

Type	Member Function
int	InitTCPServer(uint16 port, uint8 automataType, char *ipAddress=0, unsigned char *parm=0, int length=0)
	This function initializes the TCP server. Once initialized, the server waits for a request to establish the TCP connection with a remote client.

TABLE 6.6

FiniteStateMachine Constructor Summary

```
FiniteStateMachine(uint16 numOfTimers=DEFAULT_TIMER_NO, uint16
numOfState=DEFAULT_STATE_NO, uint16
maxNumOfProceduresPerState=DEFAULT_PROCEDURE_NO_PE_STATE, bool
getMemory=true)
```
  This constructor initializes the object that represents the instance of a given automata type along with the data structures needed for its proper operation.

The following sections contain a detailed description of *FSMSystem* library API functions.

6.8.1 *FSMSystem*

Function prototype:
```
FSMSystem(
  uint8 numOfAutomata,
  uint8 numberOfMbx)
```

Function description: This constructor initializes the object that represents the FSM system together with the data structures needed for its proper operation.

Parameters:

numOfAutomata: the number of various automata types to be added to the FSM system

numberOfMbx: the number of mailboxes to be used by the FSM system

Note: Typically, a single mailbox is assigned to each automata type but other arrangements are also allowed. Normally, an automata type corresponds to a protocol. For example, the IP protocol may be implemented as one automata type and the TCP protocol may be implemented as another automata type. A typical arrangement would be to assign one mailbox to IP and one to TCP. Another arrangement would be to assign two mailboxes to each protocol. For example, in this arrangement, IP would use the first

TABLE 6.7

FiniteStateMachine Member Functions Summary

Type	Member Function
uint8*	AddParam(uint16 paramCode, uint32 paramLength, uint8 *param)
	This function is used to add the given parameter of the given length to the new message.
uint8*	AddParamByte(uint16 paramCode, BYTE param)
	This function is used to add the given parameter of length 1 byte to the new message.
uint8*	AddParamDWord(uint16 paramCode, DWORD param)
	This function is used to add the given parameter of length 4 bytes to the new message.
uint8*	AddParamWord(uint16 paramCode, WORD param)
	This function is used to add the given parameter of length 2 bytes to the new message.
virtual void	CheckBufferSize(uint32 paramLength)
	This function provides a new message buffer with the size sufficient to accept the parameter of the given length.
virtual void	ClearMessage()
	This function returns the buffer allocated for the current message to the kernel and assigns value *NULL* to the internal pointer to the current message. The current message is the last message received by the automata instance.
virtual void	CopyMessage()
	This function makes a copy of the current message and assigns that copy to the new message.
virtual void	CopyMessage(uint8 *msg)
	This function makes a copy of the given message and assigns that copy to the new message.
virtual void	CopyMessageInfo(uint8 infoCoding, uint16 lengthCorrection=0)
	This function copies the part of the message containing the useful information, referred to as a payload (message without its header), from the current to the new message.
virtual void	Discard(uint8* buff)
	This function deletes the message placed in the given buffer and returns the buffer to the kernel.
void	DoNothing()
	This function performs no operation. It is called when the automata receives an unexpected message unless a new function is provided to handle unexpected messages.
void	Free FSM()
	This function reports to the FSM system that the automata instance has finished its current assignment and is free for further assignments.
virtual uint8	GetAutomata()=0
	This function returns the identification of the automata type for this automata instance.

FSM Library

TABLE 6.7 (CONTINUED)

FiniteStateMachine Member Functions Summary

Type	Member Function
uint8	GetBitParamByteBasic(uint32 offset, uint32 mask=MASK_32_BIT)
	This function returns the value of the current message parameter of length 1 byte masked with the given mask.
uint16	GetBitParamWordBasic(uint32 offset, uint32 mask=MASK_32_BIT)
	This function returns the value of the current message parameter of length 2 bytes masked with the given mask.
uint32	GetBitParamDWordBasic(uint32 offset, uint32 mask=MASK_32_BIT)
	This function returns the value of the current message parameter of length 4 bytes masked with the given mask.
virtual uint8*	GetBuffer(uint32 length)
	This function returns the buffer whose size is not less than the size given by the value of its parameter.
uint32	GetBufferLength(uint8 *buff)
	This function returns the size of the given buffer in bytes.
virtual inline uint32	GetCallId()
	This function returns the identification of the communication process in which this instance is currently involved, e.g., the call ID.
uint32	GetCount(uint8 mbx)
	This function returns the current number of messages in the given mailbox.
virtual uint8	GetGroup()
	This function returns the identification of the group of automata to which this instance belongs.
virtual uint8	GetInitialState()
	This function returns the identification of the initial state of this automata type.
virtual inline uint8	GetLeftMbx()
	This function returns the identification of the mailbox assigned to the automata instance that is logically to the left of this automata instance.
virtual inline uint8	GetLeftAutomata()
	This function returns the identification of the automata type that is logically to the left of this automata instance.
virtual inline uint8	GetLeftGroup()
	This function returns the identification of the group of automata that is logically to the left of this automata instance.
virtual inline uint32	GetLeftObjectId()
	This function returns the identification of the automata instance that is logically to the left of this automata instance.
virtual uint8	GetMbxId()
	This function returns the identification of the mailbox assigned to this automata instance.

TABLE 6.7 (CONTINUED)
FiniteStateMachine Member Functions Summary

Type	Member Function
virtual MessageInterface*	GetMessageInterface(uint32 id)
	This function returns the object that governs the coding of messages used by this automata instance. The returned object is an instance of the class derived from the class *MessageInterface*.
uint8*	GetMsg()
	This function returns the first unread message from the mailbox assigned to this automata instance.
static uint8*	GetMsg(uint8 mbx)
	This function returns the first unread message from the mailbox identified by the value of its parameter.
inline uint32	GetMsgCallId()
	This function returns the identification of the communication process (e.g., call ID) from the current message.
inline uint16	GetMsgCode()
	This function returns the message code from the current message header.
inline uint8	GetMsgFromAutomata()
	This function returns the identification of the originating automata type from the current message.
inline uint8	GetMsgFromGroup()
	This function returns the identification of the group of the originating automata instance for the current message.
inline uint8	GetMsgInfoCoding()
	This function returns the identification of the information coding scheme used for the current message.
inline uint16	GetMsgInfoLength()
	This function returns the payload length of the current message in bytes.
inline uint16	GetMsgInfoLength(uint8 *msg)
	This function returns the payload length of the given message in bytes. The message is specified by its pointer.
inline uint32	GetMsgObjectNumberFrom()
	This function returns the identification of the originating automata instance from the current message.
inline uint32	GetMsgObjectNumberTo()
	This function returns the identification of the destination automata instance from the current message.
inline uint8	GetMsgToAutomata()
	This function returns the identification of the destination automata type from the current message.
inline uint8	GetMsgToGroup()
	This function returns the identification of the type of group of the destination automata from the current message.
inline uint8*	GetNewMessage()
	This function returns the address of the buffer that contains the new message.

TABLE 6.7 (CONTINUED)

FiniteStateMachine Member Functions Summary

Type	Member Function
inline uint8	GetNewMsgInfoCoding()
	This function returns the identification of the information coding scheme used for the new message.
inline uint16	GetNewMsgInfoLength()
	This function returns the payload length of the new message in bytes.
uint8*	GetNextParam(uint16 paramCode)
	This function returns the address of the next instance of the given type of message parameter within the current message.
bool	GetNextParamByte(uint16 paramCode, BYTE &param)
	This function searches for the next instance of the given type of the single-byte parameter in the current message. If the instance is found, the function copies it into its parameter specified by the reference and returns the value *true*; otherwise, it returns the value *false*.
bool	GetNextParamDWord(uint16 paramCode, DWORD &param)
	This function searches for the next instance of the given type of the 4-byte parameter in the current message. If the instance is found, the function copies it into its parameter specified by the reference and returns the value *true*; otherwise, it returns the value *false*.
bool	GetNextParamWord(uint16 paramCode, WORD &param)
	This function searches for the next instance of the given type of the 2-byte parameter in the current message. If the instance is found, the function copies it into its parameter specified by the reference and returns the value *true*; otherwise, it returns the value *false*.
virtual uint32	GetObjectId()
	This function returns the unique identification of this automata instance.
uint8*	GetParam(uint16 paramCode)
	This function returns the address of the first instance of the given type of the message parameter within the current message.
bool	GetParamByte(uint16 paramCode, BYTE &param)
	This function searches for the first instance of the given type of single-byte parameter in the current message. If the instance is found, the function copies it into its parameter specified by the reference and returns the value *true*; otherwise, it returns the value *false*.

TABLE 6.7 (CONTINUED)
FiniteStateMachine Member Functions Summary

Type	Member Function
bool	GetParamDWord(uint16 paramCode, DWORD &param)
	This function searches for the first instance of the given type of 4-byte parameter in the current message. If the instance is found, the function copies it into its parameter specified by the reference and returns the value *true*; otherwise, it returns the value *false*.
bool	GetParamWord(uint16 paramCode, WORD &param)
	This function searches for the first instance of the given type of 2-byte parameter in the current message. If the instance is found, the function copies it into its parameter specified by the reference and returns the value *true*; otherwise, it returns the value *false*.
PROC_FUN_PTR	GetProcedure(uint16 event)
	This function returns the pointer to the event handler for the given event identifier and the current state of automata.
virtual inline uint8	GetRightMbx()
	This function returns the identification of the mailbox assigned to the automata instance that is logically to the right of this automata instance.
virtual inline uint8	GetRightAutomata()
	This function returns the identification of the automata type that is logically to the right of this automata instance.
virtual inline uint8	GetRightGroup()
	This function returns the identification of the type of the group of automata that is logically to the right of this automata instance.
virtual inline uint32	GetRightObjectId();
	This function returns the identification of the automata instance that is logically to the right of this automata instance.
virtual inline uint8	GetState()
	This function returns the identification of the current state of this automata instance.
virtual bool	IsBufferSmall(uint8 *buff, uint32 length)
	This function returns the value *true* if the size of the given buffer is not greater than the given size specified as the value of its second parameter; otherwise, it returns the value *false*.
virtual void	Initialize()
	This function defines the automata state transition event handlers and timers used by this automata type.
void	InitEventProc(uint8 state, uint16 event, PROC_FUN_PTR fun)
	This function defines the given state transition event handler for the given automata state and the given event (message code).

TABLE 6.7 (CONTINUED)

FiniteStateMachine Member Functions Summary

Type	Member Function
void	`InitTimerBlock(uint16 tmrId, uint32 count, uint16 signalId)`
	This function initializes the given timer by the given duration and the timer expiration message code.
void	`InitUnexpectedEventProc(uint8 state, PROC_FUN_PTR fun)`
	This function defines the given state transition event handler for unexpected events in the given automata state.
bool	`IsTimerRunning(uint16 id)`
	This function returns the value *true* if the given timer is active (running); otherwise, it returns the value *false*.
void	`NoFreeObjectProcedure(uint8 *msg)`
	This function defines the behavior of this automata type if the list of free automata of this type is used and if it is empty at the moment when a free instance is requested.
virtual void	`NoFreeInstances()`
	This function defines the behavior of the FSM system if a list of free automata is used and if it is empty at the moment when a free instance is requested.
virtual bool	`ParseMessage(uint8 *msg)`
	This function checks if the given message is coded properly and if it is, it becomes the current message (its pointer is assigned to the internal variable *CurrentMessage*).
virtual void	`PrepareNewMessage(uint8 *msg)`
	This function defines the given buffer as the new message buffer by assigning the given pointer to the internal variable *NewMessage*. The buffer is used as a working area for the construction of the new message.
virtual void	`PrepareNewMessage(uint32 length, uint16 code, uint8 infoCode = LOCAL_PARAM_CODING)`
	This function creates the new message of the given length with the given message code and the given type of information coding.
virtual void	`Process(uint8 *msg)`
	This function performs the preparations for the message processing and selects the state transition event handler based on the message code and current state of this automata instance.
void	`PurgeMailBox()`
	This function purges all the messages from the mailbox assigned to this automata type and releases all the buffers assigned to the messages.
bool	`RemoveParam(uint16 paramCode)`
	This function removes the given type of message parameter from the new message.
virtual void	`Reset()`
	This function resets this automata instance by returning it to its initial state and stopping all its active timers.

TABLE 6.7 (CONTINUED)
FiniteStateMachine Member Functions Summary

Type	Member Function
void	ResetTimer(uint16 id)
	This function resets the internal timer block object and returns the buffer allocated by the *StartTimer* primitive to the FSM library kernel.
void	RestartTimer(uint16 tmrId)
	This function restarts the given timer. It is logically equivalent to a sequence of *StopTimer* and *StartTimer* primitives.
virtual void	RetBuffer(uint8 *buff)
	This function returns the given buffer to the FSM library kernel. Normally, each memory buffer is returned at the end of its life cycle. The failure to do so leads to the memory leak problem.
void	ReturnMsg(uint8 mbxId)
	This function makes a copy of the current message and sends it to the given mailbox. This primitive is used frequently for message forwarding. On many occasions, the communication process must react in this simple way.
void	SetBitParamByteBasic(BYTE param, uint32 offset, uint32 mask=MASK_32_BIT)
	This function sets the given single byte parameter of the new message to the result of the bit-wise inclusive OR operation applied to the given parameter and its previous value masked (bit-wise AND operation) with the given bit-mask.
void	SetBitParamDWordBasic(DWORD param, uint32 offset, uint32 mask=MASK_32_BIT)
	This function sets the given 4-byte parameter of the new message to the result of the bit-wise inclusive OR operation applied to the given parameter and its previous value masked (bit-wise AND operation) with the given bit-mask.
void	SetBitParamWord(WORD param, uint32 offset, uint32 mask=MASK_32_BIT)
	This function sets the given 2-byte parameter of the new message to the result of the bit-wise inclusive OR operation applied to the given parameter and its previous value masked (bit-wise AND operation) with the given bit-mask.
inline void	SetCallId()
	This function sets the default value of the attribute *CallId* of this automata instance.
inline void	SetCallId(uint32 id)
	This function sets the given value of the attribute *CallId* of this automata instance.
inline void	SetCallIdFromMsg()
	This function sets the attribute *CallId* of this automata instance to the value of the parameter *CallId* of the current message. This primitive is used to store the reference number specific to the communication protocol.

FSM Library

TABLE 6.7 (CONTINUED)

FiniteStateMachine Member Functions Summary

Type	Member Function
virtual void	SetDefaultFSMData()
	This function sets the automata specific data to their default values. It is typically used before the normal operation phase.
virtual void	SetDefaultHeader(uint8 infoCoding)
	This function sets the default header field values for the given type of the message information coding.
inline void	SetGroup(uint8 id)
	This function sets the identification of the group of automata for this automata type to the given value. This primitive is used to declare the group membership.
virtual void	SetInitialState()
	This function sets the current state of this automata instance to its initial state.
static void	SetKernelObjects(TPostOffice *postOffice, TBuffers *buffers, CTimer *timer)
	This function sets the *FSMSystem* library kernel objects (post office, buffers, and timers), which are common for all of the automata in the FSM system.
inline void	SetLeftMbx(uint8 mbx)
	This function sets the identification of the mailbox assigned to the automata instance that is logically to the left of this automata instance.
inline void	SetLeftAutomata(uint8 automata)
	This function sets the identification of the automata type that is logically to the left of this automata instance.
inline void	SetLeftObject(uint8 group)
	This function sets the identification of the type of the group of automata that is logically to the left of this automata instance.
inline void	SetLeftObjectId(uint32 id)
	This function sets the identification of the automata instance that is logically to the left of this automata instance.
static void	SetLogInterface(LogInterface *logingObject)
	This function defines the object responsible for message logging. The object is an instance of a class derived from the class *LogInterface*.
inline void	SendMessage(uint8 mbxId)
	This function sends the new message to the given mailbox. The mailbox is specified by its identification.
inline void	SendMessage(uint8 mbxId, uint8 *msg)
	This function sends the given message to the given mailbox.
void	SetMessageFromData()
	This function sets the header fields of the new message related to the originating automata instance to the values specific to this automata instance.
inline void	SetMsgCallId(uint32 id)
	This function sets the call ID parameter of the new message to the given value.

TABLE 6.7 (CONTINUED)

FiniteStateMachine Member Functions Summary

Type	Member Function
inline void	`SetMsgCallId(uint32 id, uint8 *msg)`
	This function sets the call ID parameter of the given message to the given value.
inline void	`SetMsgCode(uint16 code)`
	This function sets the message code parameter of the new message to the given value.
inline void	`SetMsgCode(uint16 code, uint8 *msg)`
	This function sets the message code parameter of the given message to the given value.
inline void	`SetMsgFromAutomata(uint8 from)`
	This function sets the type of the originating automata parameter of the new message to the given value.
inline void	`SetMsgFromAutomata(uint8 from, uint8 *msg)`
	This function sets the type of the originating automata parameter of the given message to the given value.
inline void	`SetMsgFromGroup(uint8 from)`
	This function sets the type of the originating group of automata parameter of the new message to the given value.
inline void	`SetMsgFromGroup(uint8 from, uint8 *msg)`
	This function sets the type of the originating group of automata parameter of the given message to the given value.
inline void	`SetMsgInfoCoding(uint8 codingType)`
	This function sets the message information coding parameter of the new message to the given value.
inline void	`SetMsgInfoCoding(uint8 codingType, uint8 *msg)`
	This function sets the message information coding parameter of the given message to the given value.
inline void	`SetMsgInfoLength(uint16 length)`
	This function sets the message payload (useful information) length parameter of the new message.
inline void	`SetMsgInfoLength(uint16 length, uint8 *msg)`
	This function sets the message payload (useful information) length parameter of the given message.
inline void	`SetMsgObjectNumberFrom(uint32 from)`
	This function sets the originating automata instance identification parameter of the new message to the given value.
inline void	`SetMsgObjectNumberFrom(uint32 from, uint8 *msg)`
	This function sets the originating automata instance identification parameter of the given message to the given value.
inline void	`SetMsgObjectNumberTo(uint32 to)`
	This function sets the destination automata instance identification parameter of the new message to the given value.

FSM Library

TABLE 6.7 (CONTINUED)

FiniteStateMachine Member Functions Summary

Type	Member Function
inline void	SetMsgObjectNumberTo(uint32 to, uint8 *msg)
	This function sets the destination automata instance identification parameter of the given message to the given value.
inline void	SetMsgToAutomata(uint8 to)
	This function sets the destination automata type identification parameter of the new message to the given value.
inline void	SetMsgToAutomata(uint8 to, uint8 *msg)
	This function sets the destination automata type identification parameter of the given message to the given value.
inline void	SetMsgToGroup(uint8 to)
	This function sets the destination automata group identification parameter of the new message to the given value.
inline void	SetMsgToGroup(uint8 to, uint8 *msg)
	This function sets the destination automata group identification parameter of the given message to the given value.
void	SendMessageLeft()
	This function sends the new message to the mailbox assigned to the automata instance that is logically to the left of this automata instance.
void	SendMessageRight()
	This function sends the new message to the mailbox assigned to the automata instance that is logically to the right of this automata instance.
inline void	SetNewMessage(uint8 *msg)
	This function sets the new message to the given message by assigning the given message pointer to the internal pointer to the new message.
inline void	SetObjectId(uint32 id)
	This function sets the identification of this automata instance to the given value.
inline void	SetRightMbx(uint8 mbx)
	This function sets the identification of the mailbox assigned to the automata instance that is logically to the right of this automata instance.
inline void	SetRightAutomata(uint8 automata)
	This function sets the identification of the automata type that is logically to the right of this automata instance.
inline void	SetRightObject(uint8 group)
	This function sets the identification of the type of the group of automata that is logically to the right of this automata instance.

TABLE 6.7 (CONTINUED)

FiniteStateMachine Member Functions Summary

Type	Member Function
`inline void`	`SetRightObjectId(uint32 id)`
	This function sets the identification of the automata instance that is logically to the right of this automata instance.
`inline void`	`SetState(uint8 state)`
	This function sets the identification of the current state of this automata instance.
`void`	`StartTimer(uint16 tmrId)`
	This function starts the given timer. The timer is specified by its identification.
`void`	`StopTimer(uint16 tmrId)`
	This function stops the given timer. The timer is specified by its identification.
`static void`	`SysClearLogFlag()`
	This function stops the logging of the messages exchanged by the automata.
`static void`	`SysStartAll()`
	This function starts the logging of the messages exchanged by the automata.

TABLE 6.8

NetFSM Constructor Summary

`NetFSM(uint16 numOfTimers=DEFAULT_TIMER_NO, uint16 numOfState=DEFAULT_STATE_NO, uint16 maxNumOfProceduresPerState=DEFAULT_PROCEDURE_NO_PER_STATE, bool getMemory=true)`
The constructor initializes the object that represents an instance of the given automata type along with the data structures needed for its proper operation.

mailbox to receive the messages from network interfaces (drivers) and the second to receive the messages from TCP. Yet another arrangement would be to assign a single mailbox to all the protocols. Finally, a set of mailboxes can be used to prioritize the messages. For example, three mailboxes may be used to distinguish high, middle, and low priority messages.

6.8.2 *Add(ptrFiniteStateMachine, uint8, uint32, bool)*

Function prototype:

```
void Add(
 ptrFiniteStateMachine object,
 uint8 automataType,
 uint32 numberOfObjects,
 bool useFreeList = false)
```

FSM Library

TABLE 6.9

NetFSM Member Functions Summary

Type	Member Function
virtual void	convertFSMToNetMessage()
	This function converts the internal message format into the external message format appropriate for the transmission over the TCP/IP network.
virtual uint16	convertNetToFSMMessage()
	This function converts the external message format into the internal message format appropriate for the communication within the FSM system.
void	establishConnection()
	This function establishes the TCP connection between two geographically distributed FSM systems.
virtual uint8	getProtocolInfoCoding()
	This function returns the identification of the type of the external message coding.
void	sendToTCP()
	This function sends the new message to the remote FSM system over the previously established TCP connection.

Function description: This function adds the first instance of each automata type to the FSM system. At the same time, this function defines the unique identification of this automata type and the number of instances of this automata type that will be subsequently added to the FSM system. It also declares a group of instances of this automata type as either a set of resources to be used individually or as a pool of resources of the same type available for dynamic allocation.

Function parameters:

object: the pointer to the first instance of this automata type to be added to the FSM system

automataType: the unique identification of this type of automata

numberOfObjects: the total number of instances of this type to be added to the FSM system.

useFreeList: the indicator selecting the mode of usage of individual instances of this type.

Note: Typically, the FSM system is created at system startup and then groups of various automata types are added to it. As a rule, the first instance of the given automata type is added by this function. Its parameters specify, in order from left to right, the pointer to the first object of this type, the identification of this automata type, the total number of instances that will be added to the FSM system, and the mode of individual instance allocation. This last parameter has a default value *false*, which means that each automata instance represents an individual resource. If this default is overridden by the value *true*, the group of instances of this automata type represents a pool of resources of the same type. The individual instances from this pool are allocated dynamically and on-demand, based on the use of the internal FSMSystem library kernel list of resources of the given type. (This is the origin of the name of the last parameter of this function, *useFreeList*.) This dynamic allocation is requested by sending a message to an unknown automata, which is identified by the instance identification set to the value –1 (see function *SetMsgObjectNumberTo*).

6.8.3   Add(ptrFiniteStateMachine, uint8)

Function prototype:
```
void Add(
  ptrFiniteStateMachine object,
  uint8 automataType)
```

Function description: This function adds all the automata instances except the first instance of the given type to the FSM system. It assumes that the first instance of this automata type has been added previously to the FSM system by calling the overloaded function *Add* with four parameters in its signature.

Function parameters:

object: the pointer to the instance of this automata type to be added to the FSM system

automataType: the unique identification of this automata type

Note: As already mentioned, after the FSM system is created at system startup, the groups of various automata types are added to it. As a rule, the first instance of the given automata type is added by the overloaded function *Add* with four parameters in its signature (see the previous section for more details on its parameters). All the other instances of the given automata type are added to the FSM system by this overloaded function *Add*. An advantage of differentiating these two functions becomes obvious in a dynamic environment where objects are created on-demand and added to the FSM system. If the given automata type already exists and a need arises for another instance of it, this overloaded *Add* function is sufficient.

6.8.4   *InitKernel*

Function prototype:
```
void InitKernel(
  uint8 buffClassNo,
  uint32 *buffersCount,
  uint32 *buffersLength,
  uint8 numOfMbxs=0,
```

FSM Library

```
TimerResolutionEnum timerRes = Timer1s)
```

Function description: This function initializes the elements of the kernel responsible for time, buffer, and message management. The parameters of this function specify the number of buffer types, the number of instances per each buffer type and their lengths, the number of mailboxes to be used by the automata added to the FSM system, and the basic timer resolution. The default value of the basic timer resolution is 1 sec, which is defined by the symbolic constant *Timer1s*.

Function parameters:

buffClassNo: the number of buffer types

buffersCount: the pointer to the array of the numbers of instances of the corresponding buffer types

buffersLength: the pointer to the array of the sizes of the corresponding buffer types

numOfMbxs: the number of the mailboxes

timerRes: the basic timer resolution

Note: This function essentially initializes the *FSMSystem* library kernel. It must be called after the FSM system has been created and before it can be started. It also assumes that the arrays of the cardinal numbers and the sizes of individual buffer types are already created and filled by the programmer. Because the specification of the buffers to be provided by the kernel may look cumbersome, we provide the following example. Suppose that a need arises for three buffer types, namely, small, medium, and large. The programmer should set the first parameter of this function to the number 3. Next, suppose that the programmer needs 300 small buffers, 200 medium buffers, and 100 large buffers, and that their sizes should be 64, 128, and 256 bytes, respectively. Before calling this function, the programmer should create the following two arrays:

- Array of cardinal numbers = [300, 200, 100]
- Array of sizes = [64, 128, 256]

Finally, the programmer should specify the pointers to these two arrays as the second and the third parameter of this function.

6.8.5 *Remove(uint8)*

Function prototype:
```
void Remove(unit8 automataType)
```

Function description: This function removes all instances of the given automata type from the FSM system.

Function parameters:
automataType — the type of automata to be removed from the system

Note: First, the FSM system removes all instances of the given automata type from the FSM system. Next, the kernel frees all the memory zones occupied by the internal data structures used by the automata of this type.

6.8.6 Remove(uint8, uint32)

Function prototype:
```
ptrFiniteStateMachine Remove(
  uint8 automataType
  uint32 object)
```

Function description: This function removes the given instance of the given automata type. The parameters of this function specify the identification of the automata type and the identification of the automata instance.

Function parameters:
automataType: the identification of the automata type
object: the identification of the instance of the given automata type

Function returns: This function returns the pointer to the automata instance removed from the FSM system.

6.8.7 Start

Function prototype:
```
virtual void Start()
```

Function description: This function starts the FSM system and is the main function of the FSM system. In this function, the FSM system thread enters a loop in which it reads the kernel mailboxes and distributes the messages to the destination automata.

Note: The FSM system thread remains in the loop while the internal attribute *SystemWorking* is set to the value *true*. A typical implementation of the FSM system thread is shown in the example in Section 6.2.1.2.

6.8.8 StopSystem

Function prototype:
```
void StopSystem()
```

FSM Library   345

Function description: This function stops the FSM system. It sets the internal attribute *SystemWorking* to the value *false*, thus causing the FSM system thread to exit its loop and stop the FSM system.

Note: If the function *Start* has been called from the separate operating system thread, the call to the function *StopSystem* will cause the termination of that thread.

6.8.9 *FSMSystemWithTCP*

Function prototype:
```
FSMSystemWithTCP(
  uint8 numOfAutomata,
  uint8 numberOfMbx)
```

Function description: This constructor initializes the object that represents the FSM system supporting communication over TCP/IP network, along with the data structures needed for its proper operation. Its parameters specify the number of automata types to be added to the FSM system and the number of mailboxes.

Function parameters:

numOfAutomata: the number of automata types that will be added to the FSM system

numberOfMbx: the number of mailboxes that will be used by the automata added to the FSM system

Note: Typically, a single mailbox is assigned to each automata type included in the FSM system but other arrangements are also allowed. For example, a single mailbox may be assigned to all the automata types included in the FSM system. Also allowed is to assign an arbitrary number of mailboxes to each automata type, e.g., to enable message prioritization.

6.8.10 *InitTCPServer*

Function prototype:
```
int InitTCPServer(
  uint16 port,
  unit8 automataType,
  char *ipAddress = 0,
  unsigned char *parm = 0,
  int length = 0)
```

Function description: This function initializes the TCP server. Once initialized, the server waits for a request to establish the TCP connection with a remote client. The parameters of this function specify the number of the TCP port on which the server awaits the connection request, the automata type included in the FSM system engaged in the communication, the server IP address, the pointer to the area where the connection parameters should be passed to the specified automata type, and the parameter lengths in bytes. After the reception of the request, the server allocates an instance of the given automata type and passes the connection together with the received parameters to the allocated automata instance. Further communication continues directly between the remote client and the allocated automata instance, i.e., the server is completely isolated from it.

Function parameters:

Port: the number of the TCP port on which the server awaits a connection request

automataType: the automata type included in the FSM system that is engaged in the communication. This automata type must be derived from the class *NetFSM*. After the connection has been initially estbalished, the server transfers it to the allocated instance of this automata type.

ipAddress: the pointer to the server IP address

parm: the pointer to the area where the parameters received while establishing the connection should be passed and subsequently taken by to the specified automata type

length: the parameter lengths specified by the previous pointer, in bytes

Function returns: If the TCP server awaiting a request from a remote client is successfully started, this function returns the value 0. Otherwise, it returns the value –1.

Note: This function should be called only once, just initially to start the TCP server.

6.8.11 *FiniteStateMachine*

Function prototype:
```
FiniteStateMachine(
  unit16 numOfTimers = DEFAULT_TIMER_NO,
  uint16 numOfState = DEFAULT_STATE_NO,
  uint16 maxNumOfProceduresPerState = DEFAULT_PROCEDURE_NO_PER_STATE,
  bool getMemory = true)
```

Function description: This constructor initializes the object that represents the instance of a given automata type together with the data structures needed for its proper operation. Its parameters specify the number of the timers to be used by this automata type, the number of the states that this automata type has, the maximal number of state transitions per state, and the indicator specifying whether this constructor should reserve the memory for the objects that represent the states and state transitions of this automata

FSM Library

type or not. The default value of this indicator is *true*, which means that this constructor is responsible for memory allocation.

Function parameters:
numOfTimers: the number of the timers to be used by this automata type
numOfState: the number of the states that this automata type has
maxNumOfProceduresPerState: the maximal number of state transitions per state
getMemory: the memory allocation indicator (by default, its value is *true*)

Note: This constructor may be called either with some or without any of the parameters. If the parameter is not specified, the constructor will use its default value. The indicator *getMemory* may be set to the value *false* when the programmer wants to do manual memory allocation to optimize overall memory consumption.

6.8.12 *AddParam*

Function prototype:
```
uint8 *AddParam(
  uint16 paramCode,
  uint32 paramLength,
  uint8 *param)
```

Function description: This function is used to add the given parameter of the given length to the new message. The parameters of this function specify the unique identification of the parameter type, the parameter length in bytes, and the pointer to the parameter itself. If the parameter to be added to the message is too large to fit in the buffer that is assigned to the new message, this function will get a bigger buffer, copy the new message into it, add the parameter, and release the old buffer.

Function parameters:
paramCode: the parameter type
paramLength: the parameter length, in bytes
param: the pointer to the parameter

Function returns: This function returns the pointer to the buffer that contains the new message.

Note: This function enables the programmer to add a parameter of an arbitrary size to the new message with the limitation that it must not exceed the maximal parameter length specified for the given type of message coding (e.g., for the type *StandardMessage*, the maximal parameter length is 256 bytes). The message parameters in *StandardMessage* are sorted in the ascending order of their corresponding type identifiers.

6.8.13 *AddParamByte*

Function prototype:
```
uint8 *AddParamByte(
  uint16 paramCode,
  BYTE param)
```

Function description: This function is used to add the given parameter of length 1 byte to the new message. The parameters of this function specify the unique identification of the parameter type and the parameter value.

Function parameters:
paramCode: the parameter type
param: the parameter value

Function returns: This function returns the pointer to the buffer that contains the new message.

Note: The total message length must not exceed the limit specified for the given type of message coding. In any case, it must not exceed 8G bytes.

6.8.14 *AddParamDWord*

Function prototype:
```
uint8 *AddParamDWord(
  uint16 paramCode,
  DWORD param)
```

Function description: This function is used to add the given parameter of length 4 bytes to the new message. The parameters of this function specify the unique identification of the parameter type and the parameter value.

Function parameters:
paramCode: the parameter type
param: the parameter value

Function returns: This function returns the pointer to the buffer that contains the new message.

Note: The total message length must not exceed the limit specified for the given type of message coding. In any case, it must not exceed 232 bytes.

6.8.15 *AddParamWord*

Function prototype:
```
uint8 *AddParamDWord(
  uint16 paramCode,
  WORD param)
```

Function description: This function is used to add the given parameter of length 2 bytes to the new message. The parameters of this function specify the unique identification of the parameter type and the parameter value.

Function parameters:
paramCode: the parameter type
param: the parameter value

Function returns: This function returns the pointer to the buffer that contains the new message.

Note: The total message length must not exceed the limit specified for the given type of message coding. In any case, it must not exceed 8G bytes.

6.8.16 *CheckBufferSize*

Function prototype:
```
uint8 *CheckBufferSize(uint32 paramLength)
```

Function description: This function provides a new message buffer with the size sufficient to accept the parameter of the given length. The parameter of this function specifies the parameter length in bytes.

Function parameters:
paramLength: the parameter length

Function returns: This function returns the pointer to the new message.

Note: This function is obsolete. In the previous version of the FSM library, this function ensured the new message buffer management was transparent to the programmer. Typically, the programmer would call this function before calling some of the *AddParam* functions to ensure that the new message is stored in a buffer of a sufficient size. This means that the buffer is large enough to accept a new parameter of the given size in addition to the current content of the new message. Behind the scenes, this function checked the current size of the new message. If it was not sufficient, the function allocated a new, larger buffer, copied the current new message into it, released the old buffer, and returned the pointer to the newly allocated buffer containing the new message. In the current version of the FSM library, all the *AddParam* functions call this function internally at their very beginning and the programmer need no longer call it explicitly.

6.8.17 *ClearMessage*

Function prototype:
```
virtual void ClearMessage()
```

Function description: This function returns the buffer allocated for the current message to the kernel and assigns the value *NULL* to the internal pointer to the current message. The current message is the last message received by the automata instance.

Note: If the *FSMSystem* library has been compiled for the debug mode, this function will additionally verify that the return value of the function is *NULL*.

6.8.18 *CopyMessage()*

Function prototype:
```
virtual void CopyMessage()
```

Function description: This function makes a copy of the current message and assigns that copy to the new message. By definition, a current message is the last received message and a new message is the message under construction to be subsequently sent. The value of the pointer to the current message copy is assigned to the internal pointer to the new message.

Note: This function first checks if the new message already exists by checking the internal pointer to the new message. If the new message has already been defined or is under construction (the internal pointer is not equal to the value *NULL*), the function releases the buffer that contains the new message and assigns the value *NULL* to the internal pointer. Next, the function makes a copy of the current message and assigns its address to the pointer to the new message. This function is typically used for message forwarding. The protocol *A* sends a message to the protocol *B*, which in turn forwards the copy of the same message to the protocol *C*.

6.8.19 *CopyMessage(uint*)*

Function prototype:
```
virtual void CopyMessage(uint8 *msg)
```

Function description: This function makes a copy of the given message and assigns that copy to the new message. The parameter of this function specifies the pointer to the original message.

Function parameters:
msg: the pointer to the original message

Note: This function assumes that the new message does not exist, i.e., the internal pointer to the new message should contain the value *NULL* before this function is called. However, if the new message already exists, this function will return its buffer and get a fresh buffer for the new message before copying the given message into it.

FSM Library 351

6.8.20 *CopyMessageInfo*

Function prototype:
```
virtual void CopyMessageInfo(
  uint8 infoCoding,
  uint16 lengthCorrection = 0)
```

Function description: This function copies the part of the message containing the useful information, referred to as a payload (message without its header), from the current into the new message stored in a newly allocated buffer. The parameters of this function specify the type of the information coding that governs the formatting and length correction of the message.

Function parameters:
infoCoding: the identification of the type of the information coding
lengthCorrection: the message length correction

Note: The message length correction depends on the type of applied information coding. If the new message buffer does not exist, this function will get a buffer, assign it to the new message, and make the required copy.

6.8.21 *Discard*

Function prototype:
```
virtual void Discard(uint8* buff)
```

Function description: This function deletes the message placed in the given buffer and returns the buffer to the kernel. The parameter of this function specifies the buffer to be cleared and released.

Function parameters:
buff: the pointer to the buffer

6.8.22 *DoNothing*

Function prototype:
```
void DoNothing()
```

Function description: This function performs no operation. It is called when automata receives an unexpected message unless a new function to handle unexpected messages is defined. By definition, an unexpected message is any type of message that has not been defined as a legal type of message in the current automata state.

Note: This function may be redefined by calling the function *InitUnexpectedEventProc* if a need exists for concrete functionality handling unexpected messages.

6.8.23 *FreeFSM*

Function prototype:
```
void FreeFSM()
```

Function description: This function reports to the FSM system that an automata instance has finished its current assignment and is free for further assignments. If the first instance of this automata type has been added to the FSM system with the parameter *useFreeList* set to the value *true*, the group of the instances of this automata type is viewed as a pool of resources. In that case, this function returns the resource to the corresponding pool by queuing it to the internal list of the resources of the same type.

Note: If a group of instances of this automata type is used as a set of individual resources rather than as a pool of resources (the parameter *useFreeList* has been set to the value *false* when the first automata instance has been added to the FSM system), this function has no effect.

6.8.24 *GetAutomata*

Function prototype:
```
virtual uint8 GetAutomata() = 0
```

Function description: This function returns the identification of the automata type for this automata instance.

Function returns: This function returns the unique ID of the automata type.

Note: This function is a pure virtual function, which means that it must be defined in the class that models some concrete automata type. Typically, this function returns the constant value that represents the required identification. It finds this constant by looking up the table of identifications created by reading the file of all the known automata types at the FSM system startup time.

6.8.25 *GetBitParamByteBasic*

Function prototype:
```
unit8 GetBitParamByteBasic(
  uint32 offset,
  uint32 mask=MASK_32_BIT)
```

Function description: This function returns the value of the current message parameter of length 1 byte masked with the given mask. The parameters

FSM Library 353

of this function specify the offset of the original parameter of the message and the value of the mask.

> Function parameters:
> *offset*: the offset of the original parameter of the message
> *mask*: the value of the mask

Function returns: This function returns the result of the bit-wise AND operation between the value of the message parameter at the given message *offset* and the given value of the parameter *mask*.

Note: Normally, depending on the value of the parameter mask, testing the value of a single bit, or of a group of bits simultaneously, is possible in the parameter of size 1 byte that is at a given distance from the beginning of the message.

6.8.26 *GetBitParamWordBasic*

Function prototype:
```
unit8 GetBitParamWordBasic(
  uint32 offset,
  uint32 mask=MASK_32_BIT)
```

Function description: This function returns the value of the current message parameter of length 2 bytes masked with the given mask. The parameters of this function specify the offset of the original parameter of the message and the value of the mask.

> Function parameters:
> *offset*: the offset of the original parameter of the message
> *mask*: the value of the mask

Function returns: This function returns the result of the bit-wise AND operation between the value of the message parameter at the given message *offset* and the given value of the parameter *mask*.

Note: Normally, depending on the value of the parameter mask, testing the value of a single bit, or a group of bits simultaneously, is possible in the parameter of size 2 bytes that is at a given distance from the beginning of the message.

6.8.27 *GetBitParamDWordBasic*

Function prototype:
```
unit8 GetBitParamDWordBasic(
  uint32 offset,
  uint32 mask=MASK_32_BIT)
```

Function description: This function returns the value of the current message parameter of length 4 bytes masked with the given mask. The parameters of this function specify the offset of the original parameter of the message and the value of the mask.

Function parameters:
offset: the offset of the original parameter of the message
mask: the value of the mask

Function returns: This function returns the result of the bit-wise AND operation between the value of the message parameter at the given message *offset* and the given value of the parameter *mask*.

Note: Normally, depending on the value of the parameter mask, testing the value of a single bit, or of a group of bits simultaneously, is possible in the parameter of size 4 bytes that is at a given distance from the beginning of the message.

6.8.28 *GetBuffer*

Function prototype:
```
virtual uint8 *GetBuffer(uint32 length)
```

Function description: This function returns a buffer whose size is not less than the size given by the value of its parameter. The parameter of this message specifies the minimal buffer length in bytes.

Function parameters:
length: the buffer length

Function returns: This function returns the pointer to a newly allocated buffer.

Note: The *FSMSystem* library kernel handles a limited number of buffer types with a limited number of instances per each type defined during the kernel initialization by calling the function *InitKernel*. By definition, this function first searches for the buffer types of the size that ideally match the desired buffer. If such a type does not exist, the function searches for the next size greater buffer types. This allocation policy may yield a buffer of a size much bigger than needed, and frequent occurrence of this type of allocation may lead to inefficient memory usage. For example, suppose that the programmer has mistakenly defined only two buffer sizes, small and large, such that not a single protocol message can fit into the small buffer. In this case, only the large buffers will be consumed and the small buffers will not be used at all. Therefore, special care must be taken when defining the buffers before calling the function *InitKernel*.

FSM Library 355

Now let us go back to the buffer allocation algorithm. When this function finds a buffer type of a sufficient size, it checks for a free buffer of that type. If no such type is found, the system is badly designed and a new buffer type must be added to the system. If such a buffer type exists but no free buffers of that type are available, the function will look for the next size buffer. If all the buffers of the sufficient size are already allocated, the FSM system experiences the memory exhaustion problem. In academic examples, the system is allowed to crash under these circumstances. However, industrial-strength applications require implementation of additional mechanisms, such as overload protection and intelligent automatic restarts.

6.8.29 *GetBufferLength*

Function prototype:
```
uint32 GetBufferLength(uint8 *buff)
```

Function description: This function returns the size of the given buffer in bytes. The parameter of this function specifies the pointer to the buffer.

Function parameters:
buff: the address of the buffer

Function returns: This function returns the specified buffer length in bytes.

6.8.30 *GetCallId*

Function prototype:
```
virtual inline uint32 GetCallId()
```

Function description: This function returns the identification of the communication process that this instance is currently involved in, e.g., the call ID. The actual meaning of this identification is application specific.

Function returns: This function returns the value of the attribute *CallId*.

Note: Historically, the attribute *CallId* is tied to call processing (e.g., Q.71) and signaling (e.g., SS7, DSS1) protocols, but it also proved to be useful in modern multimedia protocols (e.g., H.323 and SIP). Generally, this attribute may be used as an identification of the process or of the transaction that engages more cooperative automata. If a single attribute is not sufficient the programmer may introduce additional attributes in classes derived from the base class *FiniteStateMachine*.

6.8.31 GetCount

Function prototype:
```
uint32 GetCount(uint8 mbx)
```

Function description: This function returns the current number of messages in the given mailbox. The parameter of this message specifies the identification of the mailbox.

Function parameters:
mbx: the mailbox identification

Function returns: This function returns the number of unread messages contained in the mailbox of interest.

6.8.32 GetGroup

Function prototype:
```
virtual uint8 GetGroup()
```

Function description: This function returns the identification of the group of automata to which this instance belongs.
Function returns: This function returns a number that uniquely identifies the group of automata which, besides other members, includes this automata instance.

6.8.33 GetInitialState

Function prototype:
```
virtual uint8 GetInitialState()
```

Function description: This function returns the identification of the initial state of this automata type.
Function returns: This function returns the number that uniquely identifies the initial state of this automata type.
Note: The default value of the initial state is 0.

6.8.34 GetLeftMbx

Function prototype:
```
virtual inline uint8 GetLeftMbx()
```

Function description: This function returns the identification of the default mailbox assigned to the automata instance that is logically to the left of this automata instance.

Function returns: This function returns the number that uniquely identifies the default mailbox assigned to the left automata instance.

Note: Historically, the terms *left* and *right* automata instance originate from SDL, where an automata instance typically communicates with its left and right neighbors. These neighbors might have their own mailboxes, sometimes briefly called left and right mailboxes.

6.8.35   GetLeftAutomata

Function prototype:
```
virtual inline uint8 GetLeftAutomata()
```

Function description: This function returns the identification of the automata type that is logically to the left of this automata instance.

Function returns: This function returns the number that uniquely identifies the left automata type.

Note: By definition, left automata are logically placed to the left of the currently observed automata instance.

6.8.36   GetLeftGroup

Function prototype:
```
virtual linline uint8 GetLeftGroup()
```

Function description: This function returns the identification of the group of automata that is logically to the left of this automata instance.

Function returns: This function returns the number that uniquely identifies the left group of automata.

Note: By definition, a left group of automata is a group that contains left automata.

6.8.37   GetLeftObjectId

Function prototype:
```
virtual inline uint32 GetLeftObjectId()
```

Function description: This function returns the identification of the automata instance that is logically to the left of this automata instance.

Function returns: This function returns the number that uniquely identifies the left automata instance.

Note: By definition, left automata are logically placed to the left of the currently observed automata instance. This function returns the identification of the particular left automata instance with which the currently observed automata instance communicates.

6.8.38 GetMbxId

Function prototype:
```
virtual uint8 GetMbxId()
```

Function description: This function returns the identification of the default mailbox assigned to this automata type. Note that an instance of a given automata type may receive its messages through any mailbox, i.e., through the default mailbox as well as through other mailboxes. Alternately, a single mailbox may by assigned to more than one automata type.

Function returns: This function returns the number that uniquely identifies the default mailbox assigned to this automata instance.

Note: This function is a pure virtual function, which means that it must be defined by the programmer when he writes a class derived from the class *FiniteStateMachine*. Typically, this function returns the constant value that represents the required mailbox identification (the content of the corresponding class field). This constant can be initially determined by looking up the table of identifications and set by calling the function *SetMbxId*. The table of identifications can be created by reading the file containing all the known automata types at the FSM system startup time. A mailbox ID is typically a record field that describes a single automata type.

6.8.39 GetMessageInterface

Function prototype:
```
virtual MessageInterface *GetMessageInterface(uint32 id) = 0
```

Function description: This function returns the object that governs the coding of messages used by this automata instance. The parameter of this function specifies the identification of the information coding scheme. The returned object is an instance of the class derived from the class *MessageInterface*.

Function parameters:
id: the information coding scheme

Function returns: This function returns the pointer to the object responsible for parsing and coding the messages used by this automata instance.

Note: This function is a virtual function, which means that it must be defined when the programmer writes a class derived from the class *FiniteStateMachine*. The identification with the value 0 is reserved for the information coding used by the format of the class *StandardMessage*, which is a basic type of a message supported by the *FSMSystem* library.

6.8.40 GetMsg()

Function prototype:
```
uint8* GetMsg()
```

Function description: This function returns the first unread message from the mailbox assigned to this automata instance.

Function returns: This function returns a pointer to the buffer that has been removed from the head of the list, which is hidden by the abstraction of the mailbox assigned to this automata instance. If no such buffer exists, i.e., if the list is empty, the function returns the value *NULL*.

6.8.41 GetMsg(uint8)

Function prototype:
```
static uint8* GetMsg(uint8 mbx)
```

Function description: This function returns the first unread message from the given mailbox. The parameter of this function specifies the identification of the mailbox.

Function parameters:
mbx: the mailbox ID

Function returns: This function returns the pointer to the buffer that has been removed from the head of the list, which is hidden by the abstraction of the given mailbox. If no such buffer exists, i.e., if the list is empty, the function returns the value *NULL*.

Note: Although this function is defined as a static function, a call to this function is not allowed before the kernel initialization and the FSM system startup. The call to this function made before that may cause unpredictable behavior.

6.8.42 GetMsgCallId

Function prototype:
```
inline uint32 GetMsgCallId()
```

Function description: This function returns the identification of the communication process (e.g., call ID) from the current message.

Function returns: This function returns the value of the attribute *CallId*.

Note: The attribute *CallId* is application specific. It can be used to indicate a process or a transaction in which more cooperating automata are involved. The size of *CallId* is 32 bits. It is considered large enough for most of the applications. To increase the size of *CallId*, the programmer would need to modify the base class *FiniteStateMachine*.

6.8.43 GetMsgCode

Function prototype:
```
inline uint16 GetMsgCode()
```

Function description: This function returns the message code from the current message header.

Function returns: This function returns the value of the message code from the header of the current (last received) message.

6.8.44 GetMsgFromAutomata

Function prototype:
```
inline uint8 GetMsgFromAutomata()
```

Function description: This function returns the identification of the originating automata type from the current message. This value is provided from the header of the current message.

Function returns: This function returns the value of the identification of the automata type that has created and sent the current message to this automata instance.

6.8.45 GetMsgFromGroup

Function prototype:
```
inline uint8 GetMsgFromGroup()
```

Function description: This function returns the identification of the group of the originating automata instance for the current message. This value is provided from the header of the current message.

Function returns: This function returns the value of the identification of the group of automata instance that has created and sent the current message to this automata instance.

6.8.46 GetMsgInfoCoding

Function prototype:
```
inline uint8 GetMsgInfoCoding()
```

Function description: This function returns the identification of the information coding scheme used for the current message.

Function returns: This function returns the value that identifies the type of information coding that has been used to create the current message.

Note: This information is provided from the header of the current message.

FSM Library 361

6.8.47 GetMsgInfoLength()

Function prototype:
```
inline uint16 GetMsgInfoLength()
```

Function description: This function returns the payload length of the current message in bytes.

Function returns: This function returns the value of the current message payload size in bytes.

Note: The length of the message header is not included in the length returned by this message. By definition, the total message length is the sum of the length of the message header and the length of the message payload.

6.8.48 GetMsgInfoLength(uint8*)

Function prototype:
```
inline uint16 GetMsgInfoLength(uint8 *msg)
```

Function description: This function returns the payload length of the given message in bytes. The parameter of this function specifies the pointer to the message.

Function parameters:
msg: the pointer to the message

Function returns: This function returns the value of the size of the given message payload in bytes.

Note: The length of the message header is not included in the length returned by this message. By definition, the total message length is the sum of the length of the message header and the length of the message payload.

6.8.49 GetMsgObjectNumberFrom

Function prototype:
```
inline uint32 GetMsgObjectNumberFrom()
```

Function description: This function returns the identification of the originating automata instance from the current message.

Function returns: This function returns the value that identifies the automata instance that has created and sent the message.

Note: This value is provided from the header of the current (last received) message.

6.8.50 *GetMsgObjectNumberTo*

Function prototype:
```
inline uint32 GetMsgObjectNumberTo()
```

Function description: This function returns the identification of the destination automata instance from the current message. This value is actually this automata instance.

Function returns: This function returns the value that identifies the automata instance that has received the message and that must process it.

Note: This value is provided from the header of the current (last received) message.

6.8.51 *GetMsgToAutomata*

Function prototype:
```
inline uint8 GetMsgToAutomata()
```

Function description: This function returns the identification of the destination automata type from the current message. This value is actually this automata type.

Function returns: This function returns the value that identifies the automata type that should receive the message and that should process it.

Note: This value is provided from the header of the current (last received) message.

6.8.52 *GetMsgToGroup*

Function prototype:
```
inline uint8 GetMsgToGroup()
```

Function description: This function returns the identification of the type of the group of the destination automata from the current message. This value is actually the group to which this automata type belongs.

Function returns: This function returns the value that identifies the group of automata that has received the message and that must process it.

Note: This value is provided from the header of the current (last received) message.

6.8.53 *GetNewMessage*

Function prototype:
```
inline uint8 *GetNewMessage()
```

FSM Library 363

Function description: This function returns the address of the buffer that contains the new message.

Function returns: This function returns the pointer to the already defined new message or the message under construction.

Note: If the new message does not exist, this function returns the value *NULL*. This function assumes that the programmer has already allocated a buffer for the new message by previously calling the function *PrepareNewMessage* or calling the function *GetBuffer*.

6.8.54 *GetNewMsgInfoCoding*

Function prototype:
```
inline uint8 GetNewMsgInfoCoding()
```

Function description: This function returns the identification of the information coding scheme used for the new message.

Function returns: This function returns the value that uniquely identifies the type of information coding.

Note: This value is provided from the header of the new message.

6.8.55 *GetNewMsgInfoLength*

Function prototype:
```
inline uint16 GetNewMsgInfoLength()
```

Function description: This function returns the payload length of the new message in bytes.

Function returns: This function returns the value of the new message payload size in bytes.

Note: The length of the message header is not included in the length returned by this message. By definition, the total message length is the sum of the length of the message header and the length of the message payload.

6.8.56 *GetNextParam*

Function prototype:
```
uint8 *GetNextParam(uint16 paramCode)
```

Function description: This function returns the address of the next instance of the given parameter type within the current message. The parameter of this function specifies the type of the message parameter.

Function parameters:
paramCode: the identification of the type of the message parameter

Function returns: The function returns the pointer to the next instance of the message parameter. If it does not exist, the function returns the value *NULL*.

Note: This function cannot be used by the programmer to get the first instance of the message parameter of a given type. It assumes that the first instance has already been provided by calling the function *GetParam*. Typically, the function *GetParam* is called once to provide the first instance of the parameter and then called iteratively to provide the next instances of the parameter.

6.8.57 GetNextParamByte

Function prototype:
```
bool GetNextParamByte(
  uint16 paramCode,
  BYTE &param)
```

Function description: This function searches for the next instance of the given type of the single-byte parameter in the current message. If the instance is found, the function copies it into its parameter specified by the reference and returns the value *true*; otherwise, it returns the value *false*. The parameters of this function specify the identification of the type of the message parameter and the pointer to the memory area where this function should store the next instance of the message parameter.

Function parameters:
paramCode: the identification of the type of the message parameter

param: the pointer to the memory area reserved by the programmer for the next instance of the message parameter

Function returns: This function returns the value *true* if the next instance of the message parameter is found. If the instance is not found, this function returns the value *false*.

Note: The programmer cannot use this function to get the first instance of the message parameter of the given type. This function assumes that the first instance has already been provided by calling the function *GetParamByte*. Typically, the function *GetParamByte* is called once to provide the first instance of the parameter and then called iteratively to provide the next instances of the parameter.

6.8.58 GetNextParamDWord

Function prototype:
```
bool GetNextParamDWord(
  uint16 paramCode,
  DWORD &param)
```

Function description: This function searches for the next instance of the given type of parameter 4 bytes in the current message. If the instance is found, the function copies it into its parameter specified by the reference and returns the value *true*; otherwise, it returns the value *false*. The parameters of this function specify the identification of the type of the message parameter and the pointer to the memory area where this function should store the next instance of the message parameter.

Function parameters:

paramCode: the identification of the type of message parameter

param: the pointer to the memory area reserved by the programmer for the next instance of the message parameter

Function returns: This function returns the value *true* if the next instance of the message parameter is found. If the instance is not found, this function returns the value *false*.

Note: The programmer cannot use this function to get the first instance of the message parameter of the given type. This function assumes that the first instance has already been provided by calling the function *GetParamDWord*. Typically, the function *GetParamDWord* is called once to provide the first instance of the parameter and then called iteratively to provide the next instances of the parameter.

6.8.59 *GetNextParamWord*

Function prototype:
```
bool GetNextParamWord(
  uint16 paramCode,
  WORD &param)
```

Function description: This function searches for the next instance of the given type of parameter 2 bytes in the current message. If the instance is found, the function copies it into its parameter specified by the reference and returns the value *true*; otherwise, it returns the value *false*. The parameters of this function specify the identification of the type of the message parameter and the pointer to the memory area where this function should store the next instance of the message parameter.

Function parameters:

paramCode: the identification of the type of message parameter

param: the pointer to the memory area reserved by the programmer for the next instance of the message parameter

Function returns: This function returns the value *true* if the next instance of the message parameter is found. If the instance is not found, this function returns the value *false*.

Note: The programmer cannot use this function to get the first instance of the message parameter of the given type. This function assumes that the first instance has already been provided by the call to the function *GetParamWord*. Typically, the function *GetParamWord* is called once to provide the first instance of the parameter and then called iteratively to provide the next instances of the parameter.

6.8.60 *GetObjectId*

Function prototype:
```
virtual uint32 GetObjectId()
```

Function description: This function returns the unique identification of this automata instance.

Function returns: This function returns the value that uniquely identifies this particular automata instance.

Note: This value has been automatically assigned to this automata instance by the function *Add*, which is called to add this automata instance to the FSM system.

6.8.61 *GetParam*

Function prototype:
```
uint8 *GetParam(uint16 paramCode)
```

Function description: This function returns the address of the first instance of the given type of the message parameter within the current message. The parameter of this function specifies the identification of the parameter type.

Function parameters:
paramCode: the identification of the parameter type

Function returns: This function returns the pointer to the first instance of the message parameter within the current message. If no message parameters of the given type are found, this function returns the value *NULL*.

Note: This function returns the pointer to the beginning of the message parameter. The format of the message parameter is governed by the selected type of the message information coding. For example, the parameter of the message *StandardMessage* consists of three fields. These fields are the parameter type (stored in 2 bytes), the parameter length (stored in 1 byte), and the information part of the parameter (stored in the number of bytes determined by the content of the previous field of the parameter).

6.8.62 GetParamByte

Function prototype:
```
bool GetParamByte(
  uint16 paramCode,
  BYTE &param)
```

Function description: This function searches for the first instance of the given type of single-byte parameter in the current message. If the instance is found, the function copies it into its parameter specified by the reference and returns the value *true*; otherwise, it returns the value *false*. The parameters of this function specify the identification of the type of the message parameter and the pointer to the memory area where this function should store the first instance of the message parameter.

Function parameters:

paramCode: the identification of the type of message parameter

param: the pointer to the memory area reserved by the programmer for the next instance of the message parameter

Function returns: This function returns the value *true* if the first instance of the message parameter is found. If the instance is not found, this function returns the value *false*.

Note: The programmer must use this function to get the first instance of the message parameter of the given type. Typically, this function is called once to provide the first instance of the parameter and then the function *GetNextParamByte* is called iteratively to provide the next instances of the parameter.

6.8.63 GetParamDWord

Function prototype:
```
bool GetParamDWord(
  uint16 paramCode,
  DWORD &param)
```

Function description: This function searches for the first instance of the given type of parameter 4 bytes in the current message. If the instance is found, the function copies it into its parameter specified by the reference and returns the value *true*; otherwise, it returns the value *false*. The parameters of this function specify the identification of the type of message parameter and the pointer to the memory area where this function should store the first instance of the message parameter.

Function parameters:

paramCode: the identification of the type of message parameter

param: the pointer to the memory area reserved by the programmer for the next instance of the message parameter

Function returns: This function returns the value *true* if the first instance of the message parameter is found. If the instance is not found, this function returns the value *false*.

Note: The programmer must use this function to get the first instance of the message parameter of the given type. Typically, this function is called once to provide the first instance of the parameter and then the function *GetNextParamDWord* is called iteratively to provide the next instances of the parameter.

6.8.64 *GetParamWord*

Function prototype:
```
bool GetParamWord(
  uint16 paramCode,
  BYTE &param)
```

Function description: This function searches for the first instance of the given type of parameter 2 bytes in the current message. If the instance is found, the function copies it into its parameter specified by the reference and returns the value *true*; otherwise, it returns the value *false*. The parameters of this function specify the identification of the type of message parameter and the pointer to the memory area where this function should store the first instance of the message parameter.

Function parameters:
paramCode: the identification of the type of message parameter

param: the pointer to the memory area reserved by the programmer for the next instance of the message parameter

Function returns: This function returns the value *true* if the first instance of the message parameter is found. If the instance is not found, this function returns the value *false*.

Note: The programmer must use this function to get the first instance of the message parameter of the given type. Typically, this function is called once to provide the first instance of the parameter and then the function *GetNextParamWord* is called iteratively to provide the next instances of the parameter.

6.8.65 *GetProcedure*

Function prototype:
```
PROC_FUN_PTR GetProcedure(uint16 event)
```

Function description: This function returns the pointer to the event handler for the given event identifier and the current state of automata. The parameter of this function specifies the identification of the event type.

Function parameters:
event: the identification of the event type (message code)

Function returns: This function returns the pointer to the event handler. Essentially, the event handler is a C++ class function member that handles the given event type in the current state.

Note: The FSM system internal data structures contain all the necessary information about the automata states, the sets of recognizable events (messages) for all automata states, and the corresponding event handlers. This information must be defined for each automata type after it has been added to the FSM system by the function *Add*. The programmer specifies this information in the parameters of the function *Initialize*. If the event handler has not been specified by the function *Initialize* for the given event type in the current automata state, this function returns the pointer to the function *DoNothing*, which performs the default processing of the unexpected events (messages).

6.8.66 *GetRightMbx*

Function prototype:
```
virtual inline uint8 GetRightMbx()
```

Function description: This function returns the identification of the default mailbox assigned to the automata instance that is logically to the right of this automata instance.

Function returns: This function returns the number that uniquely identifies the default mailbox for the right automata instance.

Note: Historically, the terms *left* and *right* automata instance originate from SDL, where an automata instance typically communicates with its left and right neighbors. These neighbors have their own mailboxes, sometimes briefly called left and right mailboxes.

6.8.67 *GetRightAutomata*

Function prototype:
```
virtual inline uint8 GetRightAutomata()
```

Function description: This function returns the identification of the automata type that is logically to the right of this automata instance.

Function returns: This function returns the number that uniquely identifies the right automata type.

Note: By definition, right automata are logically placed to the right of the currently observed automata instance.

6.8.68 *GetRightGroup*

Function prototype:
```
virtual linline uint8 GetRightGroup()
```

Function description: This function returns the identification of the group of automata that is logically to the right of this automata instance.

Function returns: This function returns the number that uniquely identifies the right group of automata.

Note: By definition, a right group of automata is a group that contains right automata.

6.8.69 *GetRightObjectId*

Function prototype:
```
virtual inline uint32 GetRightObjectId()
```

Function description: This function returns the identification of the automata instance that is logically to the right of this automata instance.

Function returns: This function returns the number that uniquely identifies the right automata instance.

Note: By definition, right automata are logically placed to the right of the currently observed automata instance. This function returns the identification of the particular right automata instance with which the currently observed automata instance communicates.

6.8.70 *GetState*

Function prototype:
```
virtual inline uint8 GetState()
```

Function description: This function returns the identification of the current state of this automata instance.

Function returns: This function returns the value that uniquely identifies the current state of this automata instance.

6.8.71 *IsBufferSmall*

Function prototype:
```
virtual bool IsBuferSmall(
  uint8 *buff,
  uint32 length)
```

Function description: This function returns the value *true* if the size of the given buffer is not greater than the given size specified as the value of its second parameter; otherwise, it returns the value *false*. The parameters of this function specify the buffer whose size is to be checked and the size to be used as a measuring unit.

Function parameters:
buff: the pointer to the buffer whose size is to be checked
length: the value of the measuring unit

Function returns:
This function returns the value *true* if the size of the given buffer is less than or equal to the given size. If the buffer size is greater than the given size, the function returns the value *false*.

6.8.72   Initialize

Function prototype:
```
virtual void Initialize() = 0
```

Function description: This function defines the automata state transition event handlers and timers used by this automata type. State transition event handlers are essentially the C++ functions defined by the programmer, which process events (messages). Timers are primitive time mechanisms used to restrict the duration of certain communication phases.

Note: While writing the function *Initialize*, the programmer normally defines the functions that process the expected events (messages) by calling the function *InitEventProc*, the functions that process the unexpected events by calling the function *InitUnexpectedEventProc*, and the timers by calling the function *InitTimerBlock*.

6.8.73   InitEventProc

Function prototype:
```
void InitEventProc(
  uint8 state,
  uint16 event,
  PROC_FUN_PTR fun)
```

Function description: This function defines the given state transition event handler for the given automata state and the given event (message code). The parameters of this function specify the identification of the state of this automata type, the identification of the event type, and the pointer to the event handler.

Function parameters:

state: the identification of the state of this automata type

event: the identification of the event type

fun: the pointer to the event handler

Note: This function may be used only within the definition of the function *Initialize*. A sequence of calls to this function fills in the internal state table for this automata type. This table is used by the FSM system and this automata type during its normal operation to locate the event handler that corresponds to the given pair (state, event).

6.8.74 *InitTimerBlock*

Function prototype:
```
void InitTimerBlock (
  uint16 tmrId,
  uint32 count,
  uint16 signalId)
```

Function description: This function initializes the given timer by the given duration and the timer expiration message code. The parameters of this function specify the timer identification, the timer duration, and the identification of the message to be sent to this automata type when the specified timer expires.

Function parameters:

tmrId: the timer identification

count: the timer duration (in timer ticks)

signalId: the identification of the message (signal) to be sent by the specified timer

Note: The timer identification is a value selected by the programmer. This value uniquely identifies the timer to the automata type that uses it in all the timer-related primitives, namely, *InitTimerBlock*, *ResetTimer*, *RestartTimer*, *StartTimer*, and *StopTimer*. Uniqueness of identifiers is limited to the scope of a single automata type. If the timer expires, it sends a special message (referred to as a *signal*) to the automata instance that has started that timer. The code of this message is set to the value of the parameter *SignalId*. The kernel calculates the absolute timer duration in seconds by dividing the time resolution specified for automata type with the time resolution of the FSM system and by multiplying this result with the basic timer resolution specified as the parameter of the function *InitKernel*.

FSM Library 373

6.8.75 *InitUnexpectedEventProc*

Function prototype:
```
void InitUnexpectedEventProc(
  uint8 state,
  PROC_FUN_PTR fun)
```

Function description: This function defines the given state transition event handler for unexpected events in the given automata state. The parameters of the function specify the automata state and the unexpected event handler, which is essentially a C++ function that handles unexpected events (messages).

Function parameters:
state: the value that uniquely identifies the automata state
fun: the pointer to the unexpected event handler

Note: If the unexpected event (message) handler does not exist because it has not been defined by this function, the FSM system and this automata type will use the function *DoNothing* to handle unexpected messages for all the states in which the unexpected message is not defined.

6.8.76 *IsTimerRunning*

Function prototype:
```
bool IsTimerRunning(uint16 id)
```

Function description: This function returns the value *true* if a given timer is active (running); otherwise, it returns the value *false*. The parameter of this function specifies the timer identification.

Function parameters:
id: the timer identification

Function returns: This function returns the value *true* if the timer is running. If the timer is not active, this function returns the value *false*.
Note: The timer may not be active because it has not been started at all or it has been started but has expired in the meantime.

6.8.77 *NoFreeObjectProcedure*

Function prototype:
```
void NoFreeObjectProcedure(uint8 *msg)
```

Function description: This function defines the behavior of this automata type if the list of free automata of this type is used and if it is empty at the moment when a free instance is requested. The parameter of this function specifies the pending event (message).

Function parameters:
msg: the pointer to the pending message

Note: This function is used if a group of automata of this type is used as a pool of resources of the same type. This function is called if the message related to this automata type appears and no available automata instances (resources) of this type are available. The programmer should write his own function to handle this situation in an application-specific way. This situation is additionally handled at the level of the FSM system by the function *NoFreeInstances*.

6.8.78 NoFreeInstances

Function prototype:
```
virtual void NoFreeInstances() = 0
```

Function description: This function defines the behavior of the FSM system if a list of free automata is used and if it is empty at the moment when a free instance is requested.

Note: This function is used if a group of automata of this type is used as a pool of resources of the same type within the FSM system. This function is called if the message related to this automata type appears and no available automata instances (resources) of this type are available. The programmer should write his own function to handle this situation in an application-specific way. This situation is additionally handled at the level of this automata type by the function *NoFreeObjectProcedure*.

6.8.79 ParseMessage

Function prototype:
```
virtual bool ParseMessage(uint8 *msg)
```

Function description: This function checks if the given message is coded properly and, if it is, it becomes the current message (its pointer is assigned to the internal variable *CurrentMessage*). The parameter of this function specifies the message to be parsed.

Function parameters:
msg: the pointer to the message to be parsed

Function returns: This function returns the value *true* if the message syntax is correct; otherwise, it returns the value *false*.

FSM Library 375

Note: This function is called internally for each received message. Normally, this function is called after the reception of the message to check its syntax. If the message syntax is correct, further message processing functions are called. Otherwise, the FSM system reports an error and discards the syntactically incorrect message.

6.8.80 PrepareNewMessage(uint8*)

Function prototype:
```
virtual void PrepareNewMessage(uint8 *msg)
```

Function description: This function defines the given buffer as the new message buffer by assigning the given pointer to the internal variable *NewMessage*. The buffer is used by this automata instance as a working area for the construction of the new message. The parameter of this function specifies the buffer.

Function parameters:
msg: the pointer to the buffer

Note: If the programmer wants to create a new message, he would normally call the function *GetBuffer* to obtain the buffer for the construction of the message. Next, the programmer would call this function to declare the buffer provided by the kernel as the buffer that will contain the new message. After this declaration, the programmer may use all the functions from the family of functions that operate on the new message to construct the new message. Basically, these are the *AddParamX* functions.

6.8.81 PrepareNewMessage(uint32, uint16, uint8)

Function prototype:
```
  virtual void PrepareNewMessage(
    uint32 length,
    uint16 code,
 uint8 infoCode = LOCAL_PARAM_CODING)
```

Function description: This function creates the new message of the given length with the given message code and the given type of information coding. The parameters of this function specify the message length, the message code, and the identification of the type of message information coding.

Function parameters:
length: the message length

code: the value of the message code

infoCode: the identification of the type of message information coding

Note: Dealing with static messages of fixed and known sizes is easy. In this case, the programmer normally knows the size of the message he must create. The programmer creates the new message by calling this function and specifying the size as the value of the function parameter *length*. However, dealing with dynamic messages is more complicated because the message length might not be known in advance. In this case, the programmer may specify the value 0 as the value of the parameter *length*. This function in its turn will create the empty message that has its header but has no payload. Further on, the programmer typically uses functions *AddParamX* to dynamically add new parameters to the message. Whenever not enough room exists for the new parameter in the existing new message buffer, the function *AddParamX* transparently allocates a bigger buffer, moves the content of the new message into it, and releases the smaller buffer. Of course, the price paid for this flexibility is the processing overhead for the transparent buffer management.

6.8.82   *Process*

Function prototype:
```
virtual void Process(uint8 *msg)
```

Function description: This function performs the preparations for the message processing and selects the state transition event handler based on the message code and current state of this automata instance. After completion of the message processing, this function releases the buffer used by the message. The parameter of this function specifies the message to be processed.

Function parameters:
msg: the pointer to the message to be processed

Note: This function is called internally by this automata type. Because this function is virtual, the programmer may define the message handling procedure in accordance with the application-specific requirements.

6.8.83   *PurgeMailBox*

Function prototype:
```
void PurgeMailBox()
```

FSM Library

Function description: This function purges all the messages from the mailbox assigned to this automata type and releases all the buffers assigned to the messages.

Note: Notice that the mailbox is assigned to an automata type rather than to an individual instance of this type. This means that the mailbox may contain the messages addressed to different instances of this type. This function does not differentiate the messages. Instead, it simply purges all of them.

6.8.84 *RemoveParam*

Function prototype:
```
bool RemoveParam(uint16 paramCode)
```

Function description: This function removes the given type of message parameter from the new message. The parameter of this function specifies the identification of the type of message parameter.

Function parameters:
paramCode: the value that uniquely identifies the type of message parameter

Function returns: This function returns the value *true* if the given type of the message parameter is successfully found and removed. If the new message does not contain the given type, this function returns the value *false*.

Note: Removing the type of message parameter with identification 0 is not recommended because it marks the end of the parameters in the message. *FSMSystem* library debug version will report an error in that case and stop the program execution.

6.8.85 *Reset*

Function prototype:
```
virtual void Reset()
```

Function description: This function resets this automata instance by returning it to its initial state and stopping all its active timers.

Note: If the programmer wants to specify some additional actions to be undertaken during the restart operation, he may redefine this default behavior by writing the corresponding function member of a class derived from the class *FiniteStateMachine*.

6.8.86 *ResetTimer*

Function prototype:
```
void ResetTimer(uint16 id)
```

Function description: This function resets the internal timer block object and returns the buffer allocated by the *StartTimer* primitive to the FSM library kernel. The parameter of this function specifies the identification of the timer.

Function parameters:
id: the value that uniquely identifies the timer

6.8.87 *RestartTimer*

Function prototype:
```
void RestartTimer(uint16 tmrId)
```

Function description: This function restarts the given timer. It is logically equivalent to a sequence of *StopTimer* and *StartTimer* primitives. The parameter of this function specifies the identification of the timer.

Function parameters:
tmrId: the value that uniquely identifies the timer

6.8.88 *RetBuffer*

Function prototype:
```
virtual void RetBuffer(uint8 *buff)
```

Function description: This function returns the given buffer to the FSM library kernel. Normally, each memory buffer is returned at the end of its life cycle. The failure to do so leads to the memory leak problem. The parameter of this function specifies the buffer to be released.

Function parameters:
buff: the pointer to the buffer to be released

Note: The programmer must pay special attention to releasing the buffers when they are not needed anymore because the *FSMSystem* library does not include the garbage collector. Memory outage causes the exception that will stop the program execution.

6.8.89 *ReturnMsg*

Function prototype:
```
void ReturnMsg(uint8 mbxId)
```

Function description: This function makes a copy of the current message and sends it to the given mailbox. This primitive is used frequently for message forwarding. On many occasions, the communication process must react in this simple way. The parameter of this function specifies the identification of the mailbox.

Function parameters:
mbxId: the value that uniquely identifies the mailbox

6.8.90 *SetBitParamByteBasic*

Function prototype:
```
void SetBitParamByteBasic(
  BYTE param,
  uint32 offset,
  uint32 mask = MASK_32_BIT)
```

Function description: This function sets the given single-byte parameter of the new message to the result of the bit-wise inclusive OR operation applied to the given parameter and its previous value masked (bit-wise AND operation) with the given bit-mask. The parameters of this function specify the value of the single-byte parameter, the offset of the target parameter of the new message, and the value of the bit-mask.

Function parameters:

param: the value of the single-byte parameter

offset: the target parameter of the new message

mask: the value of the bit-mask

6.8.91 *SetBitParamDWordBasic*

Function prototype:
```
void SetBitParamDWordBasic(
  DWORD param,
  uint32 offset,
  uint32 mask = MASK_32_BIT)
```

Function description: This function sets the given 4-byte parameter of the new message to the result of the bit-wise inclusive OR operation applied to the given parameter and its previous value masked (bit-wise AND operation) with the given bit-mask. The parameters of this function specify the value of the 4-byte parameter, the offset of the target parameter of the new message, and the value of the bit-mask.

Function parameters:

param: the value of the 4-byte parameter

offset: the target parameter of the new message

mask: the value of the bit-mask

6.8.92 SetBitParamWordBasic

Function prototype:
```
void SetBitParamWordBasic(
  WORD param,
  uint32 offset,
  uint32 mask = MASK_32_BIT)
```

Function description: This function sets the given 2-byte parameter of the new message to the result of the bit-wise inclusive OR operation applied to the given parameter and its previous value masked (bit-wise AND operation) with the given bit-mask. The parameters of this function specify the value of the 2-byte parameter, the offset of the target parameter of the new message, and the value of the bit-mask.

Function parameters:

param: the value of the 2-byte parameter

offset: the target parameter of the new message

mask: the value of the bit-mask

6.8.93 SetCallId()

Function prototype:
```
inline void SetCallId()
```

Function description: This function sets the default value of the attribute *CallId* of this automata instance.

Note: This function automatically allocates the first available identification and assigns it to the protected class attribute *CallId*, completely transparent to the programmer.

6.8.94 SetCallId(uint32)

Function prototype:
```
inline void SetCallId(uint32 id)
```

Function description: This function sets the given value of the attribute *CallId* of this automata instance. The parameter of this function specifies the value to be assigned to the attribute *CallId*.

Function parameters:

id: the value to be assigned to the attribute *CallId*

Note: In contrast to an overloaded function without any parameters in its signature, this function enables the programmer to manually assign the value to the attribute *CallId*. However, this value must be unique. The programmer must pay special attention to the assignment of these numbers, especially if he mixes this function call with function calls to the overloaded function that assigns the default values.

6.8.95 *SetCallIdFromMsg*

Function prototype:
```
inline void SetCallIdFromMsg()
```

Function description: This function sets the attribute *CallId* of this automata instance to the value of the parameter *CallId* of the current message. This primitive is used to store the reference number specific to the communication protocol.

6.8.96 *SetDefaultFSMData*

Function prototype:
```
virtual void SetDefaultFSMData() = 0
```

Function description: This function sets the automata-specific data to their default values. It is typically used before the normal operation phase.

Note: The programmer must define this virtual function for a class derived from the class *FiniteStateMachine*. They do so by writing a C++ function that initializes the problem-specific data.

6.8.97 *SetDefaultHeader*

Function prototype:
```
virtual void SetDefaultHeader(uint8 infoCoding = 0)
```

Function description: This function sets the default header field values for the given type of the message information coding. The parameter of this function specifies the identification of the type of the message information coding.

Function parameters:
infoCoding: the type of the message information coding

Note: The programmer must define this virtual function for a class derived from the class *FiniteStateMachine*. They do so by writing a C++ function that fills in the protocol-specific data in the new message header.

6.8.98 SetGroup

Function prototype:
```
inline void SetGroup(uint8 id)
```

Function description: This function sets the identification of the group of automata for this automata type to the given value. This primitive is used to declare the group membership. The parameter of this function specifies the value to be assigned to the corresponding class attribute.

Function parameters:
id: the value that uniquely identifies the group of automata

6.8.99 SetInitialState

Function prototype:
```
virtual void SetInitialState()
```

Function description: This function sets the current state of this automata instance to its initial state.

Note: The programmer must obey the rule that the value of the identification of the initial automata state is 0.

6.8.100 SetKernelObjects

Function prototype:
```
static void SetKernelObjects(
  TPostOffice *postOffice,
  TBuffers *buffers,
  CTimer *timer)
```

Function description: This function sets the *FSMSystem* library kernel objects (post office, buffers, and timers), which are common for all the automata in the FSM system. The parameters of this function specify the post office object, the buffers object, and the timers object.

Function parameters:
postOffice: the pointer to the post office object
buffers: the pointer to the buffers object
timer: the pointer to the timers object

Note: This function is called internally by the function *InitKernel*. Remember that this function defines the kernel objects that are common for all automata types and all their instances. An accidental call to this function may cause unpredictable behavior of the FSM system.

FSM Library 383

6.8.101 *SetLeftMbx*

Function prototype:
```
inline void SetLeftMbx(uint8 mbx)
```

Function description: This function sets the default identification of the mailbox assigned to the automata instance that is logically to the left of this automata instance. The parameter of this function specifies the identification of the mailbox.

Function parameters:
mbx: the value that uniquely identifies the mailbox

6.8.102 *SetLeftAutomata*

Function prototype:
```
inline void SetLeftAutomata(uint8 automata)
```

Function description: This function sets the identification of the automata type that is logically to the left of this automata instance. The parameter of this function specifies the identification of the automata type.

Function parameters:
automata: the value that uniquely identifies the automata type

6.8.103 *SetLeftObject*

Function prototype:
```
inline void SetLeftObject(uint8 group)
```

Function description: This function sets the identification of the type of the group of automata that is logically to the left of this automata instance. The parameter of this function specifies the identification of the group of automata.

Function parameters:
group: the value that uniquely identifies the group of automata

6.8.104 *SetLeftObjectId*

Function prototype:
```
inline void SetLeftObjectId(uint32 id)
```

Function description: This function sets the identification of the automata instance that is logically to the left of this automata instance. The parameter of this function specifies the identification of the automata instance.

Function parameters:
id: the identification of the automata instance

6.8.105 *SetLogInterface*

Function prototype:
```
static void SetLogInterface(LogInterface *logingObject)
```

Function description: This function defines the object responsible for message logging. The object is an instance of a class derived from the class *LogInterface*. The parameter of this function specifies the message logging object.

Function parameters:
logingObject: the pointer to the message logging object

Note: The programmer must not call this function before the initialization of all the automata included in the FSM system has been finished. The logging object may log data to the file on the local mass memory unit (e.g., flash memory) or to the network file server. The log file is essential for debugging and test and verification purposes.

6.8.106 *SendMessage(uint8)*

Function prototype:
```
inline void SendMessage(uint8 mbxId)
```

Function description: This function sends the new message to the given mailbox. The parameter of this function specifies the identification of the mailbox.

Function parameters:
mbxId: the value that uniquely specifies the mailbox

Note: By definition, the internal pointer *NewMessage* points to the buffer that contains the new message. The programmer initializes this pointer by calling the function *PrepareNewMessage*.

6.8.107 *SendMessage(uint8, uint8*)*

Function prototype:

FSM Library

```
inline void SendMessage(
  uint8 mbxId,
  uint8 *msg)
```

Function description: This function sends the given message to the given mailbox. The parameters of this function specify the identification of the mailbox and the message to be sent to that mailbox.

Function parameters:
mbxId: the value that uniquely identifies the mailbox

msg: the pointer to the message

6.8.108 *SetMessageFromData*

Function prototype:
```
void SetMessageFromData()
```

Function description: This function sets the header fields of the new message related to the originating automata instance to the values specific to this automata instance. The data specifying the originating automata instance are its type, its group, and its identification.

Note: This function is automatically called from the function *SendMessage*.

6.8.109 *SetMsgCallId(uint32)*

Function prototype:
```
inline void SetMsgCallId(uint32 id)
```

Function description: This function sets the call ID parameter of the new message to the given value. The parameter of this function specifies the value of the call ID.

Function parameters:
id: the value of the call ID

Note: The call ID parameter has been traditionally used to identify a single telephone call. In general, it may be used to uniquely identify a communication process or a transaction that engages a group of automata that participate in its processing.

6.8.110 *SetMsgCallId(unit32, unit8*)*

Function prototype:
```
inline void SetMsgCallId(
  uint32 id,
  uint8 *msg)
```

Function description: This function sets the call ID parameter of the given message to the given value. The parameters of this function specify the value of the call ID and the target message.

Function parameters:
id: the value of the call ID
msg: the pointer to the buffer that contains the target message

Note: The value of the call ID parameter is the same for all the messages involved in a transaction or a process, e.g., a single telephone call.

6.8.111 *SetMsgCode(uint16)*

Function prototype:
```
inline void SetMsgCode(uint16 code)
```

Function description: This function sets the message code parameter of the new message to the given value. The parameter of this message specifies the message code.

Function parameters:
code: the message code

6.8.112 *SetMsgCode(uint16, uint8*)*

Function prototype:
```
inline void SetMsgCode(
  uint16 code,
  uint8 *msg)
```

Function description: This function sets the message code parameter of the given message to the given value. The parameters of this function specify the message code and the target message.

Function parameters:
code: the message code
msg: the pointer to the buffer that contains the target message

6.8.113 *SetMsgFromAutomata(uint8)*

Function prototype:
```
inline void SetMsgFromAutomata(uint8 from)
```

Function description: This function sets the type of the originating automata parameter of the new message to the given value. The parameter of this function specifies the identification of the automata type that is the message source.

Function parameters:
from: the identification of the automata type

Note: This function is automatically called by the function *SetMessageFromData*.

6.8.114 *SetMsgFromAutomata(uint8, uint8*)*

Function prototype:
```
inline void SetMsgFromAutomata(
  uint8 from,
  uint8 *msg)
```

Function description: This function sets the type of the originating automata parameter of the given message to the given value. The parameters of this function specify the type of the automata that is the message source and the target message.

Function parameters:
from: the automata type that is the message source
msg: the pointer to the buffer that contains the target message

6.8.115 *SetMsgFromGroup(uint8)*

Function prototype:
```
inline void SetMsgFromGroup(uint8 from)
```

Function description: This function sets the type of the originating group of automata parameter of the new message to the given value. The parameter of this message specifies the identification of the group of automata that is the message source.

Function parameters:
from: the identification of the group of automata that is the message source

Note: This function is automatically called by the function *SetMessageFromData*.

6.8.116 SetMsgFromGroup(uint8, uint8*)

Function prototype:
```
inline void SetMsgFromGroup(
  uint8 from,
  uint8 *msg)
```

Function description: This function sets the type of the originating group of automata parameter of the given message to the given value. The parameters of this function specify the identification of the group of automata that is the message source and the target message.

Function parameters:
from: the identification of the group of automata that is the message source
msg: the pointer to the buffer that contains the target message

6.8.117 SetMsgInfoCoding(uint8)

Function prototype:
```
inline void SetMsgInfoCoding(uint8 codingType)
```

Function description: This function sets the message information coding parameter of the new message to the given value. The parameter of this message specifies the identification of the information coding scheme.

Function parameters:
codingType: the value that uniquely specifies the information coding scheme

Note: This function is automatically called by the function *PrepareNewMessage*.

6.8.118 SetMsgInfoCoding(uint8, uint8*)

Function prototype:
```
inline void SetMsgInfoCoding(
  uint8 codingType,
  uint8 *msg)
```

Function description: This function sets the message information coding parameter of the given message to the given value. The parameters of this function specify the identification of the information coding scheme and the target message.

FSM Library 389

Function parameters:
codingType: the identification of the information coding scheme
msg: the pointer to the target message

6.8.119 SetMsgInfoLength(uint16)

Function prototype:
```
inline void SetMsgInfoLength(uint16 length)
```

Function description: This function sets the message payload (useful information) length parameter of the new message. The parameter of this function specifies the value of the payload length.

Function parameters:
length: the payload length in octets (bytes)

Note: All the *AddParamX* functions — which are responsible for adding parameters to the new message — call this function automatically to update the length of the message payload.

6.8.120 SetMsgInfoLength(uint16, uint8*)

Function prototype:
```
inline void SetMsgInfoLength(
  uint16 length,
  uint8 *msg)
```

Function description: This function sets the message payload (useful information) length parameter of the given message. The parameters of this function specify the value of the payload length and the target message.

Function parameters:
length: the payload length in octets (bytes)
msg: the pointer to the buffer that contains the target message

6.8.121 SetMsgObjectNumberFrom(uint32)

Function prototype:
```
inline void SetMsgObjectNumberFrom(uint32 from)
```

Function description: This function sets the originating automata instance identification parameter of the new message to the given value. The parameter of this function specifies the identification of the automata instance that is the message source.

Function parameters:
from: the identification of the automata instance that is the message source

Note: This function is automatically called by the function *SetMessageFromData*.

6.8.122   *SetMsgObjectNumberFrom(uint32, uint8*)*

Function prototype:
```
inline void SetMsgObjectNumberFrom(
 uint32 from,
 uint8 *msg)
```

Function description: This function sets the originating automata instance identification parameter of the given message to the given value. The parameters of this message specify the identification of the automata instance that is the message source and the target message.

Function parameters:
from: the identification of the automata instance that is the message source
msg: the pointer to the buffer that contains the target message

6.8.123   *SetMsgObjectNumberTo(uint32)*

Function prototype:
```
inline void SetMsgObjectNumberTo(uint32 to)
```

Function description: This function sets the destination automata instance identification parameter of the new message to the given value. The parameter of this function specifies the automata instance that is the message destination.

Function parameters:
to: the automata instance that is the message destination

6.8.124   *SetMsgObjectNumberTo(uint32, uint8*)*

Function prototype:
```
inline void SetMsgObjectNumberTo(uint32 to,uint8 *msg)
```

Function description: This function sets the destination automata instance identification parameter of the given message to the given value. The parameters of this function specify the automata instance that is the message destination and the target message.

Function parameters:
to: the automata instance that is the message destination
msg: the pointer to the buffer that contains the target message

6.8.125 *SetMsgToAutomata(uint8)*

Function prototype:
```
inline void SetMsgToAutomata(uint8 to)
```

Function description: This function sets the destination automata type identification parameter of the new message to the given value. The parameter of this function specifies the automata type that is the message destination.

Function parameters:
to: the automata type that is the message destination

6.8.126 *SetMsgToAutomata(uint8, uint8*)*

Function prototype:
```
inline void SetMsgToAutomata(
  uint8 to,
  uint8 *msg)
```

Function description: This function sets the destination automata type identification parameter of the given message to the given value. The parameters of this function specify the identification of the automata type that is the message destination and the target message.

Function parameters:
to: the identification of the automata type that is the message destination
msg: the pointer to the buffer that contains the target message

6.8.127 *SetMsgToGroup(uint8)*

Function prototype:
```
inline void SetMsgToGroup(uint8 to)
```

Function description: This function sets the destination automata group identification parameter of the new message to the given value. The parameter of this function specifies the identification of the group of automata that is the message destination.

Function parameters:

to: the identification of the group of automata that is the message destination

6.8.128 *SetMsgToGroup(uint8, uint8*)*

Function prototype:
```
inline void SetMsgToGroup(
  uint8 to,
  uint8 *msg)
```

Function description: This function sets the destination automata group identification parameter of the given message to the given value. The parameters of this function specify the identification of the group of automata that is the message destination and the target message.

Function parameters:

to: the identification of the group of automata that is the message destination

msg: the pointer to the buffer that contains the target message

6.8.129 *SendMessageLeft*

Function prototype:
```
void SendMessageLeft()
```

Function description: This function sends the new message to the mailbox assigned to the automata instance that is logically to the left of this automata instance.

Note: The programmer may use this function if he has already defined the left automata instance for the currently observed automata instance. This definition includes the definition of the mailbox assigned to the left automata instance. If the left automata instance and its mailbox are defined, this function automatically fills in all the data related to both source (originating) and destination automata instances within the new message and sends the new message to the left mailbox.

6.8.130 *SendMessageRight*

Function prototype:
```
void SendMessageLeft()
```

Function description: This function sends the new message to the mailbox assigned to the automata instance that is logically to the right of this automata instance.

FSM Library

Note: The programmer may use this function if he has already defined the right automata instance for the currently observed automata instance. This definition includes the definition of the mailbox assigned to the right automata instance. If the right automata instance and its mailbox are defined, this function automatically fills in all the data related to both source (originating) and destination automata instances within the new message and sends the new message to the right mailbox.

6.8.131 SetNewMessage

Function prototype:
```
inline void SetNewMessage(uint8 *msg)
```

Function description: This function sets the new message to the given message by assigning the given message pointer to the internal pointer to the new message. The parameter of this function specifies the target message.

Function parameters:
msg: the pointer to the buffer that contains the target message

6.8.132 SetObjectId

Function prototype:
```
inline void SetObjectId(uint32 id)
```

Function description: This function sets the identification of this automata instance to the given value. The parameter of this function specifies the identification of this automata instance.

Function parameters:
id: the value that uniquely identifies this automata instance

6.8.133 SetRightMbx

Function prototype:
```
inline void SetRightMbx(uint8 mbx)
```

Function description: This function sets the identification of the mailbox assigned to the automata instance that is logically to the right of this automata instance. The parameter of this message specifies the identification of the right mailbox for this automata instance.

Function parameters:
mbx: the identification of the right mailbox for this automata instance

6.8.134 *SetRightAutomata*

Function prototype:

```
inline void SetRightAutomata(uint8 automata)
```

Function description: This function sets the identification of the automata type that is logically to the right of this automata instance. The parameter of this function specifies the automata type that is to the right of this automata instance.

Function parameters:
automata: the identification of the automata type

6.8.135 *SetRightObject*

Function prototype:

```
inline void SetRightObject(uint8 group)
```

Function description: This function sets the identification of the type of the group of automata that is logically to the right of this automata instance. The parameter of this function specifies the type of the group of automata that is to the right of this automata instance.

Function parameters:
group: the identification of the group of automata

6.8.136 *SetRightObjectId*

Function prototype:

```
inline void SetRightObjectId(uint32 id)
```

Function description: This function sets the identification of the automata instance that is logically to the right of this automata instance. The parameter of this function specifies the identification of the automata instance that is to the right of this automata instance.

Function parameters:
id: the identification of the automata instance

6.8.137 *SetState*

Function prototype:

```
inline void SetState(uint8 state)
```

FSM Library

Function description: This function sets the identification of the current state of this automata instance. The parameter of this function specifies the identification of the state.

Function parameters:
state: the value that uniquely identifies the particular state of automata

6.8.138 *StartTimer*

Function prototype:
```
void StartTimer(uint16 tmrId)
```

Function description: This function starts the given timer. The parameter of this function specifies the identification of the timer.

Function parameters:
tmrId: the value that uniquely identifies the particular timer

Note: Uniqueness of the timer identifier is limited to the scope of a single automata type that uses it.

6.8.139 *StopTimer*

Function prototype:
```
void StopTimer(uint16 tmrId)
```

Function description: This function stops the given timer. The parameter of this function specifies the identification of the timer.

Function parameters:
tmrId: the value that uniquely identifies the particular timer

Note: Uniqueness of the timer identifier is limited to the scope of a single automata type that uses it.

6.8.140 *SysClearLogFlag*

Function prototype:
```
static void SysClearLogFlag()
```

Function description: This function stops the logging of the messages exchanged by the automata.

6.8.141 *SysStartAll*

Function prototype:
```
Static void SysStartAll()
```

Function description: This function starts the logging of the messages exchanged by the automata.

Note: Normally, the programmer should start the logging of messages before he starts the individual automata included in the FSM system.

6.8.142 *NetFSM*

Function prototype:
```
NetFSM(
    uint16 numOfTimers = DEFAULT_TIMER_NO,
    uint16 numOfState = DEFAULT_STATE_NO,
    uint16 maxNumOfProceduresPerState = DEFAULT_PROCEDURE_NO_PER_STATE,
    bool getMemory = true)
```

Function description: This constructor initializes the object that represents an instance of the given automata type together with the data structures needed for its proper operation. The parameters of this function specify the number of timers to be used by this automata type, the total number of states for this automata type, the maximal number of state transitions per state for this automata type, and the memory allocation indicator. All the parameters have their default values as shown in the function prototype declaration above.

Function parameters:

numOfTimers: the number of timers to be used by this automata type

numOfState: the total number of states for this automata type

maxNumOfProceduresPerState: the maximal number of state transitions per state

getMemory: the memory allocation indicator

Note: The programmer may call this constructor without parameters. In that case, the parameters will be set to their corresponding default values. The value of the fourth parameter *getMemory* regulates memory allocation. By default, this indicator is set to the value *true*, which means that the constructor will take care of memory allocation. Default memory allocation is not optimal because it is based on the maximal number of transitions per state. This compromise has been made intentionally because it leads to a very simple FSM definition API. If the programmer wants to optimize memory allocation, he may build the data structure describing the FSM by allocating necessary memory blocks from the memory heap, linking them together, and storing the pointer to this data structure in the protected class field member *States* before this function is called. In that case, the programmer would set the fourth parameter *getMemory* to the value *false*.

6.8.143 convertFSMToNetMessage

Function prototype:
```
virtual void convertFSMToNetMessage() = 0
```

Function description: This function converts the internal message format into the external message format appropriate for the transmission over the TCP/IP network.

Note: The programmer must define this virtual function by writing the corresponding function member of a class derived from the class *NetFSM*.

6.8.144 convertNetToFSMMessage

Function prototype:
```
virtual uint16 convertNetToFSMMessage() = 0
```

Function description: This function converts the external message format into the internal message format appropriate for the communication within the FSM system.

Function returns: This function returns the code of the received message.

Note: The programmer must define this virtual function by writing the corresponding function member of a class derived from the class *NetFSM*.

6.8.145 establishConnection

Function prototype:
```
void establishConnection()
```

Function description: This function establishes the TCP connection between two geographically distributed FSM systems.

Note: The programmer must call this function before he can call the function *sendToTCP* to send the message to the remote FSM system.

6.8.146 getProtocolInfoCoding

Function prototype:
```
virtual uint8 getProtocolInfoCoding() = 0
```

Function description: This function returns the identification of the type of external message coding.

Function returns: This function returns the value that uniquely identifies the type of the coding of the external message.

6.8.147 *sendToTCP*

Function prototype:
```
void sendToTCP()
```

Function description: This function sends the new message to the remote FSM system over the previously established TCP connection.

Note: The programmer must call the function *establishConnection* before he can call this function.

6.9 A Simple Example with Three Automata Instances

This section shows how the programmer can construct the FSM system and how he can add individual automata instances to it. To keep the example simple, we include only one use case, *Show Simple Demo* (Figure 6.1). The realization of this use case is a simple collaboration that comprises three instances (*instance_1*, *instance_2*, and *instance_3*) of the same automata type (*Automata*), which are added to the FSM system (Figure 6.2). These three automata instances have a trivial task to exchange the given number of messages in a "round robin" fashion.

At the beginning, the main thread calls the function *StartDemo* of *instance_1*, which in turn asynchronously sends itself the message *IDLE_START*. Upon reception of this message, *instance_1* sends the message *IDLE_MSG* to *instance_2*, which increments the message sequence number and forwards the message to *instance_3*, and the latter translates it to the message *MSG_MSG* and sends it back to *instance_1*. This message then makes two full circles around the collaborating objects. Finally, *instance_1* translates it to the message *MSG_STOP* and sends it to *instance_2*, which in turn forwards it to *instance_3*. The corresponding sequence diagram is shown in Figure 6.3. The conditions A, B, and C regulate the already mentioned translations of the messages.

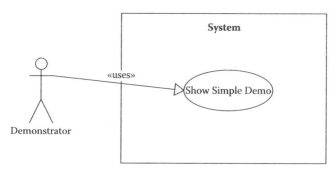

FIGURE 6.1
The simple use case diagram for the example with three automata instances.

FSM Library 399

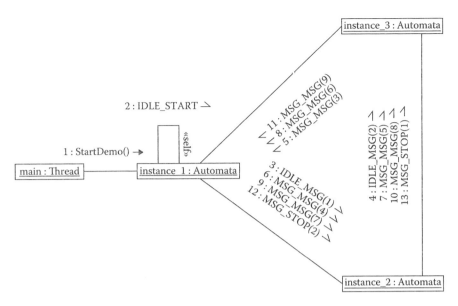

FIGURE 6.2
The collaboration diagram for the example with three automata instances.

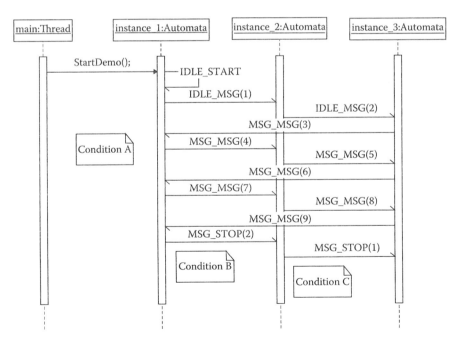

FIGURE 6.3
The sequence diagram for the example with three automata instances.

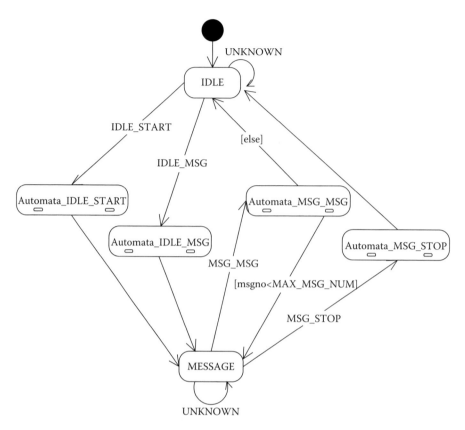

FIGURE 6.4
The statechart diagram for the example with three automata instances.

The statechart diagram that describes the behavior of a single automata instance is organized into two hierarchical levels. The top level comprises two simple states, *IDLE* and *MESSAGE*, and four composite states, *Automata_IDLE_START*, *Automata_IDLE_MSG*, *Automata_MSG_MSG*, and *Automata_MSG_STOP* (Figure 6.4). The symbolic constant *MAX_MSG_NUM* is defined to have the value 10 in this example. The variable *msgno* is the message sequence number, whose values are shown in parenthesis in Figure 6.2 and Figure 6.3. Later in the program text, this short variable name suitable for figures is replaced with the longer self-documenting name *msgNumber*.

The individual composite states *Automata_IDLE_START*, *Automata_IDLE_MSG*, *Automata_MSG_MSG*, and *Automata_MSG_STOP* are shown in Figure 6.5, Figure 6.6, Figure 6.7, and Figure 6.8, respectively. These have been made rather detailed to show how to provide the mapping from the UML model to the corresponding program code by the application of forward engineering. Essentially, the state transition actions are the sequences of calls to functions provided by the FSM library, such as *PrepareNewMessage*, *AddParamDWord*, *SendMessage*, and so on.

FSM Library

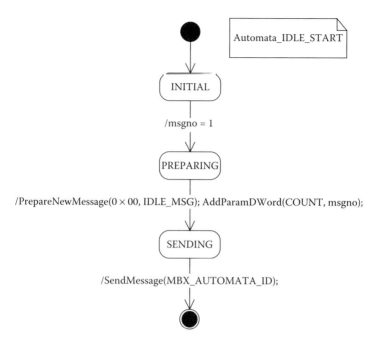

FIGURE 6.5
The statechart diagram for the composite state *Automata_IDLE_START*.

Each of the composite states can be modeled as an operation by the corresponding activity diagram. The activity diagrams for the operations *Automata_IDLE_START*, *Automata_IDLE_MSG*, *Automata_MSG_MSG*, and *Automata_MSG_STOP* are shown in Figure 6.9, Figure 6.10, Figure 6.11, and Figure 6.12, respectively. Again, these diagrams have been made by applying forward engineering but on a slightly higher abstraction level, by using informal text statements instead of explicit functions calls. Essentially, composite statecharts and activity diagrams have the same semantics in this example.

The third and semantically equivalent method of modeling the behavior of individual automata instances is by using the domain-specific SDL model. This model comprises state transitions triggered by the reception of the corresponding messages. The same names are used again so that the reader can easily follow the correspondence between the SDL state transitions and the UML composite states and activity diagrams. The SDL state transitions *Automata_IDLE_START*, *Automata_IDLE_MSG*, *Automata_MSG_MSG*, and *Automata_MSG_STOP* are shown in Figure 6.13, Figure 6.14, Figure 6.15, and Figure 6.16, respectively.

As already mentioned, all three automata instances in this example are of the same type, i.e., class. The class *Automata* is a specialization of the FSM library class *FiniteStateMachine* and is used by the FSM library class *FSM-System* (see the corresponding UML class diagram in Figure 6.17). The class *Automata* inherits all the members from its parent class and adds some field

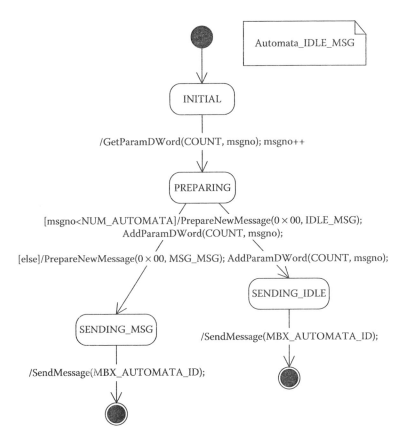

FIGURE 6.6
The statechart diagram for the composite state *Automata_IDLE_MSG*.

members, such as *msgno*, and function members, such as *Automata_IDLE_START*, *Automata_IDLE_MSG*, *Automata_MSG_MSG*, *Automata_MSG_STOP*, *Initialize*, and *StartDemo*. The first four correspond to the composite states from the previous UML statechart model.

An object diagram such as the one shown in Figure 6.18 helps us to understand better the structural relationships among objects. A collaboration diagram (Figure 6.2) shows the logical communication of automata instances over their virtual peer-to-peer connections. On a more detailed level of abstraction, we see that the real communication is governed by the FSM system, which is the owner of the mailboxes (not shown in the figure) used for storing the messages, e.g., *StandardMessage* (shown in the figure). This particular message shown in one snapshot of object collaboration is the first message sent from *instance_1* to *instance_2*. The message code is *IDLE_MSG* and the value of the message sequence parameter is 1.

The program project in this example comprises the files *Automata.h*, *Automata.cpp*, *Constants.h*, *Main.cpp*, and the FSM library (see the corresponding component diagram in Figure 6.19). Building this project in Microsoft Visual

FSM Library

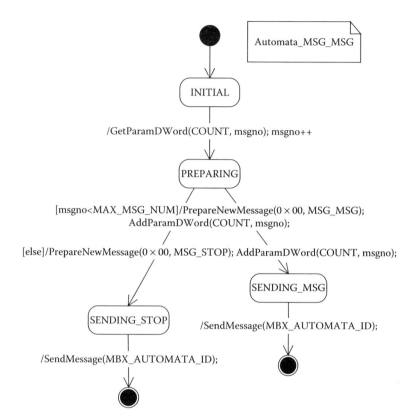

FIGURE 6.7
The statechart diagram for the composite state *Automata_MSG_MSG*.

Studio 6.0 yields a single executable, which is executed on a single PC machine (see the corresponding deployment diagram in Figure 6.20).

The rest of this section is devoted to the program implementation of the previous models. The content of the corresponding program files is as follows.

File *Automata.h*:

```
#ifndef __AUTOMATA__
#define __AUTOMATA__

#include <stdio.h>
#include "stdlib.h"
#include "kernel\fsm.h"
#include "kernel\errorObject.h"
#include "Constants.h"

class Automata: public FiniteStateMachine {
  private:
    StandardMessage StandardMsgCoding;
    MessageInterface *GetMessageInterface(uint32 id);
```

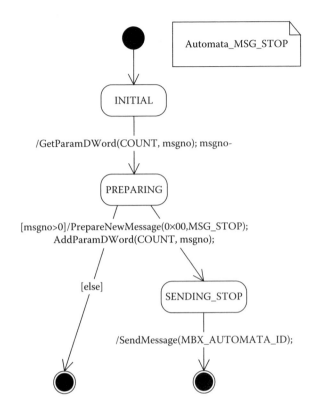

FIGURE 6.8
The statechart diagram for the composite state *Automata_MSG_STOP*.

```
        void SetDefaultHeader(uint8 infoCoding);
        uint8 GetMbxId();
        uint8 GetAutomata();
        void SetDefaultFSMData();
        void NoFreeInstances();

        uint8 text[20];
        uint32 msgNumber;
        uint32 idToMsg;

        // State transition functions for the state IDLE
        void Automata_IDLE_START();
        void Automata_IDLE_MSG();
        // State transition functions for the state MSG
        void Automata_MSG_MSG();
        void Automata_MSG_STOP();
        // Unexpected event handlers for the states IDLE and    MSG
        void Automata_UNEXPECTED_IDLE();
        void Automata_UNEXPECTED_MSG();

public:
    Automata();
    ~Automata(){};
```

FSM Library

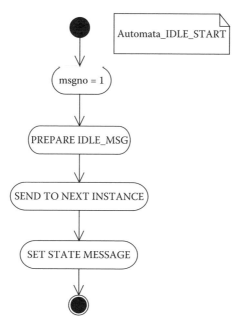

FIGURE 6.9
The activity diagram for the operation *Automata_IDLE_START*.

```
    void Initialize();
    void StartDemo();
};
#endif
```

The file *Automata.h* contains the declaration of the class *Automata* derived from the class *FiniteStateMachine*. This declaration has its private and public parts. The private field members are the message interface object *StandardMsg-Coding*, the text work area *text*, the message sequence number *msgNumber*, and the identification of the message destination automata *idToMsg*.

The common private function members are the following functions:

- *GetMessageInterface*: returns the message interface object
- *SetDefaultHeader*: sets the message header in accordance with the specified information coding
- *GetMbxId*: returns the identification of the mailbox assigned to this automata type
- *GetAutomata*: returns the identification of this automata type
- *SetDefaultFSMData*: sets the data specific for this automata type (*msgNumber* and *idToMsg*)
- *NoFreeInstances*: handles the situation when no more free instances of this type are found

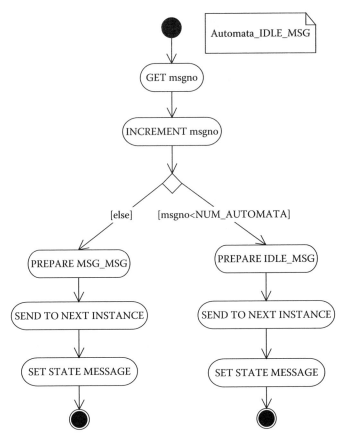

FIGURE 6.10
The activity diagram for the operation *Automata_IDLE_MSG*.

The application-specific private function members are the following state transition functions:

- *Automata_IDLE_START*: handles the message *IDLE_START* in the state *IDLE*
- *Automata_IDLE_MSG*: handles the message *IDLE_MSG* in the state *IDLE*
- *Automata_MSG_MSG*: handles the message *MSG_MSG* in the state *MESSAGE*
- *Automata_MSG_STOP*: handles the message *MSG_STOP* in the state *MESSAGE*
- *Automata_UNEXPECTED_IDLE*: handles unexpected messages in the state *IDLE*
- *Automata_UNEXPECTED_MSG*: handles unexpected messages in the state *MESSAGE*

FSM Library

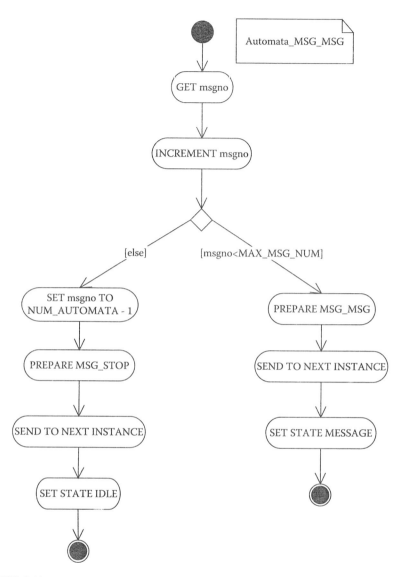

FIGURE 6.11
The activity diagram for the operation *Automata_MSG_MSG*.

The public function members are the class constructor, the class destructor, the initialization function *Initialize*, and the startup function *StartDemo*.

File *Automata.cpp*:

```
#include "kernel/LogFile.h"
#include "Automata.h"

Automata::Automata() : FiniteStateMachine(
  0,   // uint16 numOfTimers = DEFAULT_TIMER_NO,
```

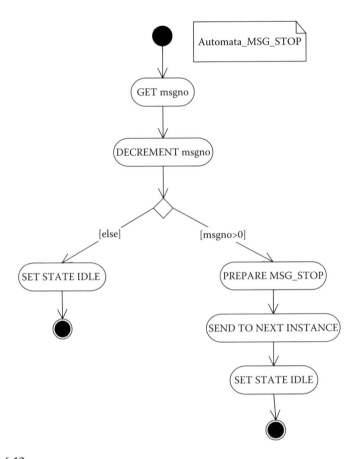

FIGURE 6.12
The activity diagram for the operation *Automata_MSG_STOP*.

```
 2,   // uint16 numOfState = DEFAULT_STATE_NO,
 3)   // uint16 maxNumOfProceduresPerState = DEFAULT_PROCEDURE_NO_PER_STATE
{
SetDefaultFSMData();
}

// This function returns the pointer to the object that governs the
// message information coding (the pointer to the message interface).
// This automata instance works only with the standard messages
// (ID 0x00). If the caller specifies another type of coding,
// this function throws the exception TErrorObject. The message
// interface is defined in Automata.h
MessageInterface *Automata::GetMessageInterface(uint32 id){
  switch(id){
    case 0x00:
      return &StandardMsgCoding;
  }
  throw TErrorObject(__LINE__,__FILE__,0x01010400);
}
```

FSM Library

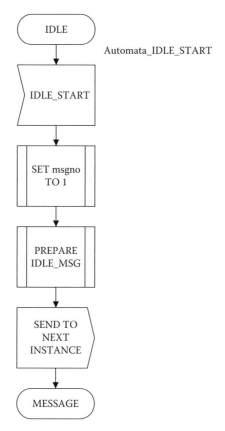

FIGURE 6.13
The SDL diagram for the transition *Automata_IDLE_START*.

```
// This function fills in the message header.
void Automata::SetDefaultHeader(uint8 infoCoding){
  SetMsgInfoCoding(infoCoding);
  SetMessageFromData();
}

// This function returns the identification of the mailbox that is
// assigned to this automata type.
uint8 Automata::GetMbxId(){
  return MBX_AUTOMATA_ID;
}

// This function returns the identification of this automata type.
uint8 Automata::GetAutomata(){
  return FSM_TYPE_AUTOMATA;
}

// This function initializes the data specific to individual
// instance of this automata type.
```

410 Communication Protocol Engineering

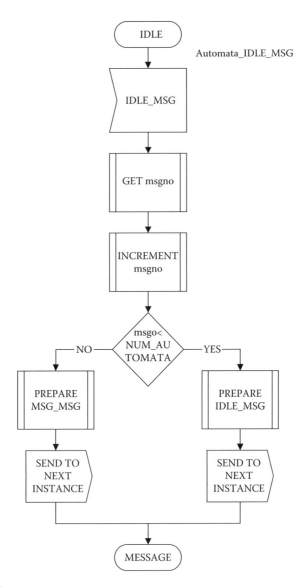

FIGURE 6.14
The SDL diagram for the transition *Automata_IDLE_MSG*.

```
void Automata::SetDefaultFSMData(){
  msgNumber = 0;
  idToMsg   = INVALID_32;
}

// This function is called if there are no free instances of this
// automata type. If the programmer wants to use this option, they must
// add the first automata instance of this type to the parameter
// useFreeList of the function Add set to true. In this example, it
```

FSM Library

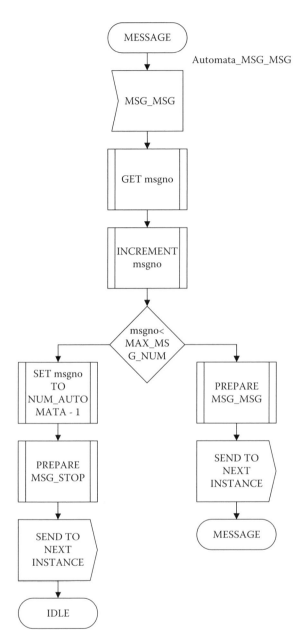

FIGURE 6.15
The SDL diagram for the transition *Automata_MSG_MSG*.

```
// is empty. In real applications, the programmer should provide
// some recovery mechanism, such as overload protection or restart.
void Automata::NoFreeInstances(){
}
```

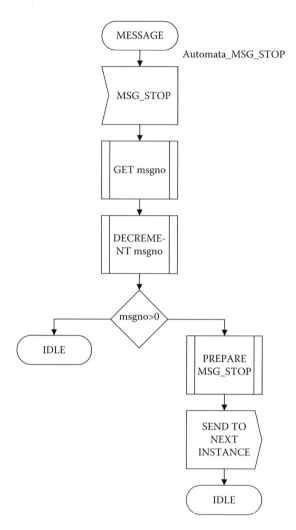

FIGURE 6.16
The SDL diagram for the transition *Automata_MSG_STOP*.

```
// This function initializes the state transition functions and the
// timers that are used by this automata type. This function is
// called implicitly by the function Add, which is responsible for
// adding individual automata instances to the FSM system.
// Each state transition function is separately declared and defined.
void Automata::Initialize(){
  // Here the programmer does the following initializations:
  // InitEventProc(uint8 state, uint16 event, PROC_FUN_PTR fun);
  // InitUnexpectedEventProc(uint8 state, PROC_FUN_PTR fun);
  // InitTimerBlock(uint16 timerId, uint32 timerCount, uint16 signalId);
  InitEventProc(IDLE,IDLE_START,(PROC_FUN_PTR)
    &Automata::Automata_IDLE_START);
  InitEventProc(IDLE,IDLE_MSG,(PROC_FUN_PTR)
    &Automata::Automata_IDLE_MSG);
```

FSM Library

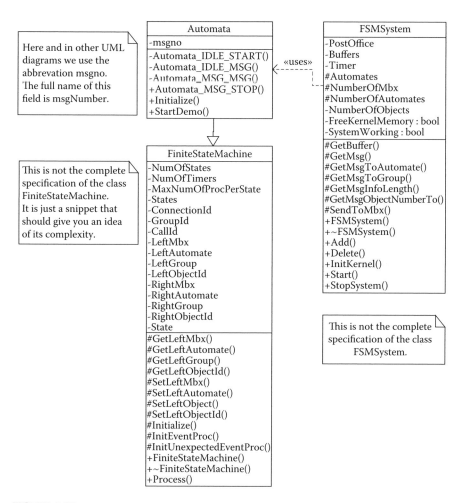

FIGURE 6.17
The class diagram for the example with three automata instances.

```
InitEventProc(MESSAGE,MSG_MSG,(PROC_FUN_PTR)
  &Automata::Automata_MSG_MSG);
InitEventProc(MESSAGE,MSG_STOP,(PROC_FUN_PTR)
  &Automata::Automata_MSG_STOP);

InitUnexpectedEventProc(IDLE,(PROC_FUN_PTR)
  &Automata::Automata_UNEXPECTED_IDLE);
InitUnexpectedEventProc(MESSAGE,(PROC_FUN_PTR)
  &Automata::Automata_UNEXPECTED_MSG);
}

// State transition functions for the state IDLE.
void Automata::Automata_IDLE_START(){
  msgNumber = 1;
  idToMsg   = GetObjectId()+1;
```

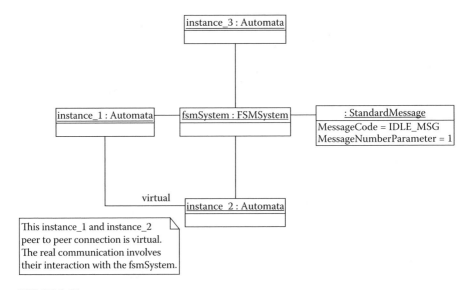

FIGURE 6.18
The object diagram for the example with three automata instances.

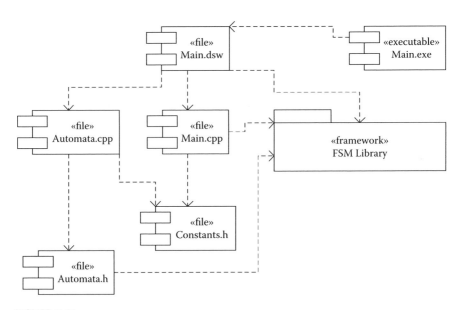

FIGURE 6.19
The component diagram for the example with three automata instances.

FSM Library

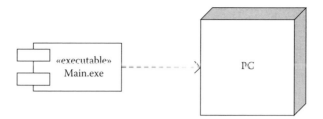

FIGURE 6.20
The deployment diagram for the example with three automata instances.

```
  // Round Robin message transfer among automata instances 0-2
  if(idToMsg == 3)
    idToMsg = 0;

  // The automata instance prepares and sends the message,
  // and changes its state to MESSAGE.
  PrepareNewMessage(0x00,IDLE_MSG);

  char text[] = "THIS IS THE FIRST MESSAGE";
  AddParam(PARAM_TEXT,strlen(text),(unsigned char *)text);
  AddParamDWord(COUNT,msgNumber);

  SetMsgToAutomata(FSM_TYPE_AUTOMATA);
  SetMsgToGroup(INVALID_08);
  SetMsgObjectNumberTo(idToMsg);
  SendMessage(MBX_AUTOMATA_ID);
  SetState(MESSAGE);
}

void Automata::Automata_IDLE_MSG(){
  idToMsg = GetObjectId()+1;

  // Round Robin message transfer among automata instances 0-2
  if((idToMsg == 3)
    idToMsg = 0;
  // Get parameters from the message
  unsigned char *tmp;
  tmp = GetParam(PARAM_TEXT);
  assert(tmp);
  memcpy(text,tmp+2,*(tmp+1));
  memset(text+(*(tmp+1)),0x00,1);// make the string
  GetParamDWord(COUNT,msgNumber);

  // Round Robin - this instance receives the message from the previous one
  uint32 idFromMsg = GetObjectId()-1;
  if(idFromMsg == -1)
    idFromMsg = 2;

  printf("Text received: %s\n from automata:%u \n",text,idFromMsg);

  // If the message sequence number is less than NUM_AUTOMATA,
```

```
      // send IDLE_MSG. If not, send MSG_MSG.
      msgNumber++;
      if(msgNumber < NUM_AUTOMATA){
        // Prepare and send the message.
        // Change automata state to MESSAGE.
        PrepareNewMessage(0x00,IDLE_MSG);

        char text[] = "THIS IS THE SECOND MESSAGE";
        AddParam(PARAM_TEXT,strlen(text),(unsigned char *)text);
        AddParamDWord(COUNT,msgNumber);

        SetMsgToAutomata(FSM_TYPE_AUTOMATA);
        SetMsgToGroup(INVALID_08);
        SetMsgObjectNumberTo(idToMsg);
        SendMessage(MBX_AUTOMATA_ID);
      }
      else {
        // Prepare and send the message.
        // Change automata state to MESSAGE.
        PrepareNewMessage(0x00,MSG_MSG);
        AddParamDWord(COUNT,msgNumber);
        SetMsgToAutomata(FSM_TYPE_AUTOMATA);
        SetMsgToGroup(INVALID_08);
        SetMsgObjectNumberTo(idToMsg);
        SendMessage(MBX_AUTOMATA_ID);
      }
      SetState(MESSAGE);
    }

    void Automata::Automata_MSG_MSG(){
      GetParamDWord(COUNT,msgNumber);
      msgNumber++;
      if(msgNumber < MAX_MSG_NUM){
        // Forward the message to the next automata    instance.
        PrepareNewMessage(0x00,MSG_MSG);
        AddParamDWord(COUNT,msgNumber);
        SetMsgToAutomata(FSM_TYPE_AUTOMATA);
        SetMsgToGroup(INVALID_08);
        SetMsgObjectNumberTo(idToMsg);
        SendMessage(MBX_AUTOMAT_ID);
      }
      else {
        printf("Stop automata:%with message:%u\n",GetObjectId(),msgNumber);

        // Prepare and send the message.
        // Change automata state to IDLE.
        PrepareNewMessage(0x00,MSG_STOP);
        AddParamDWord(COUNT,NUM_AUTOMATA-1);
        SetMsgToAutomata(FSM_TYPE_AUTOMATA);
        SetMsgToGroup(INVALID_08);
        SetMsgObjectNumberTo(idToMsg);
        SendMessage(MBX_AUTOMATA_ID);
        SetState(IDLE);
      }
    }
```

FSM Library

```
void Automata::Automata_MSG_STOP(){
  printf("Stop automata instance: %u\n",GetObjectId());

  GetParamDWord(COUNT,msgNumber);
  msgNumber--;
  if(msgNumber > 0){
    // Prepare and send the message.
    // Change automata state to IDLE.
    PrepareNewMessage(0x00,MSG_STOP);
    AddParamDWord(COUNT,msgNumber);
    SetMsgToAutomata(FSM_TYPE_AUTOMATA);
    SetMsgToGroup(INVALID_08);
    SetMsgObjectNumberTo(idToMsg);
    SendMessage(MBX_AUTOMATA_ID);
  }
  SetState(IDLE);
}

void Automata::Automata_UNEXPECTED_IDLE(){
  printf("Unexpected message in the state IDLE \n");
}

void Automata::Automata_UNEXPECTED_MSG(){
  printf("Unexpected message in the state MESSAGE \n");
}

void Automata::StartDemo(){
  uint8 *msg = GetBuffer(MSG_HEADER_LENGTH);

  SetMsgFromAutomata(FSM_TYPE_AUTOMATA,msg);
  SetMsgFromGroup(INVALID_08,msg);
  SetMsgObjectNumberFrom(0,msg);

  SetMsgToAutomata(FSM_TYPE_AUTOMATA,msg);
  SetMsgToGroup(INVALID_08,msg);
  SetMsgObjectNumberTo(0,msg);

  SetMsgInfoCoding(0,msg);  // 0 = StandardMessage
  SetMsgCode(IDLE_START,msg);
  SetMsgInfoLength(0,msg);
  SendMessage(MBX_AUTOMATA_ID,msg);
}
```

The file *Automata.cpp* contains the definition of the class *Automata*. This definition starts with the class constructor that first calls the base class constructor specifying no timers, two states, and the maximum of three state transitions per state for this automata type. After that, the constructor calls the function *SetDefaultFSMData*, which sets the data specific for this automata type.

The function *GetMessageInterface* returns the pointer to the message interface object for the given type of information coding. This class operates with only standard messages (the corresponding ID is 0x00). If the caller of this function specifies the identification of the standard message as its parameter, the function returns the pointer to the object *StandardMsgCoding*. If the caller specifies some other message type, this function throws the exception *TErrorObject*.

The function *SetDefaultHeader* sets the message information coding by calling the function *SetMsgInfoCoding* and the automata specific data by calling the function *SetMessageFromData*. The function *GetMbxId* returns the value *MBX_AUTOMATA_ID* as the identification of the mailbox assigned to this automata type. The function *GetAutomata* returns the value *FSM_TYPE_AUTOMATA* as the identification of this automata type. The function *SetDefaultFSMData* sets the field *msgNumber* to the value 0 and the field *idToMsg* to the value *INVALID_32*. The function *NoFreeInstances* is empty in this simple example. In real-world projects, it would be used to trigger some higher-level protection or recovery mechanism.

The function *Initialize* defines the event handlers by calling the function *InitEventProc* and the unexpected events handlers by calling the function *InitUnexpectedEventProc*. More precisely, this function defines the event handlers for the messages *IDLE_START* and *IDLE_MSG* in the state *IDLE* and for the messages *MSG_MSG* and *MSG_STOP* in the state *MESSAGE*. It also defines the handlers for unexpected messages in both states.

The function *Automata_IDLE_START* handles the message *IDLE_START* in the state *IDLE*. First, it sets the message sequence number *msgNumber* to the value 1. It then determines the identification of the destination automata instance by incrementing its own identification by modulo 3. (This means that the destination of the messages created and sent by *instance_0* is *instance_1*, the destination for *instance_1* is *instance_2*, and the destination for *instance_2* is *instance_0*.) Next, this function prepares and sends the message, "THIS IS THE FIRST MESSAGE". At the end, it performs the state transition from *IDLE* to *MESSAGE* by calling the function *SetState* and specifying the value *MESSAGE* as its parameter.

The function *Automata_IDLE_MSG* handles the message *IDLE_MSG* in the state *IDLE*. First, it determines the identifications of the source and destination automata instances for the received message and prints them to the monitor. It then increments the message sequence numbers and checks if they are less than the number of communicating automata instances *NUM_AUTOMATA* (value 3). If yes, the function prepares and sends the message *IDLE_MSG* with the text, "THIS IS THE SECOND MESSAGE". If not, the function prepares and sends the message *MSG_MSG* without any text. In both cases, it sets the current state of this automata instance to the value *MESSAGE*.

The function *Automata_MSG_MSG* handles the message *MSG_MSG* in the state *MESSAGE*. First, it gets the message sequence number from the received message and increments that number. It then checks if the new value of the message sequence number has reached the given limit. If not, this function prepares and sends the message *MSG_MSG* to the next automata instance in the chain. If it has, this function prepares and sends the message *MSG_STOP* to the next automata instance in the chain and sets the current state of this automata instance to *IDLE*.

The function *Automata_MSG_STOP* handles the message *MSG_STOP* in the state *MESSAGE*. First, it decrements the message sequence number and checks its new value. If the value is positive, the function prepares and sends the message *MSG_STOP* to the next automata instance in the chain and sets the current state of this automata instance to *IDLE*.

The unexpected event handlers in this example just print the warning messages. In real applications, these functions would trigger some higher-level recovery mechanisms. The function *StartDemo* creates the first message in the system. It fills in its header as if the automata instance with the identification 0 had sent that message to itself and sends the message to the mailbox assigned to this automata type.

File *Constants.h*:

```
// FSM
#define FSM_TYPE_AUTOMATA 0

// MBX
#define MBX_AUTOMATA_ID 0

#define MAX_MSG_NUM 10
#define NUM_AUTOMATA 3
#define COUNT 1
#define PARAM_TEXT 2

enum AutomataStates{
  IDLE = 0,
  MESSAGE,
};

enum Messages{
  IDLE_START = 0,
  IDLE_MSG,
  MSG_MSG,
  MSG_STOP
};
```

The file *Constants.h* first defines general symbolic constants. The identification of this automata type *FSM_TYPE_AUTOMATA* is assigned the value 0, the identification of the mailbox related to this automata type *MBX_AUTOMATA_ID* is assigned the value 0, the maximal message sequence number *MAX_MSG_NUM* is assigned the value 10, the number of automata instances of this type *NUM_AUTOMATA* is assigned the value 3, the identification of the message parameter that contains the messages sequence number *COUNT* is assigned the value 1, and the identification of the message parameter that contains the text *PARAM_TEXT* is assigned the value 2.

Next, the identifications of the individual states of this automata type are enumerated. The identification of the state *IDLE* is assigned the value 0 and the identification of the state *MESSAGES* is assigned the value 1. Finally, the identifications of various message types (message codes) are enumerated.

The message types are named as *IDLE_START, IDLE_MSG, MSG_MSG,* and *MSG_STOP*. These symbols are assigned the values 0, 1, 2, and 3, respectively.

File *Main.cpp*:

```
#include "conio.h"
#include "Kernel/fsmsystem.h"
#include "Kernel/LogFile.h"
#include "Automata.h"

// Assume the following.
// The FSM system hosts a single automata type.
// The FSM system uses a single mailbox for the message exchange.
// Create the FSM system.
FSMSystem fsmSystem(1,1);

// Create three instances of the class Automata.
Automata instance_1, instance_2, instance_3;

// FSM system thread
DWORD WINAPI ThreadFunction(void* dummy){
  uint32 buffersCount[3]  = {5,3,2};
  uint32 buffersLength[3] = {128,256,512};
  uint8  buffClassNo = 3;

  // Initialize the FSM system.
  printf("Initialize the FSM system... \n");
  fsmSystem.Add(&instance_1,FSM_TYPE_AUTOMATA,3,false);
  fsmSystem.Add(&instance_2,FSM_TYPE_AUTOMATA);
  fsmSystem.Add(&instance_3,FSM_TYPE_AUTOMATA);

    fsmSystem.InitKernel(buffClassNo,buffersCount,buffersLength,1);

  LogFile lf("log.log", "log.ini");
  LogAutomataNew::SetLogInterface(&lf);

  // Start the FSM system.
  printf("Start the FSM system... \n");
  try {
    fsmSystem.Start();
  }
  catch(...) {
    OutputDebugString("Exception - stop the FSM    system...\n");
    return 0;
  }
  OutputDebugString("The end of the operation.\n");
  return 0;
}

void main(int argc,char* argv[]){
  DWORD threadID;
  bool end = false;
  char ret;

  // Start the FSM system thread.
```

```
HANDLE hTemp =   CreateThread(NULL,0,ThreadFunction,NULL,0,&threadID);
Sleep(100);

// Program works until the character 'Q' or 'q' is  pressed.
while(!end) {
  if(_kbhit()) {
    ret = _getch();
    switch(ret) {
      case 'Q':
      case 'q':
        fsmSystem.StopSystem();
        end = true;
        Sleep(100);
        break;
      case 'S':
      case 's':
        instance_1.StartDemo();
        break;
      default:
        break;
    }
  }
}
CloseHandle(hTemp);
printf("The end. \n");
}
```

The file *Main.cpp* starts with the instantiation of the class *FSMSystem* by calling its constructor. The parameters used in this call specify that the instance of the *FSMSystem*, named *fsmSystem*, will include a single automata type and this automata type will use a single mailbox. Next, three instances of the class *Automata* are made, namely, *instance_1*, *instance_2*, and *instance_3*. Additionally, this file contains the definitions of the FSM system thread function *ThreadFunction* and the function *main*.

The function *ThreadFunction* first prepares the data needed to define three buffer types. The sizes and quantities of these buffers are five at 128 bytes, three at 256 bytes, and two at 512 bytes. Next, three automata instances are added to *fsmSystem*. Note that the fourth parameter of the first call to the function *Add* is set to the value *false*, which means that these three instances are to be used as three distinctive instances rather than as a pool of instances of the same type. After that, this function initializes the kernel by calling the function *InitKernel*, defines and sets the logging interface by calling the function *SetLogInterface*, and starts *fsmSystem* by calling its function *Start*.

The function *main* starts the FSM system thread (which executes the function *ThreadFunction*) and suspends itself for 100 ms. After that, it just waits for the character 'Q' or 'q' to be pressed and to subsequently terminate the program.

6.10 A Simple Example with Network-Aware Automata Instances

This section shows how the programmer can construct FSM systems with the TCP support that is able to communicate over the TCP/IP network and how he can add individual network-aware automata instances to it. Normally, the programmer creates the FSM system with TCP support by instantiating the class *FSMSystemWithTCP*. Alternately, network-aware automata types are normally derived from the base class *NetFSM*. Of course, the network-aware automata instances of a given type are then created simply by instantiating that automata type.

This example is very similar to the previous one. Actually, it has been created from it by a few rather simple modifications. Only one instance of the given automata type is added to the FSM system (now with the TCP/IP support). This automata instance has a trivial task of exchanging the given number of messages with its peer in the remote FSM system. The main difference is that the whole program is instantiated twice. These program instances run as two separate processes that communicate over the TCP/IP protocol stack (see the corresponding collaboration diagram in Figure 6.21).

At the beginning, as in the previous example, the main thread calls the function *StartDemo* of *instance_1*, which in turn sends itself asynchronously the message *IDLE_START*. Upon reception of this message, *instance_1* sends the message *IDLE_MSG* to its peer *instance_1* that resides at the remote FSM system. These two automata instances, local and remote, then exchange nine *MSG_MSG* messages (the last *MSG_MSG* message is not shown in the figure). At the end of the communication, the local instance sends the message *MSG_STOP* to the remote instance (not shown in the figure). The corresponding sequence diagram is shown in Figure 6.22. This diagram shows all the messages.

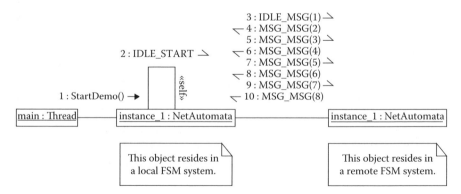

FIGURE 6.21
The collaboration diagram for the example with network-aware automata.

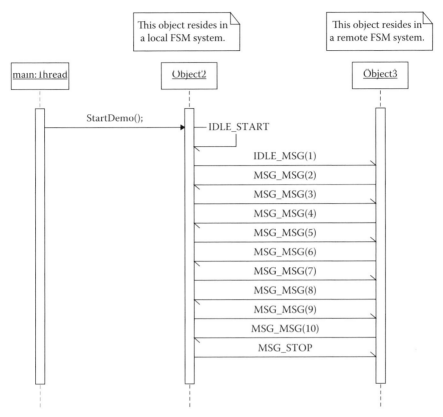

FIGURE 6.22
The sequence diagram for the example with network-aware automata.

The statechart diagram that describes the behavior of an individual automata instance is again organized into two hierarchical levels. The top level is exactly the same as the one shown in Figure 6.4. The composite states *Automata_IDLE_START*, *Automata_IDLE_MSG*, *Automata_MSG_MSG*, and *Automata_MSG_STOP* are a little simpler in this example and are shown in Figure 6.23, Figure 6.24, Figure 6.25, and Figure 6.26, respectively.

The program code given in this example assumes that both processes run on the same machine whose IP address is 192.168.0.57. To get this code running on another machine, the reader should change this parameter accordingly. If the reader wants to experiment on two different machines, he must set this parameter to the IP addresses of those machines (see the corresponding deployment diagram shown in Figure 6.27).

Before proceeding further, studying the previous example first is strongly recommended. The content of the program files is as follows.

File *NetAutomata.h*:

```
#ifndef __NET_AUTOMATA__
#define __NET_AUTOMATA__
```

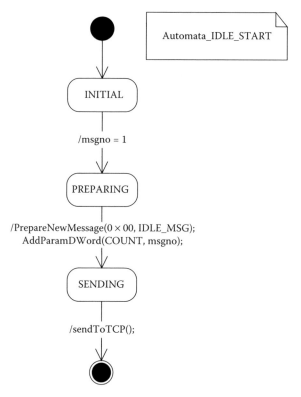

FIGURE 6.23
The composite state *Automata_IDLE_START*.


```
#include <stdio.h>
#include "stdlib.h"
#include "kernel\NetFSM.h"
#include "kernel\errorObject.h"
#include "Constants.h"

class NetAutomata: public NetFSM {
  private:
    // NetFSM
    uint16 convertNetToFSMMessage();
    void convertFSMToNetMessage();
    uint8 getProtocolInfoCoding();
    // FSM
    StandardMessage StandardMsgCoding;
    MessageInterface *GetMessageInterface(uint32 id);
    void SetDefaultHeader(uint8 infoCoding);
    uint8 GetMbxId();
    uint8 GetAutomata();
    void SetDefaultFSMData();
    void NoFreeInstances();

    uint8 text[20];
```

FSM Library

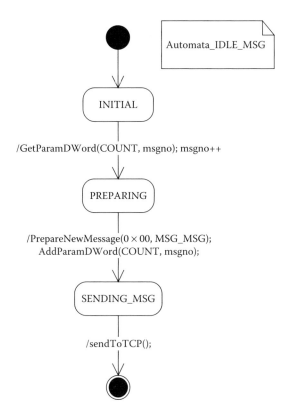

FIGURE 6.24
The composite state *Automata_IDLE_MSG*.

```
    uint32 msgNumber;
    uint32 idToMsg;

    // State transition functions for the state IDLE
    void NetAutomata_IDLE_START();
    void NetAutomata_IDLE_MSG();
    // State MSG
    void NetAutomata_MSG_MSG();
    void NetAutomata_MSG_STOP();
    // Unexpected messages in states IDLE and MSG
    void NetAutomata_UNEXPECTED_IDLE();
    void NetAutomata_UNEXPECTED_MSG();

 public:
    NetAutomata();
    ~NetAutomata(){};
    void Initialize();
    void StartDemo();

    };
#endif
```

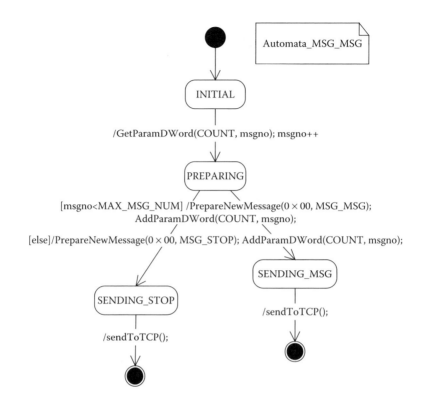

FIGURE 6.25
The composite state *Automata_MSG_MSG*.

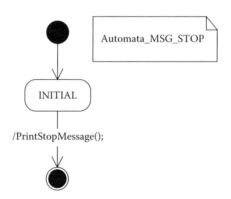

FIGURE 6.26
The composite state *Automata_MSG_STOP*.

FSM Library

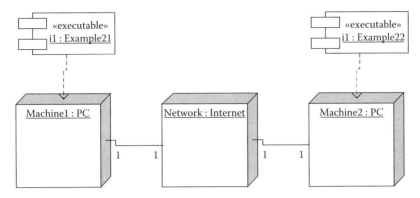

FIGURE 6.27
The deployment diagram for the example with network-aware automata.

The file *NetAutomata.h* contains the declaration of the class *NetAutomata* derived from the class *NetFSM*. This declaration has its private and public parts. The private field members are the message interface object *StandardMsgCoding*, the text work area *text*, the message sequence number *msgNumber*, and the identification of the automata instance *idToMsg*, which is the message destination.

The private function members specific to the class *NetFSM* are the following functions:

- *convertNetToFSMMessage*: converts the external message format into the internal message format appropriate for the communication within the FSM system
- *convertFSMToNetMessage*: converts the internal message format into the external message format appropriate for the transmission over the TCP/IP network
- *getProtocolInfoCoding*: returns the identification of the type of the external message coding

The private function members specific to the class *FinteStateMachine* are the following functions:

- *GetMessageInterface*: returns the message interface object
- *SetDefaultHeader*: sets the message header according to the specified information coding
- *GetMbxId*: returns the identification of the mailbox that is assigned to this automata type
- *GetAutomata*: returns the identification of this automata type
- *SetDefaultFSMData*: sets the data specific for this automata type (*msgNumber* and *idToMsg*)

- *NoFreeInstances*: handles the situation when no more free instances of this type are found

The application-specific private function members are the following state transition functions:

- *Automata_IDLE_START*: handles the message *IDLE_START* in the state *IDLE*
- *Automata_IDLE_MSG*: handles the message *IDLE_MSG* in the state *IDLE*
- *Automata_MSG_MSG*: handles the message *MSG_MSG* in the state *MESSAGE*
- *Automata_MSG_STOP*: handles the message *MSG_STOP* in the state *MESSAGE*
- *Automata_UNEXPECTED_IDLE*: handles unexpected messages in the state *IDLE*
- *Automata_UNEXPECTED_MSG*: handles unexpected messages in the state *MESSAGE*

The public function members are the class constructor, the class destructor, the initialization function *Initialize*, and the startup function *StartDemo*.

File *NetAutomata.cpp*:

```
#include "kernel/LogFile.h"
#include "NetAutomata.h"

NetAutomata::NetAutomata() : NetFSM(
  0,  // uint16 numOfTimers = DEFAULT_TIMER_NO,
  2,  // uint16 numOfState = DEFAULT_STATE_NO,
  3)  // uint16 maxNumOfProceduresPerState = DEFAULT_PROCEDURE_NO_PER_STATE
{
  SetDefaultFSMData();
}

// This function returns the pointer to the object that governs the
// message information coding (the pointer to the message interface).
// This automata instance works only with the standard messages
// (ID 0x00). If the caller specifies another type of coding,
// this function throws the exception TErrorObject.
// The message interface is defined in NetAutomata.h
MessageInterface *NetAutomata::GetMessageInterface(uint32 id){
  switch(id) {
    case 0x00:
      return &StandardMsgCoding;
  }
  throw TErrorObject(__LINE__,__FILE__,0x01010400);
}
```

FSM Library

```cpp
// This function fills in the message header.
void NetAutomata::SetDefaultHeader(uint8 infoCoding){
  SetMsgInfoCoding(infoCoding);
  SetMessageFromData();
}

// This function returns the identification of the mailbox that is
// assigned to this automata type.
uint8 NetAutomata::GetMbxId(){
  return MBX_AUTOMATA_ID;
}

// This function returns the identification of this automata type.
uint8 NetAutomata::GetAutomata(){
  return FSM_TYPE_AUTOMATA;
}

// This function initializes the data specific for individual
// instance of this automata type.
void NetAutomata::SetDefaultFSMData(){
  msgNumber = 0;
  idToMsg = INVALID_32;
}

// This function is called if there are no free instances of this
// automata type. If the programmer wants to use this option they must
// add the first automata instance of this type with the parameter
// useFreeList of the function Add set to true. In this example it is
// empty. In real applications the programmer should provide some
// recovery mechanism, such as overload protection or restart.
void NetAutomata::NoFreeInstances(){}

// This function initializes the state transition functions and the
// timers that are used by this automata type. This function is called
// implicitly by the function Add responsible for adding individual
// automata instances to the FSM system.
// Each state transition function is separately declared and defined.
void NetAutomata::Initialize(){
  // Here the programmer does the following  initializations:
  // InitEventProc(uint8 state, uint16 event,  PROC_FUN_PTR fun);
  // InitUnexpectedEventProc(uint8 state, PROC_FUN_PTR   fun);
  // InitTimerBlock(uint16 timerId, uint32 timerCount,  uint16 signalId);

  InitEventProc(IDLE,IDLE_START,(PROC_FUN_PTR)
    &NetAutomata::NetAutomata_IDLE_START);
  InitEventProc(IDLE,IDLE_MSG,(PROC_FUN_PTR)
    &NetAutomata::NetAutomata_IDLE_MSG);

  InitEventProc(MESSAGE,MSG_MSG,(PROC_FUN_PTR)
    &NetAutomata::NetAutomata_MSG_MSG);
  InitEventProc(MESSAGE,MSG_STOP,(PROC_FUN_PTR)
    &NetAutomata::NetAutomata_MSG_STOP);

  InitUnexpectedEventProc(IDLE,(PROC_FUN_PTR)
    &NetAutomata::NetAutomata_UNEXPECTED_IDLE);
```

```
    InitUnexpectedEventProc(MESSAGE,(PROC_FUN_PTR)
       &NetAutomata::NetAutomata_UNEXPECTED_MSG);
}

// State transition functions for the state IDLE.
void NetAutomata::NetAutomata_IDLE_START(){
  msgNumber = 1;
  idToMsg = 0;

  // The automata instance prepares and sends the   message,
  // and changes its state to MESSAGE.
  PrepareNewMessage(0x00,IDLE_MSG);

  char text[] = "THIS IS THE FIRST MESSAGE";
  AddParam(PARAM_TEXT,strlen(text),(unsigned char   *)text);
  AddParamDWord(COUNT,msgNumber);

  SetMsgToAutomata(FSM_TYPE_AUTOMATA);
  SetMsgToGroup(INVALID_08);
  SetMsgObjectNumberTo(idToMsg);
  sendToTCP();
  SetState(MESSAGE);
}

void NetAutomata::NetAutomata_IDLE_MSG(){
  idToMsg = 0;

  // Get parameters from the message
  unsigned char *tmp;
  tmp = GetParam(PARAM_TEXT);
  assert(tmp);
  memcpy(text,tmp+2,*(tmp+1));
  memset(text+(*(tmp+1)),0x00,1);// make the string

  GetParamDWord(COUNT,msgNumber);
  printf("Text received: %s\n",text);

  // If the message sequence number is less than given   limit,
  // continue message counting. If not stop the program.
  msgNumber++;

  // Prepare and send the message.
  // Change automata state to MESSAGE.
  PrepareNewMessage(0x00,MSG_MSG);
  AddParamDWord(COUNT,msgNumber);
  SetMsgToAutomata(FSM_TYPE_AUTOMATA);
  SetMsgToGroup(INVALID_08);
  SetMsgObjectNumberTo(idToMsg);
  sendToTCP();
  SetState(MESSAGE);
}

void NetAutomata::NetAutomata_MSG_MSG(){
  GetParamDWord(COUNT,msgNumber);
  msgNumber++;
```

FSM Library

```
  if(msgNumber < MAX_MSG_NUM){
    // Forward the message.
    PrepareNewMessage(0x00,MSG_MSG);
    AddParamDWord(COUNT,msgNumber);
    SetMsgToAutomata(FSM_TYPE_AUTOMATA);
    SetMsgToGroup(INVALID_08);
    SetMsgObjectNumberTo(idToMsg);
    sendToTCP();
  }
  else {
    printf("Stop automata: %u\n",GetObjectId());

    // Prepare and send the message.
    // Change automata state to IDLE.
    PrepareNewMessage(0x00,MSG_STOP);
    SetMsgToAutomata(FSM_TYPE_AUTOMATA);
    SetMsgToGroup(INVALID_08);
    SetMsgObjectNumberTo(idToMsg);
    sendToTCP();
    SetState(IDLE);
  }
}

void NetAutomata::NetAutomata_MSG_STOP(){
  printf("Stop automata: %u\n",GetObjectId());
  SetState(IDLE);
}
void NetAutomata::NetAutomata_UNEXPECTED_IDLE(){
  printf("Unexpected message in the state IDLE \n");
}

void NetAutomata::NetAutomata_UNEXPECTED_MSG(){
  printf("Unexpected message in the state MESSAGE \n");
}

void NetAutomata::StartDemo(){
  uint8 *msg = GetBuffer(MSG_HEADER_LENGTH);
  SetMsgFromAutomata(FSM_TYPE_AUTOMATA,msg);
  SetMsgFromGroup(INVALID_08,msg);
  SetMsgObjectNumberFrom(0,msg);

  SetMsgToAutomata(FSM_TYPE_AUTOMATA,msg);
  SetMsgToGroup(INVALID_08,msg);
  SetMsgObjectNumberTo(0,msg);

  SetMsgInfoCoding(0,msg);   // 0 = StandardMessage
  SetMsgCode(IDLE_START,msg);
  SetMsgInfoLength(0,msg);
  SendMessage(MBX_AUTOMATA_ID,msg);
}

uint16 NetAutomata::convertNetToFSMMessage(){
  // Manipulate only data because automata sends the new
  // message to itself.
  int length = receivedMessageLength-MSG_HEADER_LENGTH;
```

```
    memcpy(fsmMessageR,    protocolMessageR+MSG_HEADER_LENGTH, length);
    fsmMessageRLength=length; // mandatory - used by   workWhenReceive()

    // Rotate bytes
    uint16 msgCode =   GetUint16((uint8*)(protocolMessageR+MSG_CODE));
    switch((msgCode)){
      case IDLE_START:
        msgCode = IDLE_START;
        break;
      case IDLE_MSG:
        msgCode = IDLE_MSG;
        break;
      case MSG_MSG:
        msgCode = MSG_MSG;
        break;
      case MSG_STOP:
        msgCode = MSG_STOP;
        break;
      default:
        msgCode = 0xffff;
    }
    return msgCode;
}

void NetAutomata::convertFSMToNetMessage(){
  // Here we send the whole message.
    memcpy(protocolMessageS,fsmMessageS,fsmMessageSLength);
  sendMsgLength = fsmMessageSLength;
}

uint8 NetAutomata::getProtocolInfoCoding(){
  // Standard msg info coding
  return 0;
}
```

The file *NetAutomata.cpp* contains the definition of the class *NetAutomata*. This definition starts with the class constructor that first calls the base class constructor specifying no timers, two states, and the maximum of three state transitions per state for this automata type. After this, the constructor calls the function *SetDefaultFSMData*, which sets the data specific for this automata type.

The functions *GetMessageInterface*, *SetDefaultHeader*, *GetMbxId*, *GetAutomata*, *SetDefaultFSMData*, *NoFreeInstances*, and *Initialize* are the same as in the previous example. The only difference is that the name of the class *Automata* has been renamed to *NetAutomata*.

The function *NetAutomata_IDLE_START* handles the message *IDLE_START* in the state *IDLE*. First, it sets the message sequence number *msgNumber* to the value 1 and the identification of the destination automata instance *idToMsg* to the value 0. Next, this function prepares and sends the message, "THIS IS THE FIRST MESSAGE," to its peer in the remote FSM system by calling the function *SendToTCP*. At the end, it performs the state transition from *IDLE* to *MESSAGE* by calling the function *SetState* and specifying the value *MESSAGE* as its parameter.

The function *NetAutomata_IDLE_MSG* handles the message *IDLE_MSG* in the state *IDLE*. First, it prints the received message to the monitor. It then prepares and sends the message with the code *MSG_MSG* to its peer by calling the function *SendToTCP* and sets the current state of this automata instance to the value *MESSAGE*.

The function *NetAutomata_MSG_MSG* handles the message *MSG_MSG* in the state *MESSAGE*. First, it gets the message sequence number from the received message and increments this value. It then checks if the new value of the message sequence number has reached the given limit. If not, this function prepares and sends the message *MSG_MSG* to its peer at the remote FSM system by calling the function *SendToTCP*. If it has reached the limit, this function prepares and sends the message *MSG_STOP* to its peer at the remote FSM system and sets the current state of this automata instance to *IDLE*.

The function *NetAutomata_MSG_STOP* handles the message *MSG_STOP* in the state *MESSAGE*. It is fairly simple and just sets the current state of this automata instance to *IDLE*. The unexpected event handlers in this example just print the warning messages. In real-world applications, these functions would trigger some higher-level recovery mechanisms. The function *StartDemo* creates the first message in the system. It fills in its header as if the automata instance with the identification 0 had sent that message to itself and sends this message to the mailbox assigned to this automata type.

The function *convertNetToFSMMessage* just copies the payload of the external message received from the remote FSM system to the current FSM system internal message (the last received message), because in this simple example the two communicating instances have the same IDs and no need exists for any mappings between them. The pointer *fsmMessageR* points to the current internal message, the pointer *protocolMessageR* points to the current external message, and the variable *fsmMessageRLength* is equal to the payload size of the current external message. At the end, this function determines the message code and returns it as its return value.

The function *convertFSMToNetMessage* copies the whole new internal message to the new external message and sets the value of its length. The pointer *fsmMessageS* points to the new internal message, the pointer *protocolMessageS* points to the new external message, and the variables *fsmMessageSLength* and *sendMsgLength* contain their lengths.

The function *getProtocolInfoCoding* returns the code of the standard message coding (code 0x00) used for coding external messages. Note that in this simple example, both internal and external messages are actually standard messages.

File *Constants.h*:

```
// FSM
#define FSM_TYPE_AUTOMATA 0
```

```
// MBX
#define MBX_AUTOMATA_ID 0
#define MAX_MSG_NUM 10
#define COUNT 1
#define PARAM_TEXT 2
#define IP_ADDRESS "192.168.0.57"
#define PORT_1 7000
#define PORT_2 8000

enum AutomataStates {
  IDLE = 0,
  MESSAGE,
};

enum Messages {
  IDLE_START = 0,
  IDLE_MSG,
  MSG_MSG,
  MSG_STOP
};
```

The file *Constants.h* first defines general symbolic constants. It is very similar to the file with the same name in the previous example. The identification of this automata type *FSM_TYPE_AUTOMATA* is assigned the value 0, the identification of the mailbox related to this automata type *MBX_AUTOMATA_ID* is assigned the value 0, the maximal message sequence number *MAX_MSG_NUM* is assigned the value 10, the identification of the message parameter that contains the messages sequence number *COUNT* is assigned the value 1, and the identification of the message parameter that contains the text *PARAM_TEXT* is assigned the value 2.

The main difference with the previous example is the definition of the symbolic constants related to the communication over TCP/IP infrastructure. The IP address *IP_ADDRESS* is assigned the value 192.168.0.57, the TCP port number for the first server *PORT_1* is assigned the value 7000, and the TCP port number for the second server *PORT_2* is assigned the value 8000. Next, the identifications of the individual states of this automata type, as well as possible message codes, are enumerated. This part of the file is the same as in the previous example.

File *Main.cpp*:

```
#include "conio.h"
#include "Kernel/fsmsystem.h"
#include "Kernel/LogFile.h"
#include "NetAutomata.h"

// If the following line is not commented out we get the code for the
// server listening to the port number PORT_1.
// If the following line is commented out we get the code for the
// server listening to the port number PORT_2.
#define AUTOMATA1
```

FSM Library

```
// Assume the following.
// The FSM system hosts a single automata type.
// The FSM system uses a single mailbox for the message exchange.
// Create the FSM system
FSMSystemWithTCP fsmSystem(1,1);

// Create the instance of the class NetAutomata.
NetAutomata instance_1;

DWORD WINAPI ThreadFunction(void* dummy){
  uint32  buffersCount[3]  = {5,3,2};
  uint32  buffersLength[3] = {128,256,512};
  uint8   buffClassNo = 3;

  // Initialize the FSM system.
  printf("Initialize the FSMSystemWithTCP... \n");
  fsmSystem.Add(&instance_1,FSM_TYPE_AUTOMATA,1,true);
   fsmSystem.InitKernel(buffClassNo,buffersCount,buffersLength,1);
  LogFile lf("log.log", "log.ini");
  LogAutomataNew::SetLogInterface(&lf);

  // Server in machine number 1 will listen to the port   number PORT_1.
  // Server in machine number 2 will listen to the port   number PORT_2.
  // It does not matter which instance will establish   the TCP
  // connection by calling the function   establishConection().
#ifdef AUTOMATA1
  printf("Start server...on port:%u\n",PORT_1);
  fsmSystem.InitTCPServer(PORT_1,FSM_TYPE_AUTOMATA);
#else
  printf("Start server...on port:%u\n",PORT_2);
  fsmSystem.InitTCPServer(PORT_2,FSM_TYPE_AUTOMATA);

#endif
  // Start the FSM system.
  printf(iStart the FSM system...\n");
  try {
    fsmSystem.Start();
  }
  catch(...) {
    OutputDebugString("Exception - stop the FSM system...\n");
    return 0;
  }
  OutputDebugString("The end of the operation.\n");
  return 0;
}

void main(int argc,char* argv[]){
  DWORD threadID;
  bool end = false;
  char ret;

  // Start the FSM system thread.
  HANDLE hTemp = CreateThread(NULL,0,ThreadFunction,NULL,0,&threadID);
  Sleep(100);
```

```
  // Program works until the character 'Q' or 'q' is pressed.
  while((!end)) {
    if(_kbhit()) {
      ret = _getch();
      switch((ret)) {
        case 'Q':
        case 'q':
          fsmSystem.StopSystem();
          end = true;
          Sleep(100);
          break;
        case 'S':
        case 's':
          instance_1.StartDemo();
          break;
        case 'E':
        case 'e':
// Press 'e' to establish the connection with the remote server.
// This will enable the communication with the remote system.
#ifdef AUTOMATA1
          instance_1.setPort(PORT_2);
          instance_1.setIP((IP_ADDRESS));
          printf("establishConection on port:%u",PORT_2);
          instance_1.establishConnection();
#else
          instance_1.setPort(PORT_1);
          instance_1.setIP(IP_ADDRESS);
          printf("establishConection on port:%u",PORT_1);
          instance_1.establishConnection();
#endif
          default:
          break;
      }
    }
  }
  CloseHandle(hTemp);
  printf("The end. \n");
}
```

The file *Main.cpp* starts with the list of the necessary include files and the definition of the symbolic constant *AUTOMATA1*. This constant should be defined for the local process and not for the remote process (this is done by commenting out the source code line that defines the symbol *AUTOMATA1*).

Next, the instantiation of the class *FSMSystemWithTCP* is performed by a call to its constructor. The parameters used in this call specify that the instance of the *FSMSystemWithTCP*, named *fsmSystem*, will include a single automata type and this automata type will use a single mailbox. After that, a single instance of the class *NetAutomata* is made, *instance_1*. Additionally, this file contains the definitions of the FSM system thread function *ThreadFunction* and the function *main*.

The function *ThreadFunction* first prepares the data needed to define three buffer types. The sizes and quantities of these buffers are five at 128 bytes, three at 256 bytes, and two at 512 bytes. Next, the three automata instances are added to *fsmSystem*. Note that the fourth parameter of the first call to the function *Add* is set to the value *true*, which means that the instances are to be used as a pool of instances of the same type. After that, this function initializes the kernel by calling the function *InitKernel*, defines and sets the logging interface by calling the function *SetLogInterface*, starts the TCP server by calling the function *InitTCPServer*, and starts the *fsmSystem* by calling its function *Start*.

The function *main* starts the FSM system thread (which executes the function *ThreadFunction*) and suspends itself for 100 ms. After this, it waits for the user command. If the user presses the character 'E' or 'e', it establishes the TCP connection with the remote TCP server by calling the function *establishConnection*. If the user presses the character 'Q' or 'q', it terminates the program.

Index

A

Abstract service primitive (ASP), 127, 128
Action(s)
    definition of, 87
    state(s)
        categories of properties, 74
        examples of, 73
    types of, 88
Activity
    definition of, 87
    diagram(s), 2, 47, 71, 151, 401, 405
        definition of, 73
        e-mail operation, 86
        graphical symbols, 74, 80
        loop in, 77
        TCP, 83
        UML, 266
    state(s), 73
        attribute domain and, 82
        sequence of, 75, 89
Actor(s), 13
    generalization and specialization of, 15
    properties, categories of, 14
AddParam, 347
AddParamByte, 348
AddParamDWord, 348
AddParamWord, 348–349
Add(ptrFiniteStateMachine, uint8), 342
Add(ptrFiniteStateMachine, uint8, uint32, bool), 340–342
Address Resolution Protocol, 18
Agile programming, 1
Allocation policy, 354
Analysis, 10, 12
API, *see* Application Programming Interface
Application layer packages, 102
Application Programming Interface (API), 5, 28, 169
    functions, 319, 327
    user, example of, 170
ARQ, *see* Automatic Repeat Question
ASP, *see* Abstract service primitive
Association class, 53
Asynchronous event, 89
Atomic mailbox access, 216
Attach construct, 132
Attribute(s)
    condition, values of, 63
    default value of, 380
    domain, 82
    table of, 14
Automata, 173, 307
    control table, 180
    evolution, 179
    instances
        class diagram, 413
        collaboration diagram, 399
        component diagram, 414
        deployment diagram, 415
        object diagram, 414
        sequence diagram, 399
        statechart diagram, 400
        use case diagram, 398
    model, 55
    net, 325–326
    states, 174, 177
    type(s)
        definitions of, 225
        identification, 206
        unknown, 199
Automated theorem prover, 274, 275
Automated unit testing, 3
Automatic Repeat Question (ARQ), 76

B

Backward engineering, 165
Behavior
    model, 12
    specifications, 87
    testing, 11, 277
    tree
        evolution of, 131
        example of, 133
Binary object libraries, 167

439

Bit-mask, 379
Black box
    model, 126
    principle, 125
    testing, 249, 261
Bluetooth Host Controller Interface, 48
Boolean expressions, 78, 253
Boolean operation, 132
Branching
    coverage, 274
    flow of activities with, 76
    modeled, 22
Buffer(s), 317
    allocation, 319, 322, 355
    code, 216
    header, 215
    loss, 317
    management, 49
    types, 311, 318, 343
Bug(s)
    detection, 250
    FSM evolution-dependent, 258
    localizing, 299
    number of remaining, 278
Built-in data types, 127

C

C++, 116, 172, 175, 247
    compiler, 197–198
    FSMSystem library written in, 308
    modules, automated unit testing of, 3
Call
    establishment, 263
    event, 89
    processing, 355
Calling party telephone line, 120
Carriage-return line-feed sequences (CRLF), 31
Cartesian product, 173
CCITT, 103
Change event, 89
Channels, definition of, 107
CheckBufferSize, 349
Class(es)
    attributes, 61
    constructor, 191, 210, 232, 432
    definition of, 314
    destructor, 233
    diagram(s), 2, 10, 46, 50, 59, 195, 203, 413
        automata instances, 413
        communication protocol, 59
        domain-specific, 58

        graphical symbols, 51, 57
        vertices, 50
    difference between components and, 167
    FSM library, 203
    hierarchy, root of, 187
    N-ary association, 58
    parameterized, 58
    stereotypes of, 10
    supplementary, 297
    symbol, 51
    utility, 60
Cleanroom engineering, 1, 247
    methodology, 3
    organization of, 250
    paradigm, 11–12
ClearMessage, 349–350
Client, perspective of system, 13
Code
    buffer, 216
    generator, construction of, 178
    redundancy, 202
Coding, 166
Collaboration diagram(s), 2, 10, 61, 150, 156, 300
    automata instances, 399
    e-mail, 23, 25
    graphical symbols, 21
    links in, 22
    network-aware automata, 422
    receive e-mail application, 217
    session establishment, 157, 158
    SIP
        INVITE client transaction, 156
        session setup, 41
        softphone, 40, 154
Collaborations, end-to-end, 24
Commercial off-the-self (COTS) component, 125
Communication protocol(s), 5, 6
    class diagram, 59
    design, definition of, 46
    engineering
        generic design mechanisms in, 49
        phases of, 1, 2
    implementation, 165, 175, 307
    messages, serialization functions, 193
    modeled, 60
    notion of, 5–8
    running system for, 60
Compiler, construction of, 178
Compliance testing, 122, 248
Component(s)
    definition of, 99
    deployment, 168
    diagrams, 3, 165, 167, 168, 414

Index

differences between classes and, 167
executable, 168
most important feature of, 168
work product, 168
Composite state, 92, 93
Concrete substate class, example of, 196
Concurrent flows, 77, 79
Conformance testing, 124, 125, 248
    test case, 262
    TTCN and, 4
Connection establishment
    procedures for, 134
    scenario of successful, 136
Constant event, 281
Constraint declaration, 130
Context, user interface and, 194
Control flow transition, 74
Control predicates, 267
convertFSMToNetMessage, 397
convertNetToFSMMessage, 397
CopyMessage(), 350
CopyMessageInfo, 351
CopyMessage(uint*), 350
Copy-paste practice, 252
Coroutines, 79
COTS component, 125
CppUnit, 3, 248, 254, 298
CRLF, *see* Carriage-return line-feed sequences
Current message, 213, 320
Cyclic redundancy checking, 6

D

Daemongame, 108, 109
Data
    transfer phase, 138
    type(s)
        built-in, 127
        global, 327
        properties of, 58
Decision(s)
    properties of, 75, 89
    symbols, 90
Declaration file, 291, 292
Default memory allocation, 396
Default message processing function, 183
Deferred events, 92
Demo program code, 176
Dependency relation, 52
Deployment
    components, 168
    design, 45, 47

diagram(s), 3, 48, 99, 167, 415, 427
    automata instances, 415
    definition of, 100
    network-aware automata, 427
    symbols for, 101
model, 45
Design, 45–164
    activity diagrams, 71–85
    class diagrams, 50–60
    deployment diagrams, 99–103
    examples, 134–164
    message sequence charts, 120–124
    model, 12, 45
    object diagrams, 60–63
    sequence diagrams, 64–71
    specification and description language, 103–120
    statechart diagrams, 85–99
    for testability, 254
    tree and tabular combined notation, 124–134
Diagram(s), 2
    activity, 2, 47, 71, 151, 401, 405
        definition of, 73
        e-mail operation, 86
        graphical symbols, 74, 80
        loop in, 77
        TCP, 83
        UML, 266
    class, 2, 10, 46, 195, 203
        automata instances, 413
        communication protocol, 59
        domain-specific, 58
        graphical symbols, 51, 57
        vertices, 50
    collaboration, 2, 10, 61, 150, 300
        automata instances, 399
        e-mail, 23, 25
        graphical symbols, 21
        links in, 22
        network-aware automata, 422
        receive e-mail application, 217
        session establishment, 157, 158
        SIP, 40, 41, 154, 156
    component, 3, 165, 167, 414
    deployment, 3, 48, 99, 167
        automata instances, 415
        definition of, 100
        network-aware automata, 427
        symbols for, 101
    interaction, 46, 72
    object, 2, 46, 402
        automata instances, 414
        pure, 60
        usage of, 61

SDL, 49, 120, 136, 142
    coding, 201
    network, 143
    one-to-one mapping of, 237
    POP3 client, 222
    SIP INVITE client transaction, 161
    specification, stable states of, 135
  sequence, 2, 46, 158, 159
    appearance of, 64
    automata instances, 399
    DNS server, 68
    e-mail, 67
    error handling, 70
    graphical symbols, 66
    network-aware automata, 423
    practical detail about, 70
    SMTP server, 69
  statechart, 3, 47, 85, 135, 142, 157, 423
    automata instances, 400
    definition of, 88
    DNS server, 95
    e-mail operation, 98
    modeling concurrency, 93
    modulo 2, 174
    network, 141
    SIP INVITE client transaction, 160
    TCP, 97
  symbols common for all, 80
  UML, 46
    activity, 84
    interaction, 64
  use case, 2, 10, 13, 35, 71, 398
Dialing mode, 119
Digit event, 104
Directory tree, 171
Discard, 351
Distributed applications, 324
DLLs, *see* Dynamically linkable libraries
DNS, *see* Domain Name System
Documentation notes, 74
Domain Name System (DNS), 80
  client, 23, 80, 81
  server
    domain name and, 82
    executable, 101–102
    sequence diagram, 68
    statechart diagrams, 95
    triggerless transition, 96
Domain-specific languages, 4, 46
DoNothing, 351–352
Drivers, 248
Dynamically linkable libraries (DLLs), 167
Dynamic loaders, 181

E

e-mail(s), 217
  applications, modeled, 54
  collaboration diagram, 23, 25
  delete procedure, 218
  Internet and, 22
  messages, pending, 235
  operation activity diagram, 86
  operation statechart diagram, 98
  password processing scenario, invalid, 220
  pending, 218
  read procedure, 218
  sequence diagram, 67
  server, 18, 225
  successfully sent, 85
  use case diagram, 17
Error
  handling, sequence diagrams, 70
  processing specification, 6, 7
  recovery procedures, 121
establishConnection, 397
Ethernet cards, 63
ETSI specification, 264
Event(s)
  asynchronous, 89
  automatic logging of, 202
  call, 89
  change, 89
  class(es), 281
    definitions, 283
    GTCG random selection of, 289
    operational profile, 286
  constant, 281
  deferred, 92
  definition of, 87
  digit, 104
  handler, 316
  hook-off, 104
  hook-on, 104
  interpreter, 178, 179, 180, 183
  particular, 281
  recognizable, 369
  test, 130, 131
  time, 89
  types, 88–89, 371
  unexpected, 373, 419
Executable components, 168
Executables, modeling of, 170, 171
External signal, 106
Extreme programming, 251

Index

F

Family of protocols, 107
FIFO
  memory type, 319
  message queues, 200
  queue, 107
FiniteStateMachine, 346–347
Finite state machine (FSM), 4, 8, 307, *see also* FSM library
  axiomatic specification, 266, 268, 272
  behavior, wrapped state hierarchy, 194
  branch, 266
  classes used for construction of, 184
  evolution, 187, 189
  final states, 269
  formal specification of, 8
  implementation, 56, 172
    internal class, 202
    static structure, 185, 192
  input messages, 188, 189, 193
  modeled, 104
  net, 326
  object(s)
    diagram, example of, 62
    identification, 234
    initialization of, 204
  processes, dysfunctional behaviors, 200
  signaling between, 269
  stable states, classes that model, 190
  statecharts, 271
  state transition, 198, 269, 307
    graph, 258–259, 274
    table, 209
  state updating, 202
  structure
    event interpreter and, 179
    software maintenance and, 181
  system
    automata added to, 313
    construction of, 398
    initialization, 309, 312
    logging functionality, 310
    member functions summary, 330–340
    startup, 312
    thread function, 421
  theoretical test case for, 267
  timers, 209
  unstable states, classes that model, 190
Flat message, 65
Formal system design verification, 3
Forward engineering, 165, 400
Frame retransmission, 153, 156
FreeFSM, 352
FSM, *see* Finite state machine

FSM library, 307–437
  API functions, 327–398
    AddParam, 347
    AddParamByte, 348
    AddParamDWord, 348
    AddParamWord, 348–349
    Add(ptrFiniteStateMachine, uint8), 342
    Add(ptrFiniteStateMachine, uint8, uint32, bool), 340–342
    CheckBufferSize, 349
    ClearMessage, 349–350
    convertFSMToNetMessage, 397
    convertNetToFSMMessage, 397
    CopyMessage(), 350
    CopyMessageInfo, 351
    CopyMessage(uint*), 350
    Discard, 351
    DoNothing, 351–352
    establishConnection, 397
    FiniteStateMachine, 346–347
    FreeFSM, 352
    FSMSystem, 329–340
    FSMSystemWithTCP, 345
    GetAutomata, 352
    GetBitParamByteBasic, 352–353
    GetBitParamDWordBasic, 353–354
    GetBitParamWordBasic, 353
    GetBuffer, 354–355
    GetBufferLength, 355
    GetCallId, 355
    GetCount, 356
    GetGroup, 356
    GetInitialState, 356
    GetLeftAutomata, 357
    GetLeftGroup, 357
    GetLeftMbx, 356–357
    GetLeftObjectId, 357
    GetMbxId, 358
    GetMessageInterface, 358
    GetMsg(), 359
    GetMsgCallId, 359
    GetMsgCode, 360
    GetMsgFromAutomata, 360
    GetMsgFromGroup, 360
    GetMsgInfoCoding, 360
    GetMsgInfoLength(), 361
    GetMsgInfoLength(uint8*), 361
    GetMsgObjectNumberFrom, 361
    GetMsgObjectNumberTo, 362
    GetMsgToAutomata, 362
    GetMsgToGroup, 362
    GetMsg(uint8), 359
    GetNewMessage, 362–363
    GetNewMsgInfoCoding, 363
    GetNewMsgInfoLength, 363

444 *Communication Protocol Engineering*

GetNextParam, 363–364
GetNextParamByte, 364
GetNextParamDWord, 364–365
GetNextParamWord, 365–366
GetObjectId, 366
GetParam, 366
GetParamByte, 367
GetParamDWord, 367–368
GetParamWord, 368
GetProcedure, 368–369
getProtocolInfoCoding, 397
GetRightAutomata, 369–370
GetRightGroup, 370
GetRightMbx, 369
GetRightObjectId, 370
GetState, 370
InitEventProc, 371–372
Initialize, 371
InitKernel, 342–343
InitTCPServer, 345–346
InitTimerBlock, 372
InitUnexpectedEventProc, 373
IsBufferSmall, 370–371
IsTimerRunning, 373
NetFSM, 396
NoFreeInstances, 374
NoFreeObjectProcedure, 373–374
ParseMessage, 374–375
PrepareNewMessage(uint8*), 375
PrepareNewMessage(uint32, uint16, uint8), 375–376
Process, 376
PurgeMailbox, 376–377
RemoveParam, 377
Remove(uint8), 343–344
Remove(uint8, uint32), 344
Reset, 377
ResetTimer, 377–378
RestartTimer, 378
RetBuffer, 378
ReturnMsg, 378–379
SendMessageLeft, 392
SendMessageRight, 392–393
SendMessage(uint8), 384
SendMessage(uint8, uint8*), 384–385
sendToTCP, 398
SetBitParamByteBasic, 379
SetBitParamDWordBasic, 379
SetBitParamWordBasic, 380
SetCallId(), 380
SetCallIdFromMsg, 381
SetCallId(uint32), 380–381
SetDefaultFSMData, 381
SetDefaultHeader, 381
SetGroup, 382

SetInitialState, 382
SetKernelObjects, 382
SetLeftAutomata, 383
SetLeftMbx, 383
SetLeftObject, 383
SetLeftObjectId, 383–384
SetLogInterface, 384
SetMessageFromData, 385
SetMsgCallId(uint32), 385
SetMsgCallId(uint32, uint8*), 385–386
SetMsgCode(uint16), 386
SetMsgCode(uint16, uint8*), 386
SetMsgFromAutomata(uint8), 386–387
SetMsgFromAutomata(uint8, uint8*), 387
SetMsgFromGroup(uint8), 387
SetMsgFromGroup(uint8, uint8*), 388
SetMsgInfoCoding(uint8), 388
SetMsgInfoCoding(uint8, uint8*), 388–389
SetMsgInfoLength(uint16), 389
SetMsgInfoLength(uint16, uint8*), 389
SetMsgObjectNumberFrom(uint32), 389–390
SetMsgObjectNumberFrom(uint32, uint8*), 390
SetMsgObjectNumberTo(uint32), 390
SetMsgObjectNumberTo(uint32, uint8*), 390–391
SetMsgToAutomata(uint8), 391
SetMsgToAutomata(uint8, uint8*), 391
SetMsgToGroup(uint8), 391–392
SetMsgToGroup(uint8, uint8*), 392
SetNewMessage, 393
SetObjectId, 393
SetRightAutomata, 394
SetRightMbx, 393
SetRightObject, 394
SetRightObjectId, 394
SetState, 394–395
Start, 344
StartTimer, 395
StopSystem, 344–345
StopTimer, 395
SysClearLogFlag, 395
SysStartAll, 396
Application Programming Interface, 5
automatic message buffer reallocation, 203
-based implementation paradigm, 173
basic FSM system components, 308–316
    class *FiniteStateMachine*, 313–316
    class *FSMSystem*, 308–313
classes of, 166
design artifacts, 203

Index

example with network-aware automata instances, 422–437
example with three automata instances, 398–421
function calls, 236
fundamental classes, 203
global constants, types, and functions, 326–327
implementation based on, 197
internals, 204
kernel, 294, 323, 325
logging subsystem, 202
logging system, 199
mailboxes, 200
main task of, 198
memory management, 317–318
message handling functionality, 202
message management, 318–322
message prioritization, 200
message sending functions, 210
portability, 216
static structure, internal, 208
TCP/IP support, 323–326
  class *FSMSystemWithTCP*, 324
  class *NetFSM*, 325–326
time management, 316–317
timer identification, 201
FSMSystem, 329–340
FSMSystemWithTCP, 345
Function
  prototype declaration, 396
  virtual, 397
Functional blocks, 105, 116, 118
Functional requirements, 9, 12

G

Garbage collector, 257–258
Generalization relation, 52
Generic design mechanisms, 59
Generic Modeling Environment (GME), 279
Generic test case generator (GTCG), 251, 281–282, 287
GetAutomata, 352
GetBitParamByteBasic, 352–353
GetBitParamDWordBasic, 353–354
GetBitParamWordBasic, 353
GetBuffer, 354–355
GetBufferLength, 355
GetCallId, 355
GetCount, 356
GetGroup, 356
GetInitialState, 356

GetLeftAutomata, 357
GetLeftGroup, 357
GetLeftMbx, 356–357
GetLeftObjectId, 357
GetMbxId, 358
GetMessageInterface, 358
GetMsg(), 359
GetMsgCallId, 359
GetMsgCode, 360
GetMsgFromAutomata, 360
GetMsgFromGroup, 360
GetMsgInfoCoding, 360
GetMsgInfoLength(), 361
GetMsgInfoLength(uint8*), 361
GetMsgObjectNumberFrom, 361
GetMsgObjectNumberTo, 362
GetMsgToAutomata, 362
GetMsgToGroup, 362
GetMsg(uint8), 359
GetNewMessage, 362–363
GetNewMsgInfoCoding, 363
GetNewMsgInfoLength, 363
GetNextParam, 363–364
GetNextParamByte, 364
GetNextParamDWord, 364–365
GetNextParamWord, 365–366
GetObjectId, 366
GetParam, 366
GetParamByte, 367
GetParamDWord, 367–368
GetParamWord, 368
GetProcedure, 368–369
getProtocolInfoCoding, 397
GetRightAutomata, 369–370
GetRightGroup, 370
GetRightMbx, 369
GetRightObjectId, 370
GetState, 370
Global constants, 326
Global control predicate, 267
GME, *see* Generic Modeling Environment
Golden output, 256
Graphically-oriented languages, advantages of, 105
Graphical symbols
  activity diagrams, 74, 80
  class diagrams, 51, 57
  collaboration diagrams, 21
  interfaces, 52
  meaning of, 111–112
  sequence diagrams, 66
  statecharts, 94
  use case diagrams, 14, 16
Graphical user interface (GUI), 34, 279, 281
GTCG, *see* Generic test case generator

GUI, *see* Graphical user interface

H

HCI, *see* Host Controller Interface
History states, types of, 93
Hook-off event, 104
Hook-off signal, 133
Hook-on event, 104
Host Controller Interface (HCI), 48
HTTP, *see* Hyper Text Transport Protocol
Hyper Text Transport Protocol (HTTP), 30

I

ICMP protocol, 54
Imitators, kinds of, 248
Implementation, 165–245
    component diagrams, 167–172
    examples, 217–245
    implementation based on FSM library, 197–217
        FSM library internals, 204–216
        using FSM library, 203–204
        writing FSM library-based implementations, 216–217
    models, 168
    as phase, 165
    as product, 165
    spectrum of FSM implementations, 172–193
    state design pattern, 194–197
    under test (IUT), 124, 125, 127, 261
In-field testing, 248, 249, 250
Informal specification, 7
InitEventProc, 371–372
Initialize, 371
InitKernel, 342–343
InitTCPServer, 345–346
InitTimerBlock, 372
InitUnexpectedEventProc, 373
Inopportune behavior, 263
Input signals, 108
    identical processing of, 114
    labeled, 176
    recognizable, 115
Integer variable, declared, 115
Integration testing, 299, 300
Interaction
    definition of, 64
    diagrams, 46, 72
Interfaces, graphical symbols for rendering, 52
Internal signal, 106
International Standardization Organization (ISO), 124
Internet
    e-mail and, 22, 217
    transport messages and, 38
Interruptible activities, sequence of, 75
Invalid behavior, 263
INVITE client transaction, 235, 237
Invite transaction, 31, 38
IP address, 324
IsBufferSmall, 370–371
ISO, *see* International Standardization Organization
ISO OSI ideal, 34
IsTimerRunning, 373
Iteration, modeled, 22
ITU-T, 4
    domain-specific languages, 8
    /ETSI recommendations, 10
    recommendation, 116
IUT, *see* Implementation under test

J

Java, 116, 172, 175, 247
    compilation unit, 251
    FSM structure modeling, 181
    garbage collector, 257
    JavaCompRegister, 282
    map, 193
    module, class hierarchy, 188
    modulo 2, 184
    packages, automated unit testing of, 3
    sets, updating of, 191
    state design pattern, 195
Join synchronization point, 93
JUnit, 1, 3, 248, 254, 255

K

Kernel
    developer, 199
    FSM, 294, 309, 323, 325
    initialization, 437
    internals, 214
    static structure, 215

Index 447

L

Language(s), *see also specific language*
  domain-specific, 4, 8, 46
  graphical, 105
  modeling, 279, 280
  MSC, 122
  SDL, 104, 106
  selection, 172
  TTCN, 128, 129, 261
Left mailbox, 369, 392
Libraries, modeling of, 170
Load generator, 249
Load testing, 248, 249
Location service, 31
Log file(s)
  creation, 294
  detecting bugs through analysis of, 250
  real value of, 299
  records, 305
  theoretical, 272
Logging services, 207
Logging subsystem, FSM library, 202

M

Macro instruction, 292, 294
Mailbox(es)
  access, atomic, 216
  default, 356, 383
  definition of, 319
  FSM library, 200
  identification, 356, 358, 393, 405
  implemented, 216
  left, 369, 392
  names, 235
  right, 369
  use of as queues, 312
Main branches, 121
Mandatory functions, 233
Map identification, 191
Markov process, 278, 286
Maximal Transfer Unit (MTU), 78
MDA, *see* Model-Driven Architecture
Media Gateway Control Protocol (MEGACO), 30
MEGACO, *see* Media Gateway Control Protocol
Memory
  access, double-word, 327
  allocation, 318, 396
  exhaustion problem, 355
  leak, 317

  management, 205, 211, 214, 317
  space, partitioned, 215
Message(s)
  buffer reallocation, automatic, 203
  code, 295, 386
  common way to construct, 321
  content, 299
  current, 320
  definition of, 6
  destination, 390, 391
  flow, DNS client and, 81
  format, 6
  handling functions, 199, 202
  header, 212, 321
  interface object, 417
  length
    correction, 351
    total, 363
  logging, 384, 396
  management, 49, 211, 318
  output, 294
  parameter(s), 234, 366
    definition, 213
    identification, 419, 434
  payload length, 361, 389
  POP3-related, 221
  prioritization, absence of, 200
  -processing procedures, 6, 104
  properties, 65
  purging of, 377
  queue, 179, 200
  retransmission, 146–147, 149
  sequence number, 405, 418, 433
  source, 387
  static, 376
  target, 386, 388
  timer, 213
Message sequence charts (MSC), 3, 120
  chart(s), 19, 121, 138
    example of, 123
    main advantage of using, 124
  connection establishment, 137, 140
  language, forms of, 122
  successful message delivery, 144
  unsuccessful message delivery, 144
MIC, *see* Model integrated computing
Microsoft® Visio, 14, 21, 51, 107
Microsoft Visual Studio 6.0, 402–403
Model(s)
  active states, 92
  analysis, 12
  automata, 55
  -based software development, 166
  behavior, 12
  black box, 126

deployment, 12, 45
design, 12, 45
-Driven Architecture (MDA), 166
formal verification of, 250
implementation, 168
integrated computing (MIC), 107
interpreter, 282, 286
object mutation, 65
project, 172
requirements, 10
SDL, domain-specific, 401
SIP softphone
  analysis, 39
  requirements, 34
source code, 171
TCP/IP protocol stack, 54, 61
test, 11, 12, 45
transient states, 90
UML, 11, 50
win-lose game, 111
workflows, 82
Modeling paradigm, 279
Module definition files, 171
Modulo 2
  statechart, 268
  state class hierarchy, 187
  state transition graph, 173
MSC, see Message sequence charts
MTU, see Maximal Transfer Unit
Multimedia protocols, 355

N

N-ary association class, 58
Net automata, 325–326
NetFSM, 396
Network
  configuration, example of, 102
  Interface Controller, 18
  message format, 325
  nodes, identification of, 100
  SDL diagram, 143
  statechart diagram, 141
  structure, deployment diagrams and, 99
  topology, 147
Network-aware automata
  collaboration diagram, 422
  deployment diagram, 427
  sequence diagram, 423
Node instance, properties of, 100
NoFreeInstances, 374
NoFreeObjectProcedure, 373–374
Nonfunctional requirements, 9

Non-invite transactions, 31, 32, 38
Note, definition of, 99

O

Object(s)
  collaboration, 398, 402
  diagram(s), 2, 46, 402
    automata instances, 414
    pure, 60
    usage of, 61
  flow transition, 80
  FSM, 204
  identification, 207
  libraries, 171
  lifeline, 64
  link, 53
  message
    interface, 417
    logging, 384
  mutation, model of, 65
  -oriented design, 63, 187
  state, 63
  transition, 63
Offer-answer procedure, 33
OPEN requests, types of, 82
Open-source packages, 3, 254
Operational profile, 278
  event classes, 286
  models, 250, 280
  SIP softphone, 289, 290
  working states, 284
Output message, message code of, 294
Output signals, 108
Overloaded functions, 309
Overload protection, 355

P

Package properties, categories of, 16
Packet delivery operation, 90
Parameterized class, 58
ParseMessage, 374–375
Particular events, 281
Password, 218, 220
Payload, 6, 330, 351
Payton, 11
PCO, see Point of control and observation
PDU, see Protocol data unit
Peer(s), 127
  entities, 127

Index

-to-peer protocols, 22
Perl, 11
Personal computer, modeled, 101
Pert charts, 47, 73
PICS, *see* Protocol implementation conformance statements
PIXIT, *see* Protocol implementation extra information
Point of control and observation (PCO), 127
Polymorphism, 187, 194
POP3
    client, 24, 222
    protocol, 218
PrepareNewMessage(uint8*), 375
PrepareNewMessage(uint32, uint16, uint8), 375–376
Primitive(s)
    buffer management, 318
    example of usage, 316–317
    mapping of SDL steps, 204
    operations, 6
    self-documenting names of, 204
Private function members, 427
Process(es), 376
    declared, 111
    definition of, 105, 115
    identification of, 110
Product
    quality, measures of, 278
    state transitions, 278
Project model, 172
Protocol(s)
    data unit (PDU), 127
        constraint declarations, 130, 139, 145
        type declarations, 138, 144
        type definition, 129
    definition of, 1, 5
    family of, 107
    implementation conformance statements (PICS), 126
    implementation extra information (PIXIT), 126
    peer-to-peer, 22
    stack(s), 107
        hierarchy, 126
        implemented, 248
    state transition management, 49
Proxy server, 31
Pseudocode, 15, 52, 73
Public function members, 428
Public mailbox prioritization, 204
PurgeMailbox, 376–377
Python, 264

R

Random test cases, 279
Real-Time Streaming Protocol (RTSP), 30
Real-Time Transfer Protocol (RTP), 30
Receive e-mail session establishment scenario, 219
Recovery mechanisms, 418, 433
Registration functions, 315
Regression testing, 251, 252
RemoveParam, 377
Remove(uint8), 343–344
Remove(uint8, uint32), 344
Repeat construct, 132
Request line, 31
Requirements and analysis, 9–43
    collaboration diagrams, 21–30
    engineer, use case diagram and, 16
    functional requirements, 9, 12
    model, 10
    nonfunctional requirements, 9
    requirements and analysis example, 30–42
        SIP domain specifics, 30–34
        SIP softphone analysis model, 39–42
        SIP softphone requirements model, 34–39
    use case diagrams, 13–20
Reset, 377
ResetTimer, 377–378
RestartTimer, 378
RetBuffer, 378
Retransmission
    buffer, 237
    queue, 154
    timer, 78, 153
ReturnMsg, 378–379
Reverse engineering tool, 275
Right mailbox, 369
Root test suite, 260
Router, modeled, 53
RTP, *see* Real-Time Transfer Protocol
RTSP, *see* Real-Time Streaming Protocol

S

SDL, *see* Specification and description language
SDP, *see* Session Description Protocol
SDT, *see* Software Development Tools
Send e-mail operation, 85, 86
SendMessageLeft, 392
SendMessageRight, 392–393

Send message statement, 131
SendMessage(uint8), 384
SendMessage(uint8, uint8*), 384–385
sendToTCP, 398
Sequence diagram(s), 2, 46, 158, 159
    appearance of, 64
    automata instances, 399
    DNS server, 68
    e-mail, 67
    error handling, 70
    graphical symbols, 66
    network-aware automata, 423
    practical detail about, 70
    SMTP server, 69
Sequence numbers, mismatch of, 153
Server
    DNS, 68
        domain name and, 82
        statechart diagrams, 95
        triggerless transition, 96
    e-mail, 225
    proxy, 31
    SMTP, 61, 69, 71, 102
    TCP, 324
    Telnet, 79
    user agent, 31, 34
Session Description Protocol (SDP), 30
Session establishment collaboration diagram, 157, 158
Session Initiation Protocol (SIP), 3, 12
    conformance test specification for, 262
    domain specifics, 30
    Forum Testing Framework (SFTF), 264
    invite client transaction, 5, 217
        collaboration diagram, 156
        implementation, unit testing of, 291
        operational profile, 285
        SDL diagram, 161
        statechart diagram, 160
    message parser, 238
    protocol stack, 35
    protocol torture test, 265
    session setup, 33, 41
    softphone
        analysis model, 39
        collaboration diagram, 40, 154
        integration testing of, 299
        operational profile, 289, 290
        requirements, 3, 34
        use case diagram, 35
    standards, methods of, 32
    tester, 263
    User Agent, 264
Session modification, 263
SetBitParamByteBasic, 379

SetBitParamDWordBasic, 379
SetBitParamWordBasic, 380
SetCallId(), 380
SetCallIdFromMsg, 381
SetCallId(uint32), 380–381
SetDefaultFSMData, 381
SetDefaultHeader, 381
SetGroup, 382
SetInitialState, 382
SetKernelObjects, 382
SetLeftAutomata, 383
SetLeftMbx, 383
SetLeftObject, 383
SetLeftObjectId, 383–384
SetLogInterface, 384
SetMessageFromData, 385
SetMsgCallId(uint32), 385
SetMsgCallId(uint32, uint8*), 385–386
SetMsgCode(uint16), 386
SetMsgCode(uint16, uint8*), 386
SetMsgFromAutomata(uint8), 386–387
SetMsgFromAutomata(uint8, uint8*), 387
SetMsgFromGroup(uint8), 387
SetMsgFromGroup(uint8, uint8*), 388
SetMsgInfoCoding(uint8), 388
SetMsgInfoCoding(uint8, uint8*), 388–389
SetMsgInfoLength(uint16), 389
SetMsgInfoLength(uint16, uint8*), 389
SetMsgObjectNumberFrom(uint32), 389–390
SetMsgObjectNumberFrom(uint32, uint8*), 390
SetMsgObjectNumberTo(uint32), 390
SetMsgObjectNumberTo(uint32, uint8*), 390–391
SetMsgToAutomata(uint8), 391
SetMsgToAutomata(uint8, uint8*), 391
SetMsgToGroup(uint8), 391–392
SetMsgToGroup(uint8, uint8*), 392
SetNewMessage, 393
SetObjectId, 393
SetRightAutomata, 394
SetRightMbx, 393
SetRightObject, 394
SetRightObjectId, 394
SetState, 394–395
SFTF, *see* SIP Forum Testing Framework
Shallow history state, 94
Signal(s), 372
    definition of, 106
    external, 106
    hook-off, 133
    input, 108, 114, 176
    internal, 106
    lists of, 117
    output, 108
    source of, 89

Index

Simple mail transfer protocol (SMTP), 62
    client, 24, 29, 85
    server, 61
        modeled, 102
        sequence diagram, 69
        virtual interaction, 71
SIP, *see* Session Initiation Protocol
Sliding window concept, 149
SMTP, *see* Simple mail transfer protocol
Software
    development processes, categories of, 275–276
    engineering, development phases, 250
    quality assurance, successful, 197
    reliability estimation, 250
Software Development Tools (SDT), 107
Source code, 297
    files, 171
    model, 171
Specification and description language (SDL), 3, 103, 357
    applications, main, 107
    diagram(s), 49, 120, 136, 142
        coding, 201
        network, 143
        one-to-one mapping of, 237
        POP3 client, 222
        SIP INVITE client transaction, 161
        specification, stable states of, 135
    dilemma of creators, 104
    graphical, 106, 111
    language
        basics, 107
        characteristics of, 106
        dilemma, 104
    model, domain-specific, 401
    program, 106, 107, 110, 175
    rules, 106
    state transitions, 401
    steps, mapping of, 204
    symbol, 201
    use of, 106
Stable state, definition of, 115
Star network, hypothetical, 141
Start, 344
StartTimer, 395
State(s)
    attributes, 183
    bound actions, 56
    class hierarchy, modulo 2, 187
    composite, 92
    definition of, 87
    design pattern, 194, 195
    machine, 87, 95
    objects, 63, 190
    pattern consequences, 194
    properties of, 88
    shallow history, 94
    transition(s), 55, 90
        activity diagrams, 73
        attributes, 87
        event handlers, 371
        function, 210–211, 227, 235, 304
        graph, 104
        modeling of, 187
        table, 315
    transmission function, timer expiration and, 244
    unstable, 105
Statechart(s)
    diagram(s), 3, 47, 85, 97, 135, 157, 423
        automata instances, 400
        definition of, 88
        DNS server, 95
        e-mail operation, 98
        modeling concurrency, 93
        modulo 2, 174
        network, 141
        SIP INVITE client transaction, 160
        TCP, 97
    FSM, 271
    graphical symbols, 94
    modulo 2, 268
    set of symbols for rendering, 88
StateWORKS® tool, 166
Static structure, 61, 100
Statistical testing, 3, 11, 277
Stereotypes of classes, 10
Sticky notes, 14
StopSystem, 344–345
StopTimer, 395
Stubs, 248
Subscriber A, 276
Subsystem(s)
    example of, 103
    modeling, 48
Supplementary class, 297
Supplementary introduced operations, 127
Switch-case statement, 4, 116, 175, 178, 191
SysClearLogFlag, 395
SysStartAll, 396
System
    architecture, definition of, 10
    crash, 278
    decomposition, 269
    requirements, use case diagrams and, 71
    restarting, 178
    -software layer, 47
    static design view of, 61

T

Table of attributes, 51
Table of internal transitions, 74
Table of receptions, 22
TAL, *see* Transaction layer
Target message, 386, 388
Task sharing, 181
TCP, *see* Transmission Control Protocol
TCP/IP
  infrastructure, communication over, 434
  Internet layers, 48
  protocol stack, 53, 54, 61
  support classes, 323
Telelogic® Software Development Tools, 107
Telephone call processing, 116, 117
Telnet server, 79
Test, *see also* Test and verification
  bed, 9
  case(s), 9, 124
    automatic execution of, 260
    conformance testing, 262
    dynamic behavior specification, 133
    execution, 252
    failure, 273
    function, 258
    random, 279
    registration, 252
    reporting, 252
    results, checking of, 253
    variables declarations, 129
  configuration, illustration of, 125
  design, 45
  events, 130, 131
  harness, 9, 254, 256
  model, 11, 12, 45
  suite(s), 9, 46, 124
    generation, automatic, 250
    implemented, 247
    operation definition, 130
    statistical, 279
  verdicts, 130
Testing
  activities, software engineering and, 247
  automated unit, 3
  behavioral, 11, 277
  black box, 249, 261
  compliance, 122, 248
  conformance, 4, 124, 125, 248, 262
  framework, 263
  in-field, 248, 249, 250
  integration, 299
  load, 248, 249
  regression, 251, 252
  statistical, 3, 11, 250, 277
  unit, 122, 251
Test and verification, 247–306
  conformance testing, 261–265
  examples, 291–305
  formal verification based on theorem proving, 265–277
  further reading, 305–306
  statistical usage testing, 277–291
  unit testing, 251–261
THEO, 251, 269, 272, 273, 277
Theorem prover, 251, 269, 274, 275
Theoretical log file, 272
Threads, 48
Three and tabular combined notation (TTCN), 3, 11, 124, 131, 247
  conformance testing and, 4
  declarations of types in, 127
  language, 126, 261
    test case variables, 129
    type declarations in, 128
  test suite(s), 50, 138
    specification, 126, 138
    variables used in, 127
Three-way handshake, 33
Time
  event, 89, 92
  management, 316
  -sharing operating system, 79
  stamp, 272
Timer
  blocks, 315
  expiration, 98–99, 244, 304, 316
    message code, 372
    statement, 132
  identification, 201, 243, 372, 395
  initialization, 302, 304
  management, 49, 205, 212
  message, 213
TLI, *see* Transport layer interface
Traces relation, 58
Traffic case, 250
Transaction
  layer (TAL), 34, 38, 302
  types, 31
  user (TU), 31, 34
Transient states, sequence of, 89
Transition
  control flow, 74
  definition of, 87
  key abstractions, 56
  objects, 63
  properties of, 89
  table, modulo 2, 174
  triggerless, 74, 149
Transmission Control Protocol (TCP), 18

Index

activity diagram, 83
  blocked, 96
  client, 24
  connection, Telnet server and, 79
  entities, virtual interaction between, 70
  events, modeled, 84
  port number, 324
  protocol, activity states of, 84
  server, 324
  statechart diagram, 97
  virtual collaboration and, 28, 29
Transport layer interface (TLI), 34, 38
Triggerless transition, 74, 149
TTCN, *see* Three and tabular combined notation
TU, *see* Transaction user
Type encoding information, 128

U

UA, *see* User agent
UAC, *see* User agent client
UAS, *see* User agent server
UML, *see* Unified Modeling Language
Unexpected event(s), 373
  examples of, 7
  handlers, 419
Unified Modeling Language (UML), 1
  activity diagrams, 49, 84, 266
  diagrams, 46
  history in, 93
  interaction diagrams, types of, 64
  models, 11
  paradigm, 10
  sequence diagrams, 49
  test model, 50
  use cases in, 18
Uniform Resource Identifications (URI), 31
Unit testing, 122, 251
Unknown automata instances, 199
Unstable states, 105
Upper tester, 134
URI, *see* Uniform Resource Identifications
USDP, layers recognized by, 47

Use case(s), 10, 13
  diagram(s), 2, 10, 13, 35, 71, 398
    automata instances, 398
    e-mail, 17
    graphical symbols, 14, 16
    SIP softphone, 35
    system requirements and, 71
  structuring of, 15
  UML, 18
User
  agent (UA), 264
    client (UAC), 31, 34
    server (UAS), 31, 34
  authentication procedure, 218
  interface, context and, 194
Utility, 58
  class, 60
  functions, 327

V

Valid behavior, 263
Vertical partitioning, 48
VFSMs, *see* Virtual finite state machines
Virtual collaboration, 22, 28, 29
Virtual finite state machines (VFSMs), 166
Virtual function, 325, 352, 397
Virtual peer-to-peer connections, 402

W

Warnings, 264
Win-lose game, model of, 111
Working environment
  automated, 279
  FSM library, 197
  state-of-the-art, 197
  statistical test suites, 280
Work product components, 168
World Wide Web applications, modeled, 54